WITHDRAWN

PLASTIC ANALYSIS AND DESIGN OF PLATES, SHELLS AND DISKS

NORTH-HOLLAND SERIES IN
APPLIED MATHEMATICS AND MECHANICS

EDITORS:

H. A. LAUWERIER

Institute of Applied Mathematics
University of Amsterdam

W. T. KOITER

Laboratory of Applied Mechanics
Technological University, Delft

VOLUME 15

NORTH-HOLLAND PUBLISHING COMPANY–AMSTERDAM · LONDON
AMERICAN ELSEVIER PUBLISHING COMPANY, INC.–NEW YORK

PLASTIC ANALYSIS AND DESIGN OF PLATES, SHELLS AND DISKS

by

M. A. SAVE
Institut du Génie Civil
Liège, Belgium

C. E. MASSONNET
Département d'Architecture
Faculté Polytechnique de Mons
Mons, Belgium

1972

NORTH-HOLLAND PUBLISHING COMPANY – AMSTERDAM · LONDON
AMERICAN ELSEVIER PUBLISHING COMPANY, INC. – NEW YORK

© NORTH-HOLLAND PUBLISHING COMPANY, 1972

All rights reserved. No part of this publication may be reproduced, stored in a retrieval system, or transmitted, in any form or by any means, electronic, mechanical, photocopying, recording or otherwise, without the prior permission of the copyright owner.

Based on Calcul Plastique des Constructions, Tome II, published by Centre Belgo-Luxembourgeois d'Information de l'Acier in Brussels, Belgium 1963.

Library of Congress Catalog Card Number 73-157012
North-Holland I.S.B.N. for the series: 0 7204 2350 3
North-Holland I.S.B.N. for this volume: 0 7204 2365 1
American Elsevier I.S.B.N. 0 444 10113 6

PUBLISHERS:

NORTH-HOLLAND PUBLISHING COMPANY - AMSTERDAM
NORTH-HOLLAND PUBLISHING COMPANY, LTD. - LONDON

SOLE DISTRIBUTORS FOR THE U.S.A. AND CANADA:

AMERICAN ELSEVIER PUBLISHING COMPANY, INC.
52 VANDERBILT AVENUE
NEW YORK, N.Y. 10017

PRINTED IN THE NETHERLANDS

Preface

Plastic design of steel beams and frames has been included in building codes since 1948 in Great Britain and since 1959 in the United States. The increased use of plastic design and the present trend toward applying its methods to reinforced and prestressed concrete indicate that the basic advantages of plastic analysis — accurate estimate of the collapse load, simplicity of application, and economy of structure — are more and more appreciated by practicing engineers. Codes on reinforced concrete slabs (European Committee for Concrete) and on pressure vessels (ASME) are on the way to adopt plastic limit state as one of the design criterions. Hence, it was felt urgent to provide all interested people with a synthesis work on the subject.

The present book is the English version of "Calcul plastique des Constructions", Volume 2, 2nd edition, published by the Centre Belgo-Luxembourgeois d'Information de l'Acier in Brussels, Belgium.

Together with this second edition in French, which is to appear nearly contemporarily, not only has it been brought up to date but also appreciably improved with respect to the first edition in French:
— new experimental results have been introduced, to better clarify the real physical significance of the theoretical limit load, particularly in Chapters 6 and 8
— the treatment of circular plates has been re-written in a more systematic manner

- Section 6–8 on minimum-weight design has been re-written and amplified
- in Chapter 7 on reinforced concrete plates, a more satisfactory derivation of the yield surface has been achieved, the delicate question of nodal forces has been faced in a clearer manner, statically admissible and complete solutions have been amplified, economy of reinforcement has been treated in a more rational manner, as well as the influence of the axial forces
- Chapters 8 and 9 contain many new additions of new theoretical solutions and experimental information
- only Chapter 10 is practically unchanged except for the new section on notched bars in tension

On the other hand, the main features of the first edition in French have been maintained, namely:
- the book is directed toward engineering applications; consequently, we have limited the mathematics to what was strictly needed, and we have avoided any formalism that could be unfamiliar to some engineers (as tensor notations for example). Physical significance, with reference to experimental evidence, is emphasized throughout.
- the book should serve the engineering student in plastic limit analysis: therefore, the theory is rigorously developed, as a rule from simpler to more complicated problems, though it was felt necessary to base applications to particular structures on the firm ground of a general theory.

Though it can be read completely independently, with the sole pre-requisite basis of some knowledge of the classical theory of structures, the present book can be regarded as the companion volume of "Plastic Analysis and Design of Beams and Frames" *, which is the adaptation in English of the first volume of "Calcul plastique des Constructions", devoted by the same authors to the simpler problems of beams and frames. The knowledge of the subjects treated in "Plastic Analysis and Design of Beams and Frames" is by no means necessary to undertake studying the present work. However, references may be useful. Whenever this is done, the cited book is referred to as "Companion Volume", abreviated as Com. V. throughout the text.

We believe that the false conflict between elastic design and plastic design is now superseded. Our hope is that the present book will contribute to enlarge the designer's ability by giving him one more tool, namely the theory of limit analysis and design, together with the definition of its range of applicability. Though extremely powerful, this theory remains nothing but a tool,

---

* Published in 1965 by Blaisdell Publishing Co., Waltham, Mass., U.S.A.

to be used by the design engineer within the frame of a wider and more sophisticated design philosophy.

Last but not least, we are pleased to express here our gratitude to the numerous authors whose original works have been used in this book. We have tried to give due acknowledgement to all of them. We apologize for any, involuntary, ommission.

The junior author (M.A.S.), who wrote the adaption in English, is particularly grateful to Professor W. Prager, who read the manuscript and made many valuable suggestions that resulted in appreciable improvement of the text.

List of Symbols

a	length, radius, diameter of a hole, thickness of a web, coefficient
b	length, breadth, radius, coefficient
c	parameter, coefficient
d	specific power of stresses on strain rates, coefficient, distance
e	average axial strain, coefficient, base of natural logarithms
$\dot{\mathbf{e}}$	unit strain rate vector
f	function, reduced force, coefficient
g	gravity acceleration, distance between centers of holes
h	height, distance
j	strength parameter
k	coefficient, orthotropy coefficient, economy coefficient, parameter
k'	orthotropy coefficient (in negative bending)
l, m, n	direction cosines of an outward pointing normal
l	span, length, small side of a rectangle
m	reduced bending moment
m_x, m_y	bending components of the reduced moment tensor in rectangular cartesian x, y axes

m_{xy}	twisting component of the reduced moment tensor in rectangular cartesian x, y axes
m_t	reduced twisting moment
n	reduced axial force, number
n	outward normal vector
p, q, r	direction parameters
p	pressure, distributed load, number of parameters
$\bar{p}$	line load per unit of length
p_s	shake-down pressure
p^*	reduced pressure
p^M	limit pressure of a membrane
q	reduced load
$\dot{q}$	generalized strain rate
r	generic radius
r_1, r_2	principal radii of curvature of a shell
s	safety factor, curvilinear abscissa, distance, reduced principal stress
s_x, s_y, s_z	normal components of the stress deviator
t	time, thickness
u, v, w	components of the displacement vector
v	volume
w	transversal displacement of a plate
x, y, z	orthogonal cartesian coordinates, reduced forces
A	area, coefficient
A_S	sheared area in punching, steel reinforcement cross-sectional area
B	breadth of a stiffener, coefficient
C	coefficient, constant, parameter, capacity
D	power of dissipation, diameter
D_V	power of dissipation per unit volume
E	Young's modulus, thickness
F	function, intensity of a force, bending stiffness of a plate
F	body-force vector
H	thickness, thickness of a core, height of a pressure vessel head
I	stress invariant
K	parameter, coefficient, constant, nodal force
L	length, long side of a rectangle
M	bending moment
M_x, M_y	bending components of the moment tensor
M_{xy}	twisting component of the moment tensor

LIST OF SYMBOLS

M_e	maximum elastic bending moment of a cross-section
M_p	plastic moment
M'_p	plastic moment for negative bending
M_t	twisting moment
M_{tp}	plastic twisting moment
N	axial force
N_p	plastic axial force
O	origin of axes
P	load parameter, concentrated load, point, stress point
P_-	statically admissible load parameter
P_+	kinematically admissible load parameter
P_l	limit load
P_i	limit load for inscribed yield surface
P_c	limit load for circumscribed yield surface, carrying capacity
P_s	service load
P_V	punching load
P_v	live load
Q	generalized stress, "equivalent" shear in plates, line load, point
R	radius, pole of Mohr's circle, joint, region, corner force in plates
S	surface
T	magnitude of edge force in rotating disk
T	surface traction vector
U	elastic strain energy per unit volume
U	displacement vector
V	volume, shear force
V	velocity vector
V_x, V_y, V_z	components of the velocity vector
(**V**)	velocity field
W	weight
X, Y, Z	components of the body-force vector
$\overline{X}, \overline{Y}, \overline{Z}$	components of the surface traction vector
$\mathcal{B}$	power of body forces
$\mathcal{E}$	strain energy
$\mathcal{I}$	stress intensity
$\mathcal{J}$	strain intensity
$\dot{\mathcal{J}}$	strain rate intensity
$\mathcal{P}$	power of applied forces
$\mathcal{R}$	weight ratio
$\mathcal{V}$	reinforcement volume

LIST OF SYMBOLS

$\mathcal{V}_M$	moment volume
$\mathcal{W}$	work of applied forces
div.	divergence of a vector
α, β	angle, coefficient, parameter, ratio
γ	shear strain, affinity coefficient, angle, parameter, weight per unit volume
γ_p	coefficient of plate-beam interaction for square plate
γ_b	coefficient of plate-short beams interaction for rectangular plates
γ_B	coefficient of plate-long beams interaction for rectangular plates
$\gamma_{xy}, \gamma_{yz}, \gamma_{zx}$	shear strains
$\dot{\gamma}_{xy}, \dot{\gamma}_{yz}, \dot{\gamma}_{zx}$	shear strain rates
δ	deflection, elongation, variation, parameter
Δ	modified dissipation rate, variation
ϵ	axial strain
$\dot{\epsilon}$	axial strain rate
(ϵ)	strain tensor
$\epsilon_1, \epsilon_2, \epsilon_3$	principal strains
$(\dot{\epsilon})$	strain rate tensor
$\dot{\boldsymbol{\varepsilon}}$	strain vector in $\epsilon_x, ..., \gamma_{xy}, ...$ space
ζ	parameter
η	line, reduced load, reduced abscissa
η, ξ	coordinates
θ	angle, angle of rotation
θ_{ij}	angle of relative rotation of i and j
$\dot{\theta}$	rotation rate
$\boldsymbol{\theta}$	rotation vector
κ	curvature
$\dot{\kappa}_x, \dot{\kappa}_y$	curvature components of the curvature rate tensor
$\dot{\kappa}_{xy}$	twist component of the curvature rate tensor
λ	non negative scalar, displacement, load parameter
$\dot{\lambda}$	reduced strain rate
μ	slenderness, reinforcement ratio
ν	Poisson's ratio
ξ	reduced abscissa, parameter
π	3.1416
ρ	radius of curvature, radius, reduced radius, cutout factor, efficiency

LIST OF SYMBOLS

σ	normal component of a stress vector
$\sigma_x, \sigma_y, \sigma_z$	normal components of the stress tensor
(σ)	stress tensor
$\sigma_1, \sigma_2, \sigma_3$	principal stress
$\boldsymbol{\sigma}$	stress vector in $\sigma_x, ..., \tau_{xy}, ...$ space
σ_R	reference stress
σ_Y	yield stress in tension or compression
$\sigma_c, \sigma'_{r(\text{cyl})}$	crushing stress of concrete in compression, on cylinders
σ_r	tensile rupture stress of concrete
$(\sigma_{\_})$	statically admissible stress field
Σ	summation
τ	shear stress
$\tau_{xy}, \tau_{yz}, \tau_{zx}$	shear components of the stress tensor
τ_Y	yield stress in pure shear
τ_{oct}	octahedral shear stress
φ	angle, function, shape factor
Φ	function
$\dot{\Phi}$	reduced strain rate
ψ	function, angle
ω	geometrical parameter of a circular cylindrical shell, parameter, angular velocity
$x]$	discontinuity on x
∇	laplacian

Table of Contents

PREFACE v
LIST OF SYMBOLS viii

PART ONE: GENERAL THEORY

Chapter 1: Stress and Strain
    1.1   Stress, strain and strain rate tensors 1
    1.2   Yield conditions 7
    1.3   Problems 12

Chapter 2: Fundamental Concepts and Laws
    2.1   Perfectly plastic solid 13
    2.2   Power of dissipation 14
    2.3   Geometrical representation 14
    2.4   Fundamental assumptions on the power of dissipation . 15
    2.5   Illustrating examples 18
    2.6   Problems 21

Chapter 3: Fundamental Theorems
- 3.1 Basic definitions and statements of theorems 24
- 3.2 Proofs of the theorems 25
- 3.3 Important remarks 29
- 3.4 Elastic-plastic and rigid-plastic bodies 31
- 3.5 Influence of changes of geometry 32
- 3.6 Solutions given by the deformation theory 36
- 3.7 Uniqueness 42
- 3.8 Appendix 42
- 3.9 Problems 46

Chapter 4: General Loading Case
- 4.1 Structures with nonnegligible dead load 48
- 4.2 Loading depending on several parameters 49
- 4.3 Shake-down analysis 49

Chapter 5: Generalized Variables
- 5.1 The concept of generalized variables 58
- 5.2 The general case – choice of the generalized variable . . 63
- 5.3 Eliminating the reactions 64
- 5.4 Obtaining yield conditions in generalized stresses . . . 67
- 5.5 Simplified yield surfaces 95
- 5.6 Discontinuities 104
- 5.7 Appendix 107
- 5.8 Problems 108

PART TWO: APPLICATIONS TO PLATES, SHELLS, AND DISKS

Chapter 6: Metal Plates
- 6.1 Introduction 111
- 6.2 Experimental information 113
- 6.3 Circular isotropic plates 123
- 6.4 Circular orthotropic plates 136
- 6.5 Isotropic rectangular plates 144
- 6.6 Review of other work 155
- 6.7 Deformations of metal plates 156
- 6.8 Minimum-weight design 169
- 6.9 Examples 196
- 6.10 Problems 197

Chapter 7: Reinforced Concrete Plates
- 7.1 Introduction 203
- 7.2 Yield condition and flow rule 207
- 7.3 Discussion of the yield condition 218
- 7.4 The kinematic method (Johansen's fracture line theory) . 225
- 7.5 The static method 265
- 7.6 Minimum-weight of reinforcement 284
- 7.7 Influence of axial force 296
- 7.8 Influence of shear forces. Punching 308
- 7.9 Example of application 310
- 7.10 Problems 316

Chapter 8: Metal Shells
- 8.1 Introduction 325
- 8.2 Experiments on metal shells 326
- 8.3 Circular cylindrical shells axisymmetrically loaded . . . 334
- 8.4 Rotationally symmetric shells 356
- 8.5 Torispherical and toriconical thin pressure vessel heads . 364
- 8.6 Thin-wall beam with circular axis 375
- 8.7 Indications on other problems 388
- 8.8 Minimum-weight design 388
- 8.9 Examples of application 400
- 8.10 Problems 403

Chapter 9: Reinforced Concrete Shells
- 9.1 Introduction 407
- 9.2 Circular cylindrical tank under axisymmetric loading . . 410
- 9.3 Upper bound solutions 420

Chapter 10: Plane Stress and Plane Strain
- 10.1 Introduction 434
- 10.2 Plane stress: perforated disks 439
- 10.2 Notched bars in tension 456
- 10.4 Thin rotating disks 459
- 10.5 Other plane stress problems 469
- 10.6 Plane strain: thick tube 469

SUBJECT INDEX 476

Part One

General Theory

1

Stress and Strain

1.1. Stress, strain, and strain-rate tensors

1.1.1. *Stress tensor*

Consider an elementary parallelepiped at a generic point O of a continuum referred to orthogonal cartesian axes x, y, z (fig. 1.1). Each of the three faces in the reference planes is in general subjected to one normal stress component and two shearing stress components. The state of stress at O is thus characterized by nine components. From one face to the neighboring parallel face, these components experience a small variation. This fact is indicated by the primes shown in fig. 1.1. Rotational equilibrium of the parallelepiped about its axes yields the equality of the shearing stresses :

$$\tau_{xy} = \tau_{yx} \qquad \ldots (x,y,z)^* . \tag{1.1}$$

There thus remain six independent stress components, namely the normal stresses σ_x, σ_y, σ_z and the shearing stresses τ_{xy}, τ_{yz}, τ_{zx}. The *stress tensor*, which will be symbolically denoted by (σ), is completely defined, in the chosen system of coordinates, by these six components. Translational

---

* The three dots, followed by (x, y, z), at the end of a line, mean that similar expressions or equations are obtained by cyclic permutation of x, y, z.

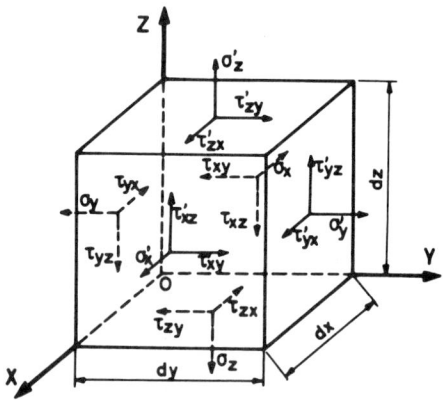

Fig. 1.1.

equilibrium in the directions of the coordinate axes furnishes the three *differential equations of equilibrium:*

$$\frac{\partial \sigma_x}{\partial x} + \frac{\partial \tau_{xy}}{\partial y} + \frac{\partial \tau_{xz}}{\partial z} + X = 0 \qquad \ldots (x, y, z), \tag{1.2}$$

where X, Y, Z are the components of the body force per unit volume.

Consider now the elementary tetrahedron of figure 1.2, with the plane BCD infinitely close to the origin O. Let l, m, n, be the direction cosines of the exterior normal to face BCD of the tetrahedron. Denote by $\overline{X}, \overline{Y}, \overline{Z}$ the components of the stress vector acting on this face. From equilibrium con-

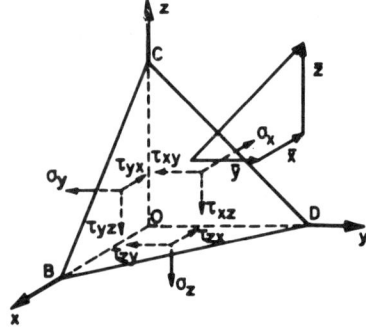

Fig. 1.2.

siderations we obtain

$$\overline{X} = \sigma_x l + \tau_{xy} m + \tau_{xz} n \qquad \begin{array}{l}\ldots(x,y,z)\\ \ldots(l,m,n).\end{array} \qquad (1.3)$$

It can be shown [1.1] that, among all plane elements such as *BCD* at point O, three are subjected solely to normal stresses σ_1, σ_2, and σ_3, respectively. The direction of each of these stresses is orthogonal to the plane determined by the other two directions. The stresses $\sigma_1, \sigma_2, \sigma_3$ are called *principal stresses* and their directions, *principal directions*. By definition, the shearing stresses associated with the principal directions vanish:

$$\tau_{12} = \tau_{23} = \tau_{31} = 0.$$

The stress tensor is completely characterized by the principal directions and the algebraic values of the principal stresses (the sign convention being defined on fig. 1.1 where all components are positive).

There exist various geometrical representations of the stress tensor [1.1], [1.2], [1.3]. Mohr's classical plane representation is particularly useful [1.4]. We assume it to be known. Functions of the components of (σ) that are not altered by rotation of coordinate axes are called invariants. The three fundamental invariants are

$$I_1 \equiv \sigma_x + \sigma_y + \sigma_z, \qquad (1.4)$$

$$I_2 \equiv \sigma_x \sigma_y + \sigma_y \sigma_z + \sigma_z \sigma_x - \tau_{xy}^2 - \tau_{yz}^2 - \tau_{zx}^2, \qquad (1.5)$$

$$I_3 \equiv \sigma_x \sigma_y \sigma_z + 2\tau_{xy} \tau_{yz} \tau_{zx} - \sigma_x \tau_{yz}^2 - \sigma_y \tau_{zx}^2 - \sigma_z \tau_{xy}^2. \qquad (1.6)$$

States of *plane stress*, where one principal stress vanishes, are frequently encountered. Let σ_3 be the vanishing principal stress. All stress vectors (with components $\overline{X}, \overline{Y}, \overline{Z}$) then lie in the plane (O, σ_1, σ_2) of the two other principal directions. With the special sign convention of fig. 1.3 where positive stress components are drawn, the equality of shearing stresses is expressed by $\tau_{xy} = -\tau_{yx}$. Consider a plane surface element with the exterior normal lying in the (O, σ_1, σ_2) plane and making an angle α (clockwise positive) with respect to the x-axis (fig. 1.3). Let σ and τ be the normal and shear components of the stress vector acting on this plane element. From equilibrium considerations, we find

$$\sigma = \sigma_x \cos^2\alpha + \sigma_y \sin^2\alpha + \tau_{xy} \sin 2\alpha, \qquad (1.7)$$

$$\tau = \tfrac{1}{2}(\sigma_y - \sigma_x) \sin 2\alpha + \tau_{xy} \cos 2\alpha. \qquad (1.8)$$

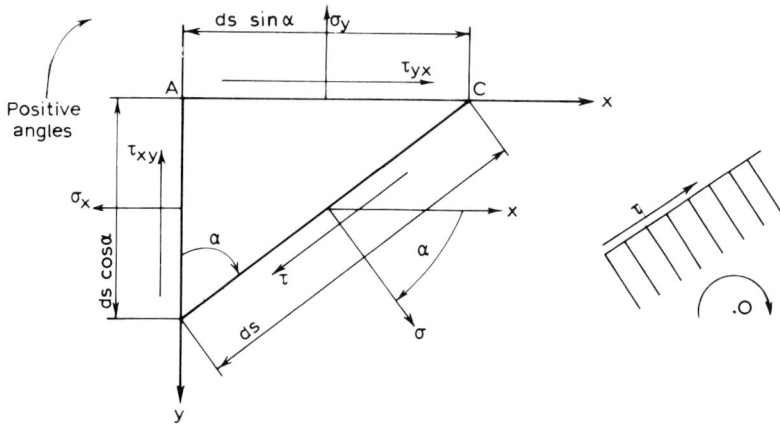

Fig. 1.3.

The principal directions are given by

$$\tan 2\alpha = \frac{2\tau_{xy}}{\sigma_x - \sigma_y}. \tag{1.9}$$

When these directions are used for coordinate axes (0,1) and (0,2), formulas (1.7) and (1.8) become

$$\sigma = \sigma_1 \cos^2\alpha + \sigma_2 \sin^2\alpha, \tag{1.10}$$

$$\tau = \frac{\sigma_2 - \sigma_1}{2} \sin 2\alpha. \tag{1.11}$$

Mohr's circle for a generic state of plane stress is shown on fig. 1.4.

Let D and D_1 be the points representing the stress vectors acting on surface elements normal to the axes Ox and Oy, respectively. Through D and D_1, draw straight lines parallel to Ox and Oy, respectively (that is, parallel to the traces, in the stress plane, of the surface elements on which the respective stress vectors act). Their intersection determines a point R on the circle. This point R is called the *pole* of the stress circle. The direction of the trace of the surface element subjected to a given stress vector is obtained by drawing the straight line through the pole and the point of the circle representing the stress vector.

For example, RA is parallel to the trace of the surface element subjected to the principal stress represented by point A, and RB is parallel to that of σ_2 (represented by point B). The principal stresses are related to the stress

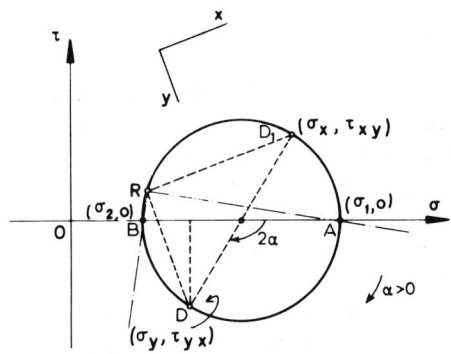

Fig. 1.4.

components $\sigma_x, \sigma_y, \tau_{xy} = -\tau_{yx}$ by*

$$\sigma_1 = \frac{\sigma_x + \sigma_y}{2} + \sqrt{\tfrac{1}{4}(\sigma_x - \sigma_y)^2 + \tau_{xy}^2}, \tag{1.12}$$

$$\sigma_2 = \frac{\sigma_x + \sigma_y}{2} - \sqrt{\tfrac{1}{4}(\sigma_x - \sigma_y)^2 + \tau_{xy}^2}, \tag{1.13}$$

and σ_1 is inclined, with respect to positive direction Ox, by an angle α given by

$$\tan 2\alpha = -\frac{2\tau_{xy}}{\sigma_x - \sigma_y}. \tag{1.14}$$

1.1.2. *Strain tensor*

When a continuum is deformed, a generic point experiences a displacement **U** with components u, v, w with respect to cartesian orthogonal axes x, y, z, respectively. For very small strains, relations between axial strains $\epsilon_x, \epsilon_y, \epsilon_z$, shear strains $\gamma_{xy}, \gamma_{yz}, \gamma_{zx}$, and displacement components are [1.1]:

$$\epsilon_x = \frac{\partial u}{\partial x} \quad \begin{matrix} \ldots (x, y, z) \\ \ldots (u, v, w), \end{matrix} \tag{1.15}$$

---

* Formulas (1.12) and (1.13) also hold when $\sigma_z \equiv \sigma_3 \neq 0$ provided σ_z is a principal stress.

$$\gamma_{xy} = \frac{\partial u}{\partial y} + \frac{\partial v}{\partial x} \qquad \begin{array}{l}\ldots (x,y,z)\\ \ldots (u,v,w).\end{array} \qquad (1.16)$$

The six components $\epsilon_x, \ldots, \gamma_{xy}, \ldots$ completely describe the state of strain at the considered point, and it is found [1.1, 1.3] that they satisfy relations formally identical to those obtained in the analysis of the stress tensor. $\epsilon_x, \epsilon_y, \epsilon_z$ are substitutes for $\sigma_x, \sigma_y, \sigma_z$, respectively, and $\gamma_{xy}/2$, $\gamma_{yz}/2$, and $\gamma_{zx}/2$ for $\tau_{xy}, \tau_{yz}, \tau_{zx}$. The state of strain is thus a tensor, with three principal directions subjected to vanishing shear strain and with three fundamental strain invariants.

Especially important is the state of *plane strain*, with one vanishing principal strain, say ϵ_3. In this case, we have [see Equation (1.12) and (1.13)]

$$\epsilon_1 = \frac{\epsilon_x + \epsilon_y}{2} + \sqrt{\tfrac{1}{4}(\epsilon_x - \epsilon_y)^2 + (\gamma_{xy}/2)^2} \qquad (1.17)$$

$$\epsilon_2 = \frac{\epsilon_x + \epsilon_y}{2} - \sqrt{\tfrac{1}{4}(\epsilon_x - \epsilon_y)^2 + (\gamma_{xy}/2)^2}. \qquad (1.18)$$

Principal directions are given by

$$\tan 2\alpha = -\frac{\gamma_{xy}}{\epsilon_x - \epsilon_y}. \qquad (1.19)$$

Formulas (1.17) to (1.19) still hold when $\epsilon_z \equiv \epsilon_3 \neq 0$, provided ϵ_z is a *principal* strain.

1.1.3. Strain-rate tensor

When plastic flow is considered, the strain components depend on time t and it proves necessary to use *strain rates* defined as follows:

$$\dot{\epsilon}_x = \frac{\partial \epsilon_x}{\partial t} \qquad \ldots (x,y,z), \qquad (1.20)$$

$$\dot{\gamma}_{xy} = \frac{\partial \gamma_{xy}}{\partial t} \qquad \ldots (x,y,z). \qquad (1.21)$$

If **V** is the velocity of a given point ($\mathbf{V} = \partial \mathbf{U}/\partial t$) with components V_x, V_y, V_z, obviously

$$\dot{\epsilon}_x = \frac{\partial}{\partial t}\frac{\partial u}{\partial x} = \frac{\partial}{\partial x}\frac{\partial u}{\partial t} = \frac{\partial \dot{u}}{\partial x} = \frac{\partial V_x}{\partial x} \qquad \begin{array}{l}\ldots (x,y,z) \\ \ldots (u,v,w),\end{array}$$

$$\dot{\gamma}_{xy} = \frac{\partial}{\partial t}\left(\frac{\partial u}{\partial y} + \frac{\partial v}{\partial x}\right) = \frac{\partial}{\partial y}\frac{\partial u}{\partial t} + \frac{\partial}{\partial x}\frac{\partial v}{\partial t} = \qquad (1.22)$$

$$= \frac{\partial \dot{u}}{\partial y} + \frac{\partial \dot{v}}{\partial x} = \frac{\partial V_x}{\partial y} + \frac{\partial V_y}{\partial x} \qquad \begin{array}{l}\ldots (x,y,z) \\ \ldots (u,v,w)\end{array}$$

because

$$V_x \equiv \dot{u} = \frac{\partial u}{\partial t} \qquad \begin{array}{l}\ldots (x,y,z) \\ \ldots (u,v,w).\end{array} \qquad (1.23)$$

The state of strain rate is also a tensor. For example, when Oz is a principal direction, we have (with $\epsilon_3 \equiv \epsilon_z$)

$$\dot{\epsilon}_1 = \frac{\dot{\epsilon}_x + \dot{\epsilon}_y}{2} + \sqrt{\tfrac{1}{4}(\dot{\epsilon}_x - \dot{\epsilon}_y)^2 + (\dot{\gamma}_{xy}/2)^2}, \qquad (1.24)$$

$$\dot{\epsilon}_2 = \frac{\dot{\epsilon}_x + \dot{\epsilon}_y}{2} - \sqrt{\tfrac{1}{4}(\dot{\epsilon}_x - \dot{\epsilon}_y)^2 + (\dot{\gamma}_{xy}/2)^2}, \qquad (1.25)$$

For purely plastic strains occurring with zero volume change, we have

$$\dot{\epsilon}_1 + \dot{\epsilon}_2 + \dot{\epsilon}_3 = 0. \qquad (1.26)$$

1.2. Yield conditions

When the state of stress is uniaxial tension or compression, the yield condition for most metals is

$$\sigma = \pm \sigma_Y. \qquad (1.27)$$

In a multiaxial state of stress, yielding will occur when a certain physical condition related to the state of stress will be satisfied. In uniaxial tension or compression, this condition must reduce to Equation (1.27). For metals, and particularly for mild steel, it has been observed that plastic deformations basically consist of slip in crystals. Hence, it was thought that the maximum shearing stress determined the onset of yielding, which was always to occur for a *fixed* value of the maximum shearing stress. This is Tresca's yield condition.

The value of the maximum shear stress at yielding can be obtained, for instance, from a tensile test where

$$\tau_{max} = \frac{\sigma}{2},$$

and hence,

$$\tau_{max,Y} = \frac{\sigma_Y}{2}. \tag{1.28}$$

In the case of plane stress defined by $\sigma_x = \sigma$, $\sigma_y = 0$, $\tau_{xy} = \tau$ (bending of beams with shear) we have

$$\tau_{max} = \frac{1}{2}\sqrt{\sigma^2 + 4\tau^2}, \tag{1.29}$$

and Tresca's yield condition

$$\tau_{max} = \tau_{max,Y} \tag{1.30}$$

becomes, using Equation (1.28),

$$\sigma^2 + 4\tau^2 = \sigma_Y^2. \tag{1.31}$$

In plane orthogonal cartesian coordinates ($O\sigma$, $O\tau$), relation (1.31) is represented by an ellipse.

In a multiaxial state of stress with principal stresses σ_1, σ_2, σ_3, the magni-

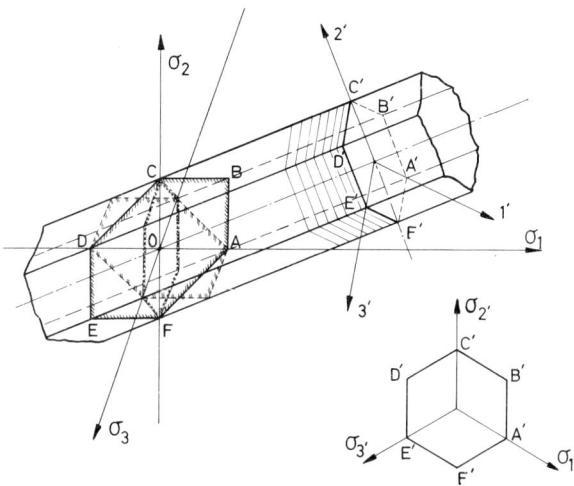

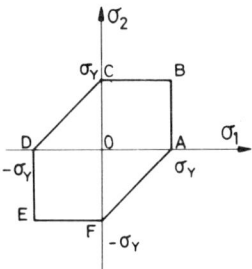

Fig. 1.6.

tude of the maximum shearing stress is the largest among the three absolute values

$$\frac{|\sigma_1 - \sigma_2|}{2}, \frac{|\sigma_2 - \sigma_3|}{2}, \frac{|\sigma_3 - \sigma_1|}{2}.$$

Consequently, condition (1.30) is represented, in cartesian orthogonal axes $(\sigma_1, \sigma_2, \sigma_3)$ by the hexagonal prism formed by the planes with equations

$$\sigma_1 - \sigma_2 = \pm \sigma_Y, \quad \sigma_2 - \sigma_3 = \pm \sigma_Y, \quad \sigma_3 - \sigma_1 = \pm \sigma_Y.$$

This prism is shown on fig. 1.5. Its axis is equally inclined with respect to the coordinate axes. When one of the principal stresses vanishes, say σ_3, the surface reduces to the hexagon obtained by intersecting the prism with the plane $\sigma_3 = 0$ (fig. 1.6). The yield condition becomes

$$\max\left[|\sigma_1|, |\sigma_2|, |\sigma_1 - \sigma_2|\right] = \sigma_Y. \tag{1.32}$$

Note finally that because the magnitude of the maximum shear stress is half the (algebraic) difference of the extreme principal stresses, the intermediate principal stress plays no role in Tresca's yield criterion.

More refined tests have however shown ([1.2, 1.4, 1.5, 1.6] and fig. 1.7) that the circular cylinder circumscribed to the considered hexagonal prism was a more exact "yield surface" for most metals. This surface represents the yield condition of Maxwell, Huber, Hencky, and von Mises, and will be simply called "the von Mises yield condition" in the following. The equation of this surface is

$$\sigma_1^2 + \sigma_2^2 + \sigma_3^2 - \sigma_1\sigma_2 - \sigma_2\sigma_3 - \sigma_3\sigma_1 = \sigma_Y^2. \tag{1.33}$$

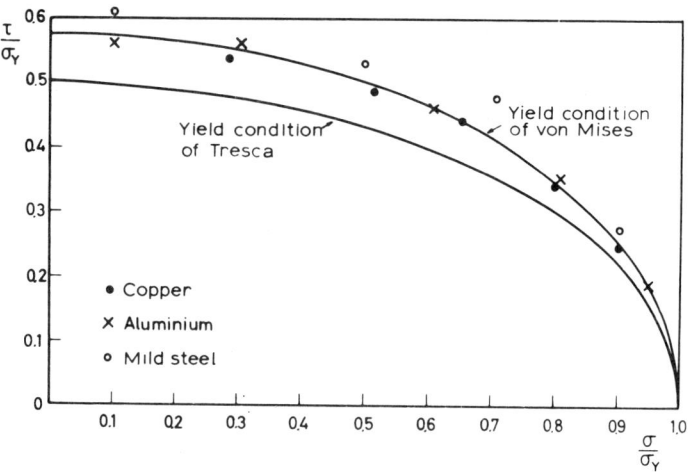

Fig. 1.7.

When the components $\sigma_x, ..., \tau_{xy}, ...$ of the stress tensor are used, condition (1.33) becomes

$$\sigma_x^2 + \sigma_y^2 + \sigma_z^2 - \sigma_x\sigma_y - \sigma_y\sigma_z - \sigma_z\sigma_x + 3\tau_{xy}^2 + 3\tau_{yz}^2 +$$
$$+ 3\tau_{zx}^2 = \sigma_Y^2. \tag{1.34}$$

For a state of plane stress ($\sigma_3 = 0$) condition (1.33) is represented by the ellipse of fig. 1.8 with the equation:

$$\sigma_1^2 + \sigma_2^2 - \sigma_1\sigma_2 = \sigma_Y^2. \tag{1.35}$$

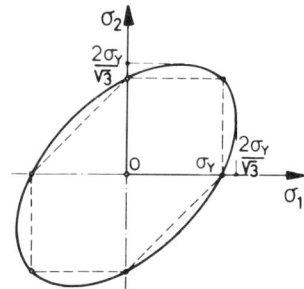

Fig. 1.8.

In the particular state of plane stress where $\sigma_x = \sigma$, $\sigma_y = 0$, $\sigma_z = 0$ (bending of beams with shear), relation (1.34) transforms into

$$\sigma^2 + 3\tau^2 = \sigma_Y^2. \tag{1.36}$$

Therefore, the yield stress in pure shear is

$$\tau_Y = \frac{\sigma_Y}{\sqrt{3}}, \tag{1.37}$$

whence it had the magnitude $\sigma_Y/2$ according to Tresca's condition.

The yield condition of von Mises sometimes is called the criterion of the octahedral shear stress. The left-hand side of Equation (1.33) [or of Equation (1.34)] is indeed proportional to the magnitude of the octahedral shear stress τ_{oct} which acts on the octahedron, the faces of which make equal intercepts on the coordinate axes (fig. 1.9).

From a more general point of view, we call the reference stress σ_R the function of the stress components that must reach the value of the yield stress in simple tension (or compression) for yielding to occur. For the von Mises yield condition, σ_R is the square root of the left-hand sides of Equations (1.33) or (1.34), and the octahedral shearing stress is $\tau_{oct} = (\sqrt{2}/3)\sigma_R$.

The two yield conditions of Tresca and of von Mises are the most commonly accepted for metals, but other conditions could be valid for other materials. Experiments will show which condition best describes a given material. Anyway, yielding can be expected to occur at a given particle when the reference stress at this particle attains the value σ_Y. This reference stress obviously is not a physical stress component but simply a reference value to compare with σ_Y. The general form of the yield condition thus is

$$\sigma_R(\sigma_x, \sigma_y, \sigma_z, \tau_{xy}, \tau_{yz}, \tau_{zx}) = \sigma_Y, \tag{1.38}$$

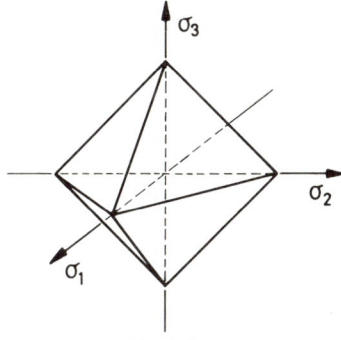

Fig. 1.9.

or, for an isotropic material,

$$\sigma_R(\sigma_1, \sigma_2, \sigma_3) = \sigma_Y.\tag{1.39}$$

1.3. Problems

1.3.1. Show that the square of the octahedral shearing-stress magnitude is proportional to the elastic-distortion energy density (that is the difference between the total elastic energy density and the elastic energy density associated with volume change). *Answer*: $\mathcal{E}_{dist} = \frac{3}{2}\left[(1+\nu)/E\right]\tau_{oct}^2$.

1.3.2. For a closed thin tube with radius R and thickness t subjected to internal pressure p, find the maximum shearing stress and corresponding surface elements. *Answer*: $\tau_{max} = pR/2t$. The corresponding surface elements are parallel to the axis of the tube and form angles of 45° with the radial direction.

1.3.3. Show that, in a state of plane stress, the expression $(\sigma_x - \sigma_y)^2 + 4\tau_{xy}^2$ is an invariant.

References

[1.1] S.P. TIMOSHENKO, J.N. GOODIER, *Theory of Elasticity*, McGraw-Hill, New-York, 1951.
[1.2] A.A. ILIOUCHINE, *Plasticité*, Eyrolles, Paris, 1956.
[1.3] L. BAES, *Resistance des Matériaux*, vol. 1, Lamertin, Brussels, 1934.
[1.4] A. NADAI, *Theory of Flow and Fracture of Solids*, vol. 1, McGraw-Hill, New York, 1950.
[1.5] W. SAUTER, A. KOCHENDÖRFER, V. DEHLINGER, "Über die Gesetzmässigkeiten der plastischen Verformung von Metallen unter einem mehrachsigen Spannungszustand", (a dissertation) Stuttgart, 1952.
[1.6] J. MARIN, A.B. WISEMAN, "Plastic Stress-Strain Relations for Combined Tension and Torsion", *NACA Technical Note* 2737, 1952.
[1.7] G.I. TAYLOR, H. QUINNEY, "The Plastic Distortion of Metals", *Phil. Trans. Roy. Soc. London*, 230: 323, 1931.

2

Fundamental Concepts and Laws

2.1. Perfectly plastic solid

We are now able to define a perfectly plastic solid as follows: solid will be called perfectly plastic if it can undergo unlimited plastic deformations under constant reference stress when it is subjected to a homogeneous state of stress with $\sigma_R = \sigma_Y$.

The value σ_Y is well defined for each material in a given environment, and is the limiting value that σ_R cannot exceed. States of stress with $\sigma_R > \sigma_Y$ are not possible.

In the following we shall frequently be interested in *purely plastic* strains and strain rates and in the corresponding displacements and velocities, and we shall disregard the corresponding elastic elements (except when the contrary is explicitly stated).

When dealing with *incipient plastic flow*, we assume that strains remain very small. Hence, strains and displacement are related through eqs. (1.15) and (1.16), whereas strain rates derive from displacement rates (or velocities) through relations (1.22) and (1.23).

2.2. Power of dissipation

During incipient plastic flow at a given particle, where the state of stress is described by $(\sigma_x,...,\tau_{xy},...)$ and the state of strain rate by $(\epsilon_x,...,\gamma_{xy},...)$, the power of the stresses per unit volume of material is

$$d = \sigma_x \dot\epsilon_x + ... + \tau_{xy}\dot\gamma_{xy} + ... , \qquad (x,y,z) . \qquad (2.1)$$

For purely plastic strains, this power is dissipated in heat during plastic flow. Therefore, it is called "power of dissipation". It is essentially *positive*.

2.3. Geometrical representation

In a six-dimensional Euclidean space, consider a rectangular cartesian coordinate system with the origin O. The vectors $\boldsymbol{\sigma}$ and $\dot{\boldsymbol{\epsilon}}$ that have the components $\sigma_x, \sigma_y, \sigma_z, \tau_{xy}, \tau_{yz}, \tau_{zx}$ and $\dot\epsilon_x, \dot\epsilon_y, \dot\epsilon_z, \dot\gamma_{xy}, \dot\gamma_{yz}, \dot\gamma_{zx}$, respectively, with respect to this coordinate system, will be called the *stress vector* and the *strain rate vector*, and the point P with the radius vector $\mathbf{OP} = \boldsymbol{\sigma}$ will be called the *stress point*. (An extension of these definitions to generalized stresses and strains will be discussed in Section 5.1.) Eq. (2.1) indicates an important property of this geometrical representation: the specific power of the stress on the strain rate is given by the scalar product of the vectors $\boldsymbol{\sigma}$ and $\dot{\boldsymbol{\epsilon}}$:

$$d = \boldsymbol{\sigma} \cdot \dot{\boldsymbol{\epsilon}} . \qquad (2.2)$$

If elastic strain rates are neglected so that $\dot{\boldsymbol{\epsilon}}$ represents the plastic strain rate, this scalar product is the *specific rate of dissipation*, which will be denoted by D.

The surface with the equation

$$\sigma_R(\sigma_x,...,\tau_{xy},...) - \sigma_Y = 0 \qquad (2.3)$$

is called the *yield surface*, because states of stress at the yield limit are represented by stress points on this surface. Note that for the perfectly plastic materials considered here, the reference stress depends only on the state of stress but not on the state of strain because these materials do not exibit work hardening. The yield surface is therefore a fixed surface in our six-dimensional space. The yield surface divides this space into two regions: the region $\sigma_R \leqslant \sigma_Y$ which consists of stress points representing attainable states of stress, and the region $\sigma_R > \sigma_Y$ which corresponds to states of stress that cannot be attained in the considered perfectly plastic material. For convenient reference, interior points of the attainable region will be described as lying *inside the yield surface*, while stress points representing unattainable states of stress will be described as being *outside the yield surface*. The coordinate origin, which represents the stress free state, must lie inside the yield surface because the material will not yield in the absence of stress.

2.4. Fundamental assumptions on the power of dissipation

As far as plastic deformations are concerned, we consider that a given material is characterized by its *constitutive equations* relating the stress tensor (σ) to the tensor $(\dot{\epsilon})$ of the plastic strain rates. Substitution of these expressions for $\sigma_x, ..., \tau_{xy}, ...,$ into relation (2.1) makes the specific power of dissipation D a function of the strain rate components only.

We further assume [2.1, 2.2] that
1. *The specific power of dissipation is a single-valued function of the strain rate components.*
2. *The dissipation function $D[(\dot{\epsilon})]$ is homogeneous of the order one.*

According to this second assumption, multiplying every strain rate component by a positive scalar λ means multiplying the power of dissipation by the factor λ:

$$D[(\lambda\dot{\epsilon})] = \lambda D[(\dot{\epsilon})], \qquad \lambda > 0.$$

Using the geometrical representation introduced in Section 2.3, we rewrite the preceding relation in the form

$$D(\lambda\dot{\epsilon}) = \lambda D(\dot{\epsilon}), \qquad \lambda > 0. \tag{2.4}$$

This second assumption expresses the inviscid nature of the considered perfectly plastic material.

If the strain rates at a particle are specified to within a common positive factor, they are said to determine a *local flow mechanism* at this particle. The strain rate vectors $\dot{\epsilon}$ and $\lambda\dot{\epsilon}$ in eq. (2.4), when λ is positive, thus determine the same local flow mechanism, and this mechanism is completely defined by the unit vector $\dot{e}$ along $\dot{\epsilon}$.

Let a local flow mechanism $\dot{e}$ be given. According to assumtion 1 above, this flow mechanism determines a unique specific rate of dissipation $D(\dot{e})$. It follows from eq. (2.2) that a state of stress σ for which

$$\sigma \cdot \dot{e} < D(\dot{e}) \tag{2.5}$$

cannot produce the flow mechanism $\dot{e}$.

Now, the stress points with radius vectors σ satisfying the relation (2.5) are interior points of the half-space that contains the origin 0 and is bounded by a plane normal to $\dot{e}$ at the distance $D(\dot{e})$ from O. As we let the vector $\dot{e}$ of the flow mechanism rotate about the origin, the interior points that all corresponding half-spaces have in common are the points inside the yield surface, and the bounding planes of the half-spaces envelope the yield surface (fig. 2.1).

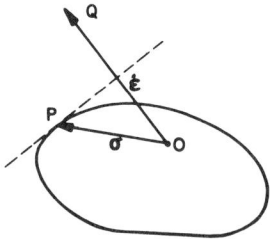

Fig. 2.1.

As the boundary of the domain common to all half-spaces, the yield surface is convex. It may:

1. be strictly convex, that is, it may have a continuously turning normal (as does the von Mises yield surface); there then is one-to-one correspondence between the stress point and the flow mechanism (fig. 2.2).
2. exhibit vertices as in fig. 2.3; at a vertex P, all outward pointing vectors that lie on or within the cone of normals define possible flow mechanisms; the vector $\dot{e}$ of a flow mechanism still determines the stress point P but the converse is no longer true.
3. exhibit flat parts as in fig. 2.4 (as does the Tresca yield surface); the stress point then determines the flow mechanism but the converse is not true when $\dot{e}$ is normal to a flat part of the yield surface.

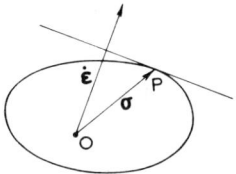

Fig. 2.2.

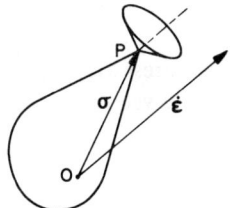

Fig. 2.3.

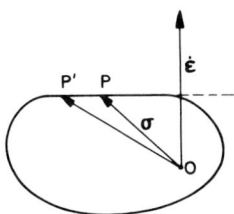

Fig. 2.4.

Despite the possible lack of one-to-one correspondence between the stress vector $\boldsymbol{\sigma}$ and the vector $\dot{\boldsymbol{e}}$ of the flow mechanism, it is readily verified that for all kinds of yield surfaces the specific dissipation $D = \boldsymbol{\sigma} \cdot \dot{\boldsymbol{\varepsilon}}$ is a single valued function of the strain rate vector $\dot{\boldsymbol{\varepsilon}}$.

The function $\sigma_R(\sigma_x,...,\tau_{xy},...)$ represented by the yield surface with equation $\sigma_R(\sigma_x,...,\tau_{xy},...) = \sigma_Y$ is a potential function for the strain rates because normality of $\dot{\boldsymbol{\varepsilon}}$ to the surface at the stress point P gives

$$\dot{\epsilon}_x = \lambda \frac{\partial \sigma_R}{\partial \sigma_x} \quad \ldots (x,y,z),$$

$$\dot{\gamma}_{xy} = \lambda \frac{\partial \sigma_R}{\partial \sigma_{xy}} \quad \ldots (x,y,z), \tag{2.6}$$

where λ is a positive scalar factor. For this reason the *normality law* just cited, with its generalizations to vertices and flats, is also called the *plastic potential flow law*. It is widely accepted, though introduced in various manners [2.4, 2.5, 2.6].

In the following, we call *flow mechanism of a body* (or structure) a field of plastic strain rate vectors $\dot{\boldsymbol{\epsilon}}$ whose magnitudes are defined to within a common positive scalar factor, whereas a vector $\dot{\boldsymbol{\epsilon}}$ (or $\dot{\mathbf{e}}$) is called *local flow mechanism*. Note that a given stress field may be related by the normality law to several fields of strain rate vectors $\dot{\boldsymbol{\epsilon}}$. These fields have in common the field of unit vectors $\dot{\mathbf{e}}$.

When the solid is isotropic, at least as far as its yield condition is concerned, the normality law, expressed in the $(\sigma_x,\ldots,\tau_{xy},\ldots)$ and $(\dot{\epsilon}_x,\ldots,\dot{\gamma}_{xy},\ldots)$ spaces, ensures that principal directions of the stress tensor and of the strain rate tensor coincide. Hence, when principal directions of these two tensors are known, or otherwise determinable, spaces of principal stresses and strain rates may as well be used. As a rule all spaces in which the dissipation is unambiguously determined may be used. All properties obtained above remain valid.

2.5. Illustrating examples

Let us first express the power of dissipation for the von Mises yield condition. We use the principal stresses and strain rates as components of the vectors $\boldsymbol{\sigma}$ and $\dot{\boldsymbol{\epsilon}}$ (fig. 2.5). The yield surface is the circular cylinder with the eq. (1.33). Let P be a generic stress point on the surface and C the foot of the perpendicular from P on the axis of the cylinder. The line CP is normal to the cylinder at P. According to relation (2.3) the power of dissipation corresponding to the stress vector $\boldsymbol{\sigma} = \mathbf{OP}$ will be given by the modulus $|\dot{\boldsymbol{\epsilon}}|$ of the strain rate vector $\dot{\boldsymbol{\epsilon}} = \mathbf{OQ}$ (parallel to line CP) multiplied by the projection of $\mathbf{OP}$ on the direction of line $\mathbf{OQ}$. This projection has the length of CP, that is the magnitude $\sqrt{2/3} \cdot \sigma_Y$ of the radius of the cylinder. Hence,

$$D = \sqrt{2/3}\sigma_Y |\mathbf{OQ}| = \sqrt{2/3}\sigma_Y \sqrt{\dot{\epsilon}_1^2 + \dot{\epsilon}_2^2 + \dot{\epsilon}_3^2}. \tag{2.7}$$

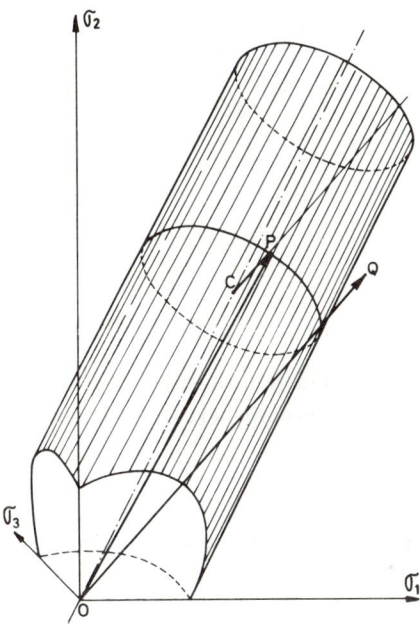

Fig. 2.5.

Consider next Tresca's yield condition. When the three principal stresses are taken as rectangular cartesian coordinates in stress space, this yield condition is represented by a hexagonal prism (fig. 1.5). Fig. 2.6 shows the normal cross section through a point P of this prism that does not lie on an edge. The normal to the prismatic surface at P is the perpendicular η in the cross-sectional plane to the corresponding side of the regular hexagon. The projection of $\boldsymbol{\sigma} = \mathbf{OP}$ on η is the segment NP. Its magnitude is that of the distance from O to any side of the hexagon, namely $\sigma_Y/\sqrt{2}$. Consequently, we have

$$D = \boldsymbol{\sigma} \cdot \dot{\boldsymbol{\varepsilon}} = \frac{\sigma_Y}{\sqrt{2}} |\dot{\varepsilon}| . \tag{2.8}$$

When the stress point lies on an edge of the yield prism (e.g., P' in fig. 2.6) and when the direction of the strain rate vector is intermediate between those of the normals to the adjacent planes (as η' on fig. 2.6), relation (2.8) is no longer valid. A strain rate vector with direction η' can be considered as re-

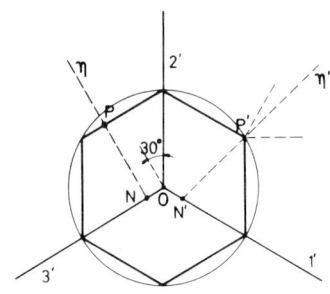

Fig. 2.6.

sulting from the linear combination of strain-rate vectors corresponding to adjacent planes [2.6, 2.7]. Hence, we have the general expression

$$\dot{\varepsilon} = \alpha \dot{e}_a + \beta \dot{e}_b \, , \tag{2.9}$$

where $\dot{e}_a$ and $\dot{e}_b$ are unit strain rate vectors for two adjacent planes, and α and β are arbitrary nonnegative scalar factors. From eqs. (2.3), (2.8), and (2.9) we obtain

$$D = \frac{\sigma_Y}{2} (\alpha + \beta) \, . \tag{2.10}$$

Clearly, from relation (2.10) it is seen that the dissipation depends on the orientation of $\dot{\varepsilon}$, except for vanishing α or β [when relation (2.8) is valid].

The preceding results can be more directly obtained when remembering that, as shown in Section 2.4, the dissipation for a unit vector $\dot{e}$ is measured by the distance from the origin to the tangent plane to the yield surface at the stress point P. This distance is $\sigma_Y \sqrt{2/3}$ for von Mises' condition. It is $\sigma_Y/\sqrt{2}$ for Tresca's, except at points on the edges where it varies with the inclination of the tangent plane.

We finally want to show that, for the yield conditions of both von Mises and Tresca, *plastic flow occurs with no volume change*. To this purpose, it is sufficient to prove that

$$\dot{\epsilon}_1 + \dot{\epsilon}_2 + \dot{\epsilon}_3 = 0 \, . \tag{2.11}$$

As we have seen, the yield condition of von Mises is represented by a cylinder in the space of principal stresses, whereas Tresca's yield condition is represented by an hexagonal prism. But in both cases, the generatrices are normal to the plane with equation

$$\sigma_1 + \sigma_2 + \sigma_3 = 0. \tag{2.12}$$

Hence, any strain rate vector must be parallel to this plane and, consequently, its components must satisfy eq. (2.11).

2.6. Problems

2.6.1. A perfectly plastic, incompressible solid obeying the von Mises yield condition is in a state of both plane stress ($\sigma_3=0$) and plane strain ($\dot{\epsilon}_3=0$). Determine its yield limit σ_1. *Answer*: $\sigma_1 = \pm(2/\sqrt{3})\sigma_Y$.

2.6.2. Apply the result of Problem 2.6.1 to find the ultimate plastic bending moment of a very long rectangular plate with thickness t, subjected to cylindrical bending. *Answer*: $M_p = 1.155 \, \sigma_Y t^2/4$.

2.6.3. A very thin cylindrical shell with radius R and thickness t (under membrane stresses only) obeys the von Mises yield condition. It is subjected to uniform internal pressure p. Find its yield pressure
  1. if it possesses heads.
  2. if it is in plane strain (vanishing axial strain, incompressible material).
  3. if the axial force vanishes.
Answer:
  1. $p = (2/\sqrt{3})(\sigma_Y t/R)$,
  2. $p = (2/\sqrt{3})(\sigma_Y t/R)$,
  3. $p = \sigma_Y t/R$.

2.6.4. Give the graphical representation of the yield conditions of von Mises and Tresca, respectively, with reference to principal axes O, σ_1, σ_2 for a perfectly plastic incompressible solid in plane strain. *Answer*: See fig. 2.7.

2.6.5. Same problem as in Problem 2.6.4, but for von Mises' condition only and with reference to cartesian orthogonal axes x, y, z such that $\dot{\epsilon}_z = \dot{\gamma}_{xz} = \dot{\gamma}_{yz} = 0$. *Answer*: $(\sigma_x - \sigma_y)^2 + 4\tau_{xy}^2 = \frac{4}{3}\sigma_Y^2$, (fig. 2.8).

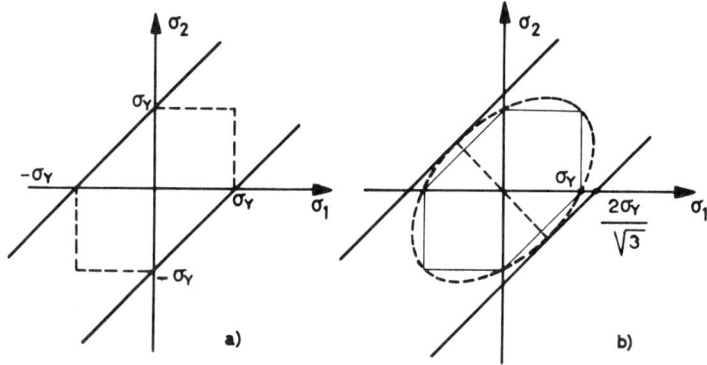

Fig. 2.7. (a) Tresca, (b) von Mises.

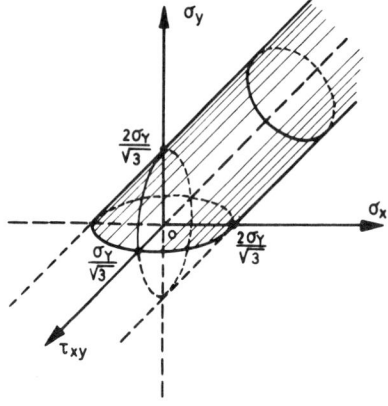

Fig. 2.8.

References

[2.1] A.A.GVOZDEV, "The Determination of the Value of the Collapse Load for Statically Indeterminate Systems Undergoing Plastic Deformation", *Proc. of the Conf. on Plastic Deformations*, December 1936, Akademia Nauk S.S.S.R., Moscow; translated by R.M.HAYTHORNTHWAITE, *Int. J. of Mech. Sci.*, **1**: 322, 1960.

REFERENCES

[2.2] W.PRAGER, "Théorie générale des états limites d'équilibre", *J. de Mathématiques pures et appliquées*, **34**: 4, 1955.

[2.3] W.PRAGER, "The General Theory of Limit Design", *Proc. 8th Int. Cong. Appl. Mech.*, Istanbul, 1952, **2**: 65, 1956.

[2.4] R.von MISES, "Mechanik der plastischen Formänderung von Kristallen", *Zeitschrift angew. Math. Mech.*, **8**: 161, 1928.

[2.5] D.C.DRUCKER, "A More Fundamental Approach to Plastic Stress-Strain Relations", *Proc. 1st U.S. Nat. Congr. Appl. Mech.*, Chicago, 1951, 487 (J.W.Edwards, Ann Arbor, Mich., 1952).

[2.6] W.T.KOITER, "Stress-Strain Relations, Uniqueness and Variational Theorems for Elastic-Plastic Materials with a Singular Yield Surface", *Quart. of Appl. Math.*, **11**: 350, 1953. See also J.L.SANDERS, "Plastic Stress-Strain Relations Based on Infinitely Many Plane Loading Surfaces", *Proc. 2nd U.S. Nat. Congr. Appl. Mech.*, Ann Arbor, Mich., 1954, 455, A.S.M.E., New York, 1955.

[2.7] W.PRAGER, "On the Use of Singular Yield Conditions and Associated Flow Rules", *J. of Appl. Mech.*, **20**: 317, 1953.

3

Fundamental Theorems

3.1. Basic definitions and statements of theorems

Let a structure be subjected to a system of loads that are increased *quasistatically* and *in proportion*, starting from zero. Here the term "quasistatic" indicates that the loading process is sufficiently slow for all dynamic effects to be disregarded. The term "in proportion" signifies that the ratio of the intensities of any two loads remains constant during the loading process*. The points of application and the lines of action of the loads, and the constant ratios of their intensities will be said to determine the *type of loading*. Choosing one of the loads, we use its magnitude P as a measure for the *intensity of loading*. The variable P will also be called the *loading parameter*.

For beams and frames the transition from purely elastic behavior to contained plastic deformation and to unrestricted plastic flow is readily studied (see Com. V, Section 3.2). For more complex structures, however, this becomes quite difficult, and the emphasis is on the direct determination of the *limit state* in which the plastic deformation in the plastic zones is no longer contained by the adjacent nonplastic zones and the structure begins to

---

* The terms "proportional loading", "simple loading", and "radial loading" are used as alternatives to "loading in proportion".

flow under constant loads. The intensity of loading for this limit state is called the *limit load*; it will be denoted by P_l.

Limit analysis is concerned with these limits states. The limit state of incipient unrestrained plastic flow is characterized by two facts:

1. The stresses are in internal equilibrium, in equilibrium with the applied loads P, and nowhere do they violate the yield inequality $\sigma_R \leqslant \sigma_Y$. A stress field of this kind is called *statically admissible*.

2. The flow mechanism satisfies the kinematical boundary conditions of the body, and for energy balance, the power of the applied loads P is equal to the (positive) power dissipated in plastic flow. A mechanism of this kind is called *kinematically admissible*.

For a given type of loading, there is an infinity of statically admissible stress fields. Each of these fields corresponds to a certain intensity of loading, which will be denoted by P_-.

Similarly, for a given kinematically admissible mechanism and a given type of loading, an intensity of loading P_+ can be defined in such a manner that the power of the loads at this intensity of loading equals the power of dissipation in the yield mechanism.

The fundamental theorems of limit analysis can then be stated as follows:
Theorem 1: *Statical (or lower bound) theorem: The limit load P_l is the largest of all loads P_- corresponding to statically admissible stress fields.*
Theorem 2: *Kinematical (or upper bound) theorem: The limit load P_l is the smallest of all loads P_+ corresponding to kinematically admissible mechanisms.*

3.2. Proofs of the theorems

We shall prove* the two fundamental theorems for the rigid perfectly plastic body (which can be regarded as an elastic perfectly plastic body with infinitely large elastic modulus). We shall then show that as far as limit loads are concerned, this idealization of the actual material is as satisfactory as the elastic perfectly plastic scheme. We assume proportional loading. Proofs are based on two preliminary theorems:
Theorem 3: *Virtual work: If a continuum is in equilibrium, the virtual work of external forces (surface tractions and body forces) on any (infinitesinal)*

---

* The following proofs are completely general, contrary to those given in Com. V., Section 3.4, for the particular case of beams and frames.

virtual displacement field compatible with the kinematic boundary conditions is equal to the virtual work of the internal forces (stresses) on the virtual strains corresponding to the considered virtual displacements.

The proof of this classical theorem is given in Section 3.8. A few remarks are worth noting:

1. If the structure under consideration is rigid, it must have at least one degree of freedom to admit nonvanishing virtual displacements. The work of the stresses on the rigid body displacements is zero, and the total work of the external forces is also equal to zero since they form an equilibrium system.
2. If the continuum is deformable, virtual deformations need only be compatible and infinitesimal; that is, to derive from a sufficiently continuous* field of infinitesimal displacements satisfying the kinematic boundary conditions.
3. For incipient flow, displacements and velocities are proportional to each other and the theorem may be rephrased as a theorem of virtual power.

Theorem 4: *Maximum dissipation: The power of dissipation $D(\dot{\varepsilon})$ associated by the normality law to a given local flow mechanism $\dot{\varepsilon}$ is larger than or at least equal to the (fictitious) power dissipated in this mechanism by any attainable state of stress.*

When referring to a local flow mechanism $\dot{\varepsilon} = \mathbf{OQ}$, (fig. 3.1), this theorem expresses the inequality

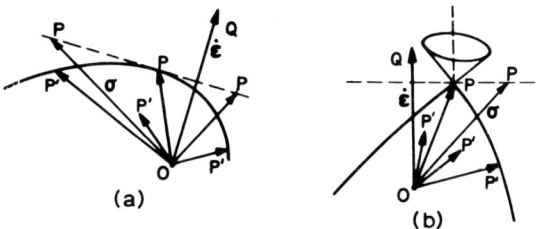

Fig. 3.1.

---

* Continuity conditions will be discussed in Section 5.6.

$$OQ \cdot OP \geqslant OQ \cdot OP' \qquad (3.1)$$

all possible stress points P' being on or within the yield surface. Inequality (3.1) is a direct consequence of convexity of the yield surface, coupled with the normality law. As inequality (3.1) is valid at every particle, it may be integrated over the entire volume of the body. Denoting by $(\dot{\varepsilon})$ the strain rate field, (σ) the stress field corresponding to $(\dot{\varepsilon})$ by the normality law* and (σ_*) any attainable stress field, we have

$$\text{Power of } (\sigma) \text{ on } (\dot{\varepsilon}) \geqslant \text{Power of } (\sigma_*) \text{ on } (\dot{\varepsilon}). \qquad (3.2)$$

For the sake of brevity, we shall symbolically write

$$\mathcal{P}(\sigma, \dot{\varepsilon})$$

to mean: power of the field (σ) on the field $(\dot{\varepsilon})$. As the left-hand side of the inequality (3.2) is the power of dissipation $\int_v D[(\dot{\varepsilon})]\, dv$ of the mechanism, we finally have

$$\int_v D[(\dot{\varepsilon})]\, dv \geqslant \mathcal{P}(\sigma, \dot{\varepsilon}). \qquad (3.3)$$

We now proceed to prove the fundamental theorems.

Theorem 1: *Statical theorem: Any load P_- corresponding to a statically admissible stress field is smaller than or at most equal to the limit load P_l.*

Let (σ_-) be a statically admissible stress field and P_- the corresponding load. In the limit state of incipient unrestrained plastic flow, let (σ) be the actual stress field, and (V) and $(\dot{\varepsilon})$ the velocity and strain rate fields. Apply the theorem of virtual power to the stress fields (σ_-) and (σ) with the deformation state described by (V) and $(\dot{\varepsilon})$. We have, respectively,

$$\mathcal{P}(\sigma_-, \dot{\varepsilon}) = \mathcal{P} \text{ of } P_- \text{ on } (V), \qquad (a)$$

$$\mathcal{P}(\sigma, \dot{\varepsilon}) = \mathcal{P} \text{ of } P_l \text{ on } (V). \qquad (b)$$

As (σ) corresponds to $(\dot{\varepsilon})$ by the normality law, the left-hand side of eq. (b) is the dissipation $\int_v D[(\dot{\varepsilon})]\, dv$. We substract from the two sides of eq. (b)

---

* It might not be unique, but the dissipation would remain uniquely determined by the $(\dot{\varepsilon})$ field.

the corresponding sides of eq. (a), note that $(\boldsymbol{\sigma}_-)$ is a possible stress field and hence apply eq. (3.2) to obtain

$$\mathcal{P} \text{ of } P_l \text{ on } (\mathbf{V}) \geq \mathcal{P} \text{ of } P_- \text{ on } (\mathbf{V}). \tag{c}$$

Because the velocity field $(\mathbf{V})$ is common and the systems of loads differ only by a positive scalar factor, the inequality (c) furnishes immediately

$$P_- \leq P_l. \tag{3.4}$$

Theorem 2: *Kinematical theorem: Any load P_+ corresponding to a kinematically admissible mechanism is larger than or at least equal to the limit load.*

Let $(\mathbf{V}_+)$ and $(\dot{\boldsymbol{\varepsilon}}_+)$ be the considered kinematically admissible velocity and strain rate fields. The corresponding load P_+ is given, by definition, by the relation

$$\mathcal{P} \text{ of } P_+ \text{ on } (\mathbf{V}_+) = \int_v D[(\dot{\boldsymbol{\varepsilon}}_+)] \, dv, \tag{a}$$

where the dissipation D is a (well defined) single valued function of $(\dot{\boldsymbol{\varepsilon}}_+)$ through the yield condition and the normality law. Let $(\boldsymbol{\sigma})$ be the actual stress field at the limit state, in equilibrium with P_l. The theorem of virtual powers, applied to the field $(\boldsymbol{\sigma})$ and the fields $(\mathbf{V}_+)$ and $(\dot{\boldsymbol{\varepsilon}}_+)$, gives

$$\mathcal{P} \text{ of } P_l \text{ on } (\mathbf{V}_+) = \mathcal{P}(\boldsymbol{\sigma}, \dot{\boldsymbol{\varepsilon}}_+). \tag{b}$$

Because (σ) is an attainable stress field that as a rule does not correspond to the field $(\dot{\varepsilon}_+)$ by the normality law, relation (3.3) gives

$$\int_v D[(\dot{\varepsilon}_+)] dv \geq \mathcal{P}(\boldsymbol{\sigma}, \dot{\varepsilon}_+). \tag{c}$$

Comparison of eqs. (a), (b), and (c) shows that

$$\mathcal{P} \text{ of } P_l \text{ on } (\mathbf{V}_+) \leq \mathcal{P} \text{ of } P_+ \text{ on } (\mathbf{V}_+). \tag{d}$$

Because the field $(\mathbf{V}_+)$ is common and the systems of loads differ only by a positive factor, the inequality (d) immediately furnishes

$$P_l \leq P_+ . \tag{3.5}$$

Note finally that it is not necessary that the sets of values of P_- and P_+ be continuous. The load P_l corresponding to a state that is simultaneously statically and kinematically admissible, belongs to both sets and is thus their unique common bound.

3.3. Important remarks

3.3.1. *Exact value of the limit load (complete solution)*

Assume that we have found a statically admissible stress field and a kinematically admissible mechanism that correspond to the same load P. According to the two fundamental theorems we have $P \leq P_l$ and $P \geq P_l$. Hence, $P = P_l$, and we have obtained the exact limit load. This situation occurs most often when it is possible to associate a statically admissible stress field and a kinematically admissible mechanism by the plastic potential flow law. The work equation defining P_+ can then be regarded as a virtual work equation expressing the equilibrium of the associated stress field. Consequently $P_+ = P_-$, and denoting by P this common value, we have $P = P_l$.

The combined theorem is thus as follows:

Theorem 5: *When it is possible to associate by the plastic potential flow law a statically admissible stress field and a kinematically admissible mechanism, the load P corresponding simultaneously to both fields is the exact limit load P_l.*

The two fields above form what is called a "complete solution" of the limit analysis of the structure. In practical problems, one starts either from a mechanism or from a statically admissible stress field, and tries to obtain the other field. Examples for this technique are given in Part II of this volume.

3.3.2. *Addition or subtraction of material*

If the dimensions of a perfectly plastic structure are increased without changing the nature of the material, and if the dead weight of the additional material is neglected, the stress field consisting of vanishing stresses in the additional material and of the stress field at the limit state in the original structure is statically admissible for the modified structure. According to the statical theorem we can hence state that *increasing the dimensions of a perfectly plastic structure cannot result in a lower limit load.*

Similarly, it could be shown that *decreasing the dimensions of a perfectly plastic structure cannot result in a larger limit load.*

3.3.3. *Residual stresses*

In the preceding sections, no assumption has been made concerning an initial stress-free state. The possible presence of residual stresses does not interfere with the proofs of the theorems, nor the existence of slight initial deformations, provided these deformations do not significantly change the geometry of the structure so that the equilibrium conditions can be set up without taking account of these deformations. Hence we can state that *unknown initial stresses and deformations have no effect on the limit load provided they do not significantly alter the geometry of the structure.*

For example, a slight settling of the supports of a continuous beam, or a small permanent twist of a beam subjected to bending, or residual stresses generated by rolling or welding, do not influence the limit load, but only the load at which the behavior of the structure ceases to be elastic (provided the plastic properties of the material are preserved).

Numerous experiments (see [3.1] and also Com. V., Section 3.3.8) have supported this property, which has been implicitly used by practicing engineers for a long time.

3.3.4. *Constancy of the stresses during plastic flow*

Considering an elastic perfectly plastic body we want to prove that *if all changes of geometry are disregarded, all stresses remain constant during the unrestrained plastic flow.*

Disregarding changes of geometry is justified because we are concerned exclusively with the *incipient* unrestrained plastic flow, of infinitesimal magnitude. This flow occurs under constant load P_l. Assume some changes $(\dot{\sigma})$ in the stress field during the flow. They are in equilibrium with vanishing load changes, $(\dot{P})$. Apply the theorem of virtual powers to these variations of loads and stresses with the velocity and strain rate fields at the limit state, (V) and $(\dot{\varepsilon})$, respectively. We have

$$\mathcal{P} \text{ of } (\dot{P}) \text{ on } (V) = \mathcal{P}(\dot{\sigma}, \dot{\varepsilon}) . \tag{3.6}$$

Because all $\dot{P}$ vanishes, the left-hand side of eq. (3.6) is equal to zero, and we obtain

$$\mathcal{P}(\dot{\sigma}, \dot{\varepsilon}) = 0 . \tag{3.7}$$

We divide the total strain rates into their elastic and plastic parts as follows:

$$\dot{\varepsilon}_x = \dot{\varepsilon}_x^e + \dot{\varepsilon}_x^p \ldots ; \qquad \dot{\gamma}_{xy} = \dot{\gamma}_{xy}^e + \dot{\gamma}_{xy}^p \ldots (x,y,z) . \tag{3.8}$$

Relations (3.8) enable us to write eq. (3.7) in the following manner:

$$\int_v (\dot{\sigma}_x \dot{\varepsilon}_x^e + ... + \dot{\tau}_{xy} \dot{\gamma}_{xy}^e + ...) \, dv + \int_v (\dot{\sigma}_x \dot{\varepsilon}_x^p + ... + \dot{\tau}_{xy} \dot{\gamma}_{xy}^p + ...) \, dv = 0. \qquad (3.9)$$

The second integral in eq. (3.9) vanishes because, for continuing flow, we must either have constant stress or a stress variation vector $\Delta\sigma$ in the tangent plane to the yield surface normal to the strain rate vector $\dot{\varepsilon}$, and therefore $\dot{\sigma}_x \dot{\varepsilon}_x^p + ... + \dot{\tau}_{xy} \dot{\gamma}_{xy}^p + ... = 0$. Because the elastic strain rates are related to the stress rates by Hooke's law, $(\dot{\sigma}_x \dot{\varepsilon}_x^e + ... + \dot{\tau}_{xy} \dot{\gamma}_{xy}^e + ...) \, dv$ will be positive unless,

$$\dot{\sigma}_x = 0 \qquad ... (x,y,z), \qquad (3.10)$$

$$\dot{\tau}_{xy} = 0 \qquad ... (x,y,z)$$

(see [3.2] for proof). As the first integral must vanish, conditions (3.10) are necessarily satisfied at all points.

3.3.5. Final remark

In certain circumstances, nonassociated flow rules are used that derive the strain rate components from a potential function not identical with the yield condition (for example, the plastic potential of von Mises with the yield condition of Tresca). Obviously, the fundamental theorems proved in Section 3.2 and their corollaries, particularly the theorem of Section 3.3.4 [3.3], are no longer valid.

3.4. Elastic-plastic and rigid-plastic bodies

The two fundamental theorems were introduced by Gvozdev [2.1], Hill [3.4], and Prager [2.2], [3.5], in the case of the rigid perfectly plastic body, and by Drucker, Prager, and Greenberg [3.6] for the elastic perfectly plastic material. Both idealizations are appropriate for the purposes of limit analysis [3.5].

When the elastic-plastic idealization is used, the limit state corresponds to incipient unrestrained plastic flow. The graph of a relevant displacement δ versus the applied load P [fig. 3.2 (a)] is initially a ray OA (elastic range), then a curve AB (elastic-plastic range: restricted plastic flow) and finally, when plastification has spread sufficiently through the body, the graph becomes a parallel to the axis, representing the unrestrained plastic flow.

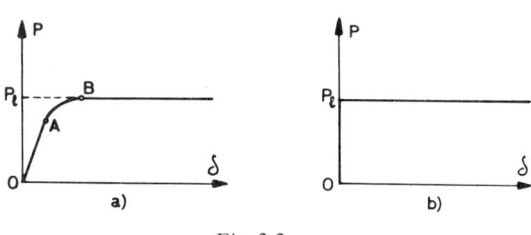

Fig. 3.2.

When using the rigid-plastic idealization, the rigid parts of the body prevent all deformations up to the onset of unrestrained plastic flow at the limit load P_l, sometimes also called the *yield-point load* [fig. 3.2 (b)].

On the other hand, *the fundamental theorems of limit analysis are absolutely identical for both idealizations.* They are based exclusively on the concepts of statically admissible stress fields and kinematically admissible plastic strain rate fields, with no reference to the elastic or rigid nature of the material not at yield. Lower bounds P_-, upper bounds P_+, and complete solutions (as described in Section 3.3.1) are valid for both idealizations.

Why the elastic-plastic idealization is not better than the rigid-plastic idealization is easily understood, if we remember that the former must assume vanishingly small elastic and elastic-plastic strains so that the known undeformed geometry of the structure can be used up to the limit state.

3.5. Influence of changes of geometry

Consider a rigid perfectly plastic structure at incipient unrestrained flow. The flow mechanism is supposed to be known. The velocity field is thus determined to within a positive scalar factor. Assuming that the flow mechanism does not change, the displacement field has the same form as the velocity field, the successive deflected shapes transforming one into the other by similarity. The magnitudes of all displacement vectors are monotonically increasing functions of a scalar parameter that can be regarded as measuring the time t. The work $\mathcal{W}$ of the applied loads and the energy $\mathcal{E}$ dissipated in plastic deformations are functions of t only, because the mechanism is given. Let $t = 0$ at impending plastic flow and consider the deflected shape at the end of the first time interval, taken as the unit of time. If this interval is small enough, we need only retain the first term in the series expansion of $\mathcal{W}(t)$. From the theorem of virtual work $\mathcal{W} = \mathcal{E}$, we obtain

$$1 \cdot k \cdot P_l = \mathcal{E}(1), \tag{3.11}$$

where k is a proportionality factor characteristic of the mechanism.

On the other hand, after a time t large enough to force us to retain two terms in the expansion of $\mathcal{W}(t)$, we have

$$\mathcal{W}(t) = (t + At^2)k \cdot P_l \tag{3.12}$$

whereas, according to eq. (2.4),

$$\mathcal{E}(t) = t\mathcal{E}(1). \tag{3.13}$$

Subtracting from eq. (3.12) the corresponding sides of eq. (3.13) and using eq. (3.11), we obtain

$$\mathcal{W}(t) - \mathcal{E}(t) = At^2 k P_l. \tag{3.14}$$

When A is positive, as in fig. 3.3 (a), the work of the loads is larger than the work dissipated and the limit state is an unstable (collapse) state [3.7]. In the case of fig. 3.3 (b), A is negative, the energy dissipated is larger than the work of the loads, and the limit state is stable. Increasing loads are needed to maintain plastic flow.

Figs. 3.4 and 3.5 are the graphs of load versus deflection for a built-in beam [3.8] and a circular simply supported plate [3.9], respectively.

Fig. 3.6 gives similar curves for a truncated conical shell subjected to a uniform load at the upper edge [3.10]. The dashed curved showing the

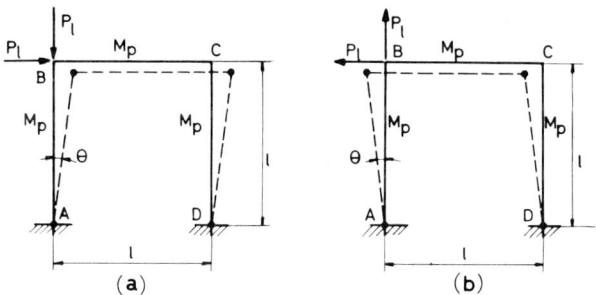

Fig. 3.3.

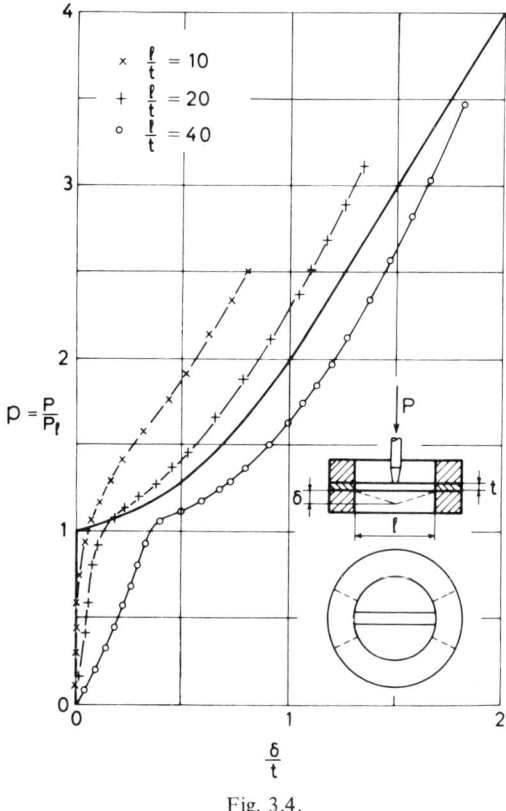

Fig. 3.4.

theoretical load versus the deflection is obtained by taking account of the changes of geometry [3.11].

In figs. 3.4 and 3.5, the changes of geometry have a favorable effect on the strength of the structure whereas in fig. 3.6 the limit state is unstable and results in plastic collapse of the structure.

Experimental and theoretical curves differ for two main reasons:

1. The solid is not rigid-plastic but elastic-plastic. If the structure is elastically very deformable and highly redundant, deformations for loads smaller than P_l may already influence the effects of the forces: examples are string and membrane effects in beams and plates primarily subjected to bending, and instability by divergence of structures under compression.

3.6]   SOLUTIONS GIVEN BY THE DEFORMATION THEORY   35

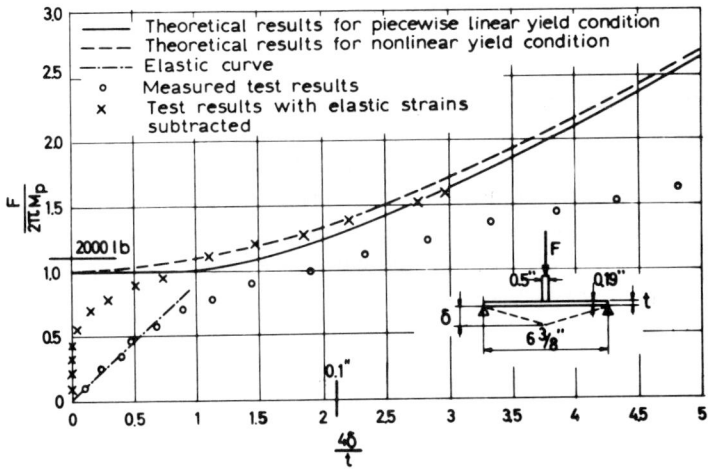

Fig. 3.5. From *Plastic Analysis of Structures* by P.G.Hodge Jr. Copyright 1959, McGraw-Hill Book Company. Used by permission of McGraw-Hill Book Company, Inc.

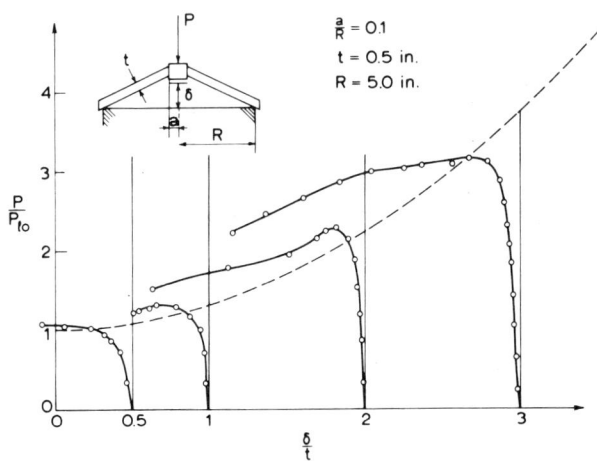

Fig. 3.6.

2. When large plastic deformations occur, the real material exhibits some work-hardening that was neglected in the theory.

3.6. Solutions given by the deformation theory

Whereas limit analysis is only concerned with the determination of the intensity of loading at the onset of unrestrained plastic flow, we may be interested in studying the elastic and elastic-plastic states that precede this onset of flow. When the plastic potential-flow law (2.6) is applied to the yield condition of von Mises,

$$\sigma_R \equiv \sigma_x^2 + \sigma_y^2 + \sigma_z^2 - \sigma_x \sigma_y - \sigma_y \sigma_z - \sigma_z \sigma_x + 3(\tau_{xy}^2 + \tau_{yz}^2 + \tau_{zx}^2) = \sigma_Y^2 , \quad (3.15)$$

we find

$$\dot{\epsilon}_x = \lambda(2\sigma_x - \sigma_y - \sigma_z) \quad \ldots (x,y,z) ,$$

$$\dot{\gamma}_{xy} = 6\lambda\tau_{xy} \quad \ldots (x,y,z) . \quad (3.16)$$

Indeed, we have

$$\frac{\partial \sigma_R}{\partial \sigma_x} = \frac{\partial \sqrt{\sigma_R^2}}{\partial \sigma_x} = \frac{1}{2\sigma_Y} \frac{\partial \sigma_R^2}{\partial \sigma_x} \quad \ldots (x,y,z) ,$$

$$\frac{\partial \sigma_R}{\partial \tau_{xy}} = \frac{1}{2\sigma_Y} \frac{\partial \sigma_R^2}{\partial \tau_{xy}} \quad \ldots (x,y,z) . \quad (3.17)$$

and the constant $1/2\sigma_Y$ may be included in the positive scalar factor λ. Introducing the stress deviator with normal components

$$s_x = \sigma_x - \frac{\sigma_x + \sigma_y + \sigma_z}{3} = \frac{2\sigma_x - \sigma_y - \sigma_z}{3} \quad \ldots (x,y,z) , \quad (3.18)$$

and the same shear components as the stress tensor, eqs. (3.16) become

$$\dot{\epsilon}_x = 3\lambda s_x \quad \ldots (x,y,z) ,$$

$$\dot{\gamma}_{xy} = 6\lambda \tau_{xy} \quad \ldots (x,y,z) . \quad (3.19)$$

SOLUTIONS GIVEN BY THE DEFORMATION THEORY

With relations (3.18), the yield condition (3.15) can be written as

$$\tfrac{3}{2}(s_x^2+s_y^2+s_z^2) + 3(\tau_{xy}^2+\tau_{yz}^2+\tau_{zx}^2) = \sigma_Y^2 . \tag{3.20}$$

Substitution of expressions (3.19) for s_x and τ_{xy} in (3.20) gives

$$\lambda^2 = \frac{1}{18\sigma_Y^2} [3(\dot{\epsilon}_x^2+\dot{\epsilon}_y^2+\dot{\epsilon}_z^2)+\tfrac{3}{2}(\dot{\gamma}_{xy}^2+\dot{\gamma}_{yz}^2+\dot{\gamma}_{zx}^2)] . \tag{3.21}$$

With the condition of plastic incompressibility,

$$\dot{\epsilon}_x + \dot{\epsilon}_y + \dot{\epsilon}_z = 0 ,$$

the bracket in eq. (3.21) can be written, after subtracting $(\dot{\epsilon}_x+\dot{\epsilon}_y+\dot{\epsilon}_z)^2$, as

$$(\dot{\epsilon}_x-\dot{\epsilon}_y)^2 + (\dot{\epsilon}_y-\dot{\epsilon}_z)^2 + (\dot{\epsilon}_z-\dot{\epsilon}_x)^2 + \tfrac{3}{2}(\dot{\gamma}_{xy}^2+\dot{\gamma}_{yz}^2+\dot{\gamma}_{zx}^2) .$$

Introducing the *strain-rate intensity*

$$\dot{g} = \frac{\sqrt{2}}{3}\sqrt{(\dot{\epsilon}_x-\dot{\epsilon}_y)^2 + (\dot{\epsilon}_y-\dot{\epsilon}_z)^2 + (\dot{\epsilon}_z-\dot{\epsilon}_x)^2 + \tfrac{3}{2}(\dot{\gamma}_{xy}^2+\dot{\gamma}_{yz}^2+\dot{\gamma}_{zx}^2)} , \tag{3.22}$$

we write eq. (3.21) in the form

$$\lambda = \frac{\dot{g}}{2\sigma_Y} . \tag{3.23}$$

Hence, eqs. (3.19) become

$$\dot{\epsilon}_x = \frac{3\dot{g}}{2\sigma_Y} s_x \quad \ldots (x,y,z) ,$$

$$\dot{\gamma}_{xy} = \frac{3\dot{g}}{\sigma_Y} \tau_{xy} \quad \ldots (x,y,z) . \tag{3.24}$$

Eqs. (3.24) are the classical Levy-von Mises equations. Note that they involve purely plastic strain rates (rigid-plastic body). To account for elastic strains, one must divide total strain rates into their elastic and plastic parts,

$$\dot{\epsilon}_x = \dot{\epsilon}_x^e + \dot{\epsilon}_x^p \qquad \ldots (x,y,z),$$

$$\dot{\gamma}_{xy} = \dot{\gamma}_{xy}^e + \dot{\gamma}_{xy}^p \qquad \ldots (x,y,z), \qquad (3.25)$$

as in Section 3.3.4. Eqs. (3.24) furnish the plastic parts and differentiation of Hooke's law with respect to time the elastic parts. In this manner, one obtains the Pràndtl-Reuss equations (see [3.13], pp. 28-29). Solutions of most practical problems with the Mises or Prandtl-Reuss equations unfortunately turn out to be very difficult [3.12, 3.13].

Some authors approach the problem from a different point of view [1.2, 3.14]. Their starting point is the experimental evidence that for any loading such that all the components of the stress tensor increase in proportion, there is a unique relation between the "intensity",

$$\mathcal{G} = \frac{\sqrt{2}}{3}\sqrt{(\epsilon_x-\epsilon_y)^2 + (\epsilon_y-\epsilon_z)^2 + (\epsilon_z-\epsilon_x)^2 + \tfrac{3}{2}(\gamma_{xy}^2+\gamma_{yz}^2+\gamma_{zx}^2)}, \qquad (3.26)$$

of the *total* strains, and the stress intensity,

$$\mathcal{G} = \frac{1}{\sqrt{2}}\sqrt{(\sigma_x-\sigma_y)^2 + (\sigma_y-\sigma_z)^2 + (\sigma_z-\sigma_x)^2 + 6(\tau_{xy}^2+\tau_{yz}^2+\tau_{zx}^2)}, \qquad (3.27)$$

as shown on figs. 3.7 and 3.8 taken from references [1.2] and [3.15], respectively. Consequently, the adopted fundamental equations are ([1.2], p. 98)

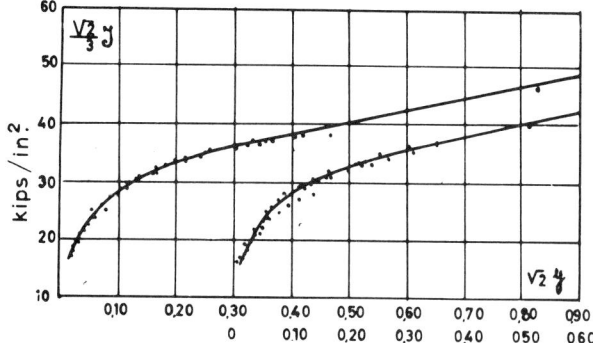

Fig. 3.7.

3.6] SOLUTIONS GIVEN BY THE DEFORMATION THEORY

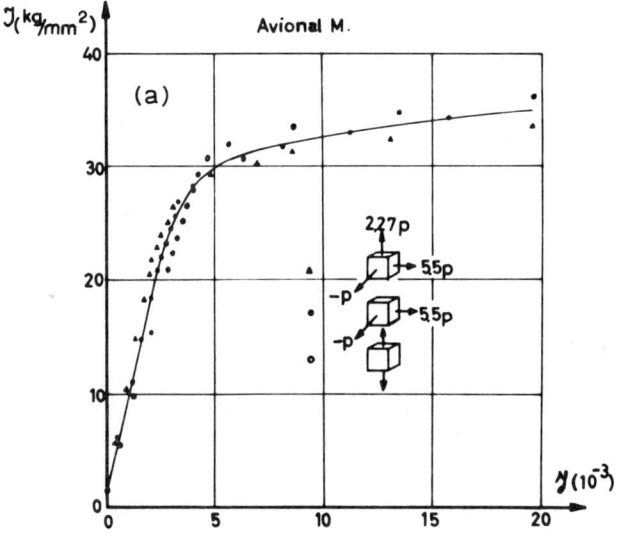

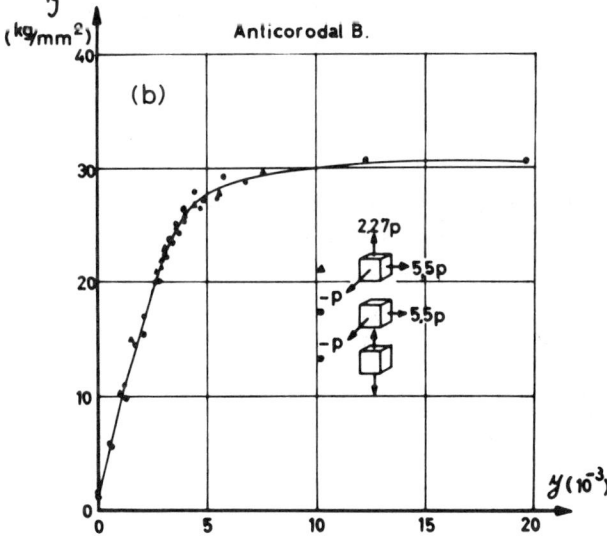

Fig. 3.8.

$$\epsilon_x - e = \frac{3}{2}\frac{\mathcal{G}}{\mathcal{G}} s_x \qquad \ldots (x,y,z) ,$$

$$\gamma_{xy} = 3\frac{\mathcal{G}}{\mathcal{G}} \tau_{xy} \qquad \ldots (x,y,z) , \tag{3.28}$$

where

$$e = \frac{\epsilon_x + \epsilon_y + \epsilon_z}{3} , \tag{3.29}$$

and

$$\mathcal{G} = \mathcal{G}(\mathcal{G}) . \tag{3.30}$$

The function $\mathcal{G}(\mathcal{G})$ is characteristic of the material. The range of applicability of this second theory of plasticity, called *deformation theory*, is narrow compared to that of the former *flow theory* ([3.16, 1.2]), but because of the simple form of eq. (3.28), many practical problems have been solved by the deformation theory. If we consider a perfectly plastic structure from the point of view of the deformation theory, and if we want to determine its "carrying capacity", we may neglect the elastic strains ([1.2], pp. 170 and 232). Hence we let $e = 0$ because plastic volume change vanishes. For perfect plasticity the relation $\mathcal{G} = \mathcal{G}(\mathcal{G})$ simply becomes (fig. 3.9)

$$\mathcal{G} = \sigma_Y . \tag{3.31}$$

From the definition (3.27) of $\mathcal{G}$, eq. (3.31) is the yield condition of von Mises. Since the strains are purely plastic, eqs. (3.28) reduce to

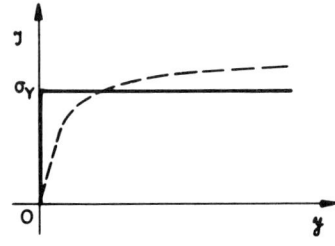

Fig. 3.9.

$$\epsilon_x = \frac{3}{2}\frac{\mathcal{G}}{\sigma_Y} s_x \qquad \ldots (x,y,z),$$

$$\gamma_{xy} = 3\frac{\mathcal{G}}{\sigma_Y} \tau_{xy} \qquad \ldots (x,y,z). \qquad (3.32)$$

Comparing eqs. (3.32) with eqs. (3.24), we note that *they are formally identical except that the former deal with plastic strains and the latter with plastic strain rates.*

Now imagine that the "carrying capacity" of a structure has been determined from eqs. (3.32). This means that there has been found:

1. an equilibrium stress field satisfying $\sigma_R \leq \sigma_Y$,
2. a field of plastic strains at impending unrestrained plastic flow.

From the point of view of limit analysis, the stress field is statically admissible. The strain field, *when regarded as a strain rate field* defined except for a constant scalar factor, *specifies a kinematically admissible mechanism that corresponds to the stress field by the plastic potential flow law* (3.24), because of the formal identity of eqs. (3.32) and (3.24). Consequently, *a "carrying capacity" determined by the deformation theory is an exact limit load for limit analysis.*

This result justifies including in our solutions those obtained by the deformation theory*.

Two final points are worth noting:

1. The "carrying capacity" is often obtained solely from a statically admissible stress field. It must then be regarded as a lower bound P_- to P_l. Nevertheless it is in some instances reasonable to expect that one obtains the "best" bound (or at least a very good bound) in this manner.

This is the case in plate problems where plastic flow occurs at all points.
2. The "carrying capacity" is sometimes obtained with the yield condition of Tresca and eqs. (3.32). That carrying capacity must be regarded exclusively as a lower bound P_- as long as a mechanism has not been found that corresponds to the stress field by the plastic potential flow law *applied to the yield condition of Tresca.*

---

* The formal analogy of eqs. (3.24) and (3.32) also explains that the yield surfaces obtained by Iliouchine [1.2] are identical to those obtained on the basis of plastic potential ([5.8, 5.9],...).

3.7. Uniqueness

As already noted at the end of Section 3.2, the limit load for proportional loading P_l is unique. Indeed, consider a load P^* that simultaneously corresponds to a statically admissible stress field (σ) and a kinematically admissible strain rate field ($\dot{\varepsilon}$). Let us assume that there exist several limit loads. The proofs of the fundamental theorems show that P^* must be equal to any one of them. Hence, P_l is unique and coincides with P^*.

Are the fields (σ) and ($\dot{\varepsilon}$) also unique? As was emphasized at the end of Section 2.4, to the field (σ) there may correspond by the normality law several mechanisms, all of them furnishing the same load P by their work equation. Moreover, it was also noted in Com. V. for beams and frames that different mechanisms could result in the same exact limit loads (e.g., Section 3.5.4). Hence, *the flow mechanism clearly need not be unique.*

When two complete solutions $(\sigma)_1$, $(\dot{\varepsilon})_1$ and $(\sigma)_2$, $(\dot{\varepsilon})_2$ are known, it can be proved [3.5, 3.17] that the stress fields of the two solutions are identical except possibly

1. in the common rigid regions,
2. where both states of stress are represented by stress points on the same flat part of the yield surface.

3.8. Appendix

Proof of the theorem of virtual works

Consider a continuous body in equilibrium under the applied forces and the reactions at the supports (fig. 3.10). Imagine a virtual field of displacements **U** of the points of the continuum. Except that **U** must be a continuous function of position satisfying the kinematic boundary conditions of the body, this displacement field may otherwise be chosen arbitrarily.

Evaluate the work done on these displacements by the surface tractions **T** per unit area (including the reactions of the supports if they do work) and the body forces **F** per unit volume. This work $\mathcal{W}_e$ is given by

$$\mathcal{W}_e = \int_A \mathbf{T} \cdot \mathbf{U} \, dA + \int_V \mathbf{F} \cdot \mathbf{U} \, dV, \qquad (a)$$

where A and V are the area and the volume of the body, respectively.

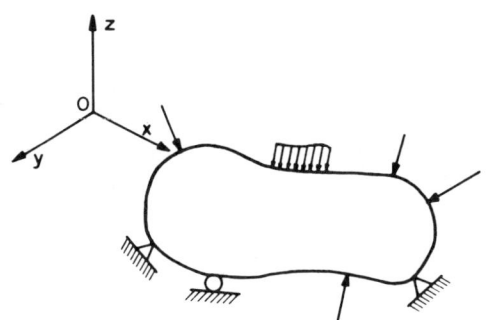

Fig. 3.10.

The assumed equilibrium (internal and at the boundary) is expressed by the following eqs. (b) and (c), respectively:

$$\frac{\partial \sigma_x}{\partial x} + \frac{\partial \tau_{xy}}{\partial y} + \frac{\partial \tau_{zx}}{\partial z} + X = 0 \qquad \ldots (x,y,z), \tag{b}$$

$$l\sigma_x + m\tau_{xy} + n\tilde{\tau}_{zx} = \overline{X} \qquad \ldots (x,y,z). \tag{c}$$

In eqs. (b) and (c), l, m, n are the direction cosines of the exterior normal at a point of the surface A of the body subjected to force $\mathbf{T}$ with components $\overline{X}, \overline{Y}, \overline{Z}$, and X, Y, Z are the components of $\mathbf{F}$.

We rewrite the first terms of eq. (a) as

$$\int_A \mathbf{T} \cdot \mathbf{U} \, dA = \int_A (\overline{X}u + \overline{Y}v + \overline{Z}w) \, dA, \tag{d}$$

where u, v, w are the components of $\mathbf{U}$. Using eq. (c) in eq. (d) we obtain

$$\int_A \mathbf{T} \cdot \mathbf{U} \, dA = \int_A (lu\sigma_x + mu\tau_{xy} + nu\tau_{zx} + lv\tau_{xy} + mv\sigma_y + nv\tau_{yz} + lw\tau_{zx} + mw\tau_{yz}$$

$$+ nw\sigma_z) \, dA$$

$$= \int_A [l(\sigma_x u + \tau_{xy} v + \tau_{zx} w) + m(\tau_{xy} u + \sigma_y v + \tau_{yz} w)$$

$$+ n(\tau_{zx} u + \tau_{yz} v + \sigma_z w)] \, dA. \tag{d'}$$

We apply the Green-Ostrogradsky formula

$$\int_A \mathbf{P} \cdot \mathbf{n}\, dA = \int_V \operatorname{div} \mathbf{P}\, dV. \tag{e}$$

Because l, m, n are the projections of the outward normal unit vector $\mathbf{n}$, relation (d') becomes

$$\int_A \mathbf{T} \cdot \mathbf{U}\, dA = \int_V \left[\frac{\partial}{\partial x}(\sigma_x u + \tau_{xy} v + \tau_{zx} w) + \frac{\partial}{\partial y}(\tau_{xy} u + \sigma_y v + \tau_{yz} w) \right.$$

$$\left. + \frac{\partial}{\partial z}(\tau_{zx} u + \tau_{yz} v + \sigma_z w) \right] dV,$$

or, more explicitly,

$$\int_A \mathbf{T} \cdot \mathbf{U}\, dA = \int_V \left[u\left(\frac{\partial \sigma_x}{\partial x} + \frac{\partial \tau_{xy}}{\partial y} + \frac{\partial \tau_{zx}}{\partial z}\right) + v\left(\frac{\partial \tau_{xy}}{\partial x} + \frac{\partial \sigma_y}{\partial y} + \frac{\partial \tau_{yz}}{\partial z}\right) \right.$$

$$+ w\left(\frac{\partial \tau_{zx}}{\partial x} + \frac{\partial \tau_{yz}}{\partial y} + \frac{\partial \sigma_z}{\partial z}\right) + \sigma_x \frac{\partial u}{\partial x} + \sigma_y \frac{\partial v}{\partial y} + \sigma_z \frac{\partial w}{\partial z}$$

$$\left. + \tau_{xy}\left(\frac{\partial v}{\partial x} + \frac{\partial u}{\partial y}\right) + \tau_{yz}\left(\frac{\partial w}{\partial y} + \frac{\partial v}{\partial z}\right) + \tau_{zx}\left(\frac{\partial u}{\partial z} + \frac{\partial w}{\partial x}\right) \right] dV. \tag{f}$$

For *very small* deformations we have

$$\epsilon_x = \frac{\partial u}{\partial x}, \qquad \epsilon_y = \frac{\partial v}{\partial y}, \qquad \epsilon_z = \frac{\partial w}{\partial z},$$

$$\gamma_{xy} = \frac{\partial u}{\partial x} + \frac{\partial v}{\partial y}, \qquad \gamma_{yz} = \frac{\partial w}{\partial y} + \frac{\partial v}{\partial z}, \qquad \gamma_{zx} = \frac{\partial u}{\partial z} + \frac{\partial w}{\partial x}. \tag{g}$$

Using eqs. (b) and (g) in eq. (f) we obtain

$$\int_A \mathbf{T} \cdot \mathbf{U}\, dA = \int_V \left[-(Xu + Yv + Zw) + \sigma_x \epsilon_x + \sigma_y \epsilon_y + \sigma_z \epsilon_z \right.$$

$$\left. + \tau_{xy} \gamma_{xy} + \tau_{yz} \gamma_{yz} + \tau_{zx} \gamma_{zx} \right] dV.$$

Because $Xu + Yv + Zw = \mathbf{F} \cdot \mathbf{U}$, we have

$$\int_A \mathbf{T} \cdot \mathbf{U} \, dA + \int_V \mathbf{F} \cdot \mathbf{U} \, dV = \int_V \sigma_x \epsilon_x + \sigma_y \epsilon_y + \sigma_z \epsilon_z$$
$$+ \tau_{xy} \gamma_{xy} + \tau_{yz} \gamma_{yz} + \tau_{zx} \gamma_{zx} \, dV. \tag{h}$$

The right-hand side of eq. (h) is the virtual work of the stresses σ_x, ..., τ_{xy}, ... on the strains ϵ_x, ..., γ_{xy}, ..., the latter being as a rule independent of the former because they are derived from an arbitrary displacement field by relations (g). We thus have proved the theorem:

If a continuum is in equilibrium under the influence of given stresses, loads, and reactions, the virtual work of the loads and reactions on an arbitrary continuously differentiable displacement field equals the virtual work of the stresses on the corresponding strain field.

When the displacement field corresponds to a rigid-body displacement, we have

$$\epsilon_x = \epsilon_y = \epsilon_z = \gamma_{xy} = \gamma_{yz} = \gamma_{zx} = 0,$$

and hence

$$\int_A \mathbf{T} \cdot \mathbf{U} \, dA + \int_V \mathbf{F} \cdot \mathbf{U} \, dV = 0. \tag{i}$$

We thus have the following theorem:

For a continuum that is in equilibrium under the influence of given stresses, loads, and reactions, the virtual work of the loads and reactions on any virtual rigid-body displacement vanishes.

3.8.1. Remarks

1. The theorems of virtual works above are simply a different way of saying that the body is in equilibrium.

2. The preceding proof only assumes that (a) continuity conditions for the applicability of the Green formula transforming a surface integral into a volume integral are fulfilled; (b) the strains are sufficiently small for eq. (g) to be valid (this condition is regarded to be included in the term "virtual"). With these two assumptions, the theorem applies to bodies of any nature: elastic, plastic, viscous, and so forth.

3.9. Problems

3.9.1. Show that, for the yield condition of von Mises, the dissipation per unit volume is $D = \sigma_Y \dot{\mathcal{G}}$.

3.9.2. Prove the following theorems (in analogy with those of Section 3.3.2).
(a) When kinematic boundary restraints are tightened, the limit load does not decrease.
(b) When kinematic boundary restraints are relaxed, the limit load does not increase. *Hint*: discuss kinematical admissibility of actual collapse mechanisms for original and modified structures.

3.9.3. Prove that, when the statical boundary conditions are unchanged (nature of the reactions of the supports) the limit load does not depend on the kinematic boundary conditions. For example, simple supports may be (slightly) deformable. *Answer*: see Section 7.5.10.

3.9.4. Show that unrestrained plastic flow under constant load is impossible for an elastically compressible perfectly plastic solid with the von Mises yield condition, when it is subjected to plane strain conditions. *Hint*: study the variation of the principal stress acting in the direction of vanishing principal strain.

References

[3.1] M.R.HORNE, "The Influence of Residual Stresses on the Behavior of Ductile Structures", *Residual Stresses*, W.R.OSGOOD, ed. Reinhold Publ. Corp, 1954.
[3.2] W.PRAGER, *Introduction to Mechanics of Continua*, Ginn and Co., Boston, 1961.
[3.3] W.T.KOITER, "On the Stress-Strain Relations and the General Theorems of Plasticity", *Lab. van toegepaste Mech.*, Techn. Hoge-school, Delft, 1953.
[3.4] R.HILL, "On the State of Stress in a Plastic-Rigid Body at the Yield Point", *Phil. Mag.*, **42**: 868, 1951.
[3.5] W.PRAGER, "Problemes de plasticite theorique", Dunod, Paris, 1958.
[3.6] D.C.DRUCKER, W.PRAGER, and H.J.GREENBERG, "Extended Limit Design Theorems for Continuous Media", *Quart. Appl. Math.*, **9**: 381, 1952.
[3.7] E.T.ONAT, "On Certain Second Order Effects in the Limit Design of Frames", *J. of Aero. Sci.*, **22**: 681, 1955.
[3.8] R.M.HAYTHORNTHWAITE, "Beams with Full End Fixity", *Engineering*, **183**: 110, 1957. See also by the same author, "Plastic Behavior of Beams with Elastic End Constraints", *Comptes rendus 9e Cong. Int. Mec. Appl.*, Bruxelles, 1956.

[3.9] P.G.HODGE Jr., *Plastic Analysis of Structures*, McGraw-Hill, 1959.
[3.10] E.T.ONAT, "On the Plastic Analysis of Shallow Conical Shells", Paper presented at the Xth Int. Congr. Appl. Mech., Stresa, 1960 (diagrams reproduced by kind permission).
[3.11] E.T.ONAT, "Plastic Analysis of Shallow Conical Shells", *Proc. A.S.C.E., J. Eng. Mech. Div.*, December, 1960.
[3.12] R.HILL, *The Mathematical Theory of Plasticity*, Clarendon Press, Oxford, 1950.
[3.13] W.PRAGER and P.G.HODGE Jr., *Theory of Perfectly Plastic Solids*, J. Wiley, New York, 1951.
[3.14] V.V.SOKOLOVSKI, *Theorie der Plastizität*, V.E.B. Verlag Technik, Berlin, 1955.
[3.15] Swiss Fed. Lab. for Testing Materials, Report 126, Zürich, Feb. 1940.
[3.16] B.BUDIANSKY, "A Reassessment of Deformation Theories of Plasticity", *J. App. Mech.*, **26**: 259, June 1959.

4

General Loading Case

4.1. Structure with nonnegligible dead load

Under the assumption of proportional loading made in the previous sections, the safety of a structure with respect to incipient unrestrained plastic flow is measured by the quotient $s_p = P_l/P_s$ of its limit load P_l over its service load P_s. It must be noted that the dead load is not likely to vary (except for corrosion, or uncertainties on specific weight and dimensions), whereas live load may vary in a wide range. This fact must be taken into account when the dead load is an important part of the total loads. Assuming the dead load to be given and fixed and the live loads to consist of a oneparameter loading with magnitude P_v, one defines statically and kinematically admissible loads P_{v-} and P_{v+} as follows:

1. P_{v-} is any intensity of the live load that, together with the fixed dead load, corresponds to a statically admissible stress field.
2. P_{v+} is any intensity of the live load that, together with the fixed dead load, corresponds by the work equation to a kinematically admissible mechanism.

If the limit live load P_{vl} is defined as the live load that produces unrestrained plastic flow when associated with the dead load, it is easily shown, along the lines of Section 3.2, that

$$P_{v-} \leqslant P_{vl} \leqslant P_{v+} .\tag{4.1}$$

Note that the point of view adopted above implies that the dead load alone can not cause the collapse of the structure.

4.2. Loading depending on several parameters

When the loading depends on several *independent* scalar loading parameters $P_1, P_2, ..., P_k$, there does not exist as a rule one limit state — but an infinity of possible combinations of the "loads" $P_1, P_2, ..., P_k$ may produce collapse.

The whole loading path must then be examined to determine *at every loading stage* whether there is unrestrained plastic flow or not.

The theorems of limit analysis that can be established along a line similar to that of Section 3.3, are then as follows:

1. If a statically admissible stress field with $\sigma_R < \sigma_Y$ can be found *at every loading stage*, plastic collapse will not occur on this loading path.
2. If at a certain loading stage, a kinematically admissible mechanism can be found in which the power of applied loads is not smaller than the power of dissipation, plastic collapse must have occurred on the considered loading path or must be impending.
3. As long as plastic collapse has not occurred, it is possible to find a statically admissible stress field *at every loading stage*.

4.3. Shake-down analysis

4.3.1. *Introduction*

To use the preceding theorems, the loading path must be completely described and every loading stage must be studied. This analysis is in most cases a very long and difficult task.

On the other hand, the loading parameters very often vary in an unknown manner and the problem is to determine *the permissible range of variation of each parameter* to avoid some kind of plastic collapse.

4.3.2. *One-parameter loading*

When the load does not increase monotonically but varies arbitrarily between prescribed limits, the structure can fail either by accumulation of plastic deformations of the same sign or by alternating plastic deformations

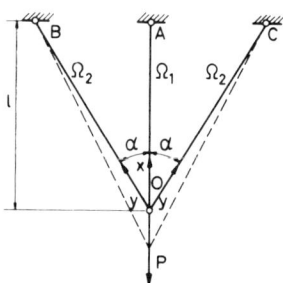

Fig. 4.1.

that eventually result in fracture. These types of failure occur, as a rule, with the highest value of the load smaller than the limit load for proportional loading. This problem was already studied in Com. V., Section 7.3.

We consider here in more detail the example of the three bar truss of fig. 4.1, in a slightly more general form than in Section 1.1 of Com. V. because the angle α is arbitrary and the cross sections of the bars OB and OC, though identical and denoted by A_2 may differ from the cross section A_1 of the bar OA. All three bars are made of the same elastic perfectly plastic material with an elastic modulus E and the stress-strain diagram shown in fig. 4.2.

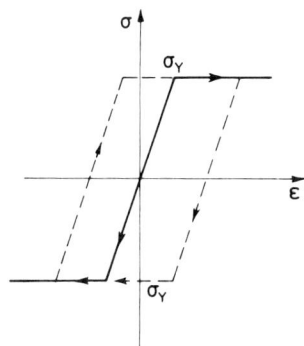

Fig. 4.2.

SHAKE-DOWN ANALYSIS

Let

$$\frac{l}{EA_1} = C_1,$$

$$\frac{l}{EA_2 \cos\alpha} = C_2, \qquad (4.2)$$

and denote by Δ the total elongation of a bar and by Δ_p its plastic elongation. We then have

$$\Delta_1 = C_1 X + \Delta_{1p},$$

$$\Delta_2 = C_2 Y + \Delta_{2p}, \qquad (4.3)$$

where X and Y are the axial forces in the bars OA and OB (or OC), respectively. With the following definitions of nondimensional forces and elongations

$$x = X\sqrt{C_1/2},$$

$$y = Y\sqrt{C_2},$$

$$\delta_1 = \frac{\Delta_1}{\sqrt{2C_1}},$$

$$\delta_2 = \frac{\Delta_2}{\sqrt{C_2}},$$

$$\delta_{1p} = \frac{\Delta_{1p}}{\sqrt{2C_1}},$$

$$\delta_{2p} = \frac{\Delta_{2p}}{\sqrt{C_2}}, \qquad (4.4)$$

the preceding expressions for the total elongations become

$$\delta_1 = x + \delta_{1p},$$

$$\delta_2 = y + \delta_{2p}. \qquad (4.5)$$

Equilibrium of joint O requires that

$$X + 2Y \cos \alpha = P. \tag{4.6}$$

With notations (4.4), eq. (4.6) can be rewritten:

$$\frac{x}{\sqrt{C_1}} + \frac{\sqrt{2}}{\sqrt{C_2}} y \cos \alpha = \frac{P}{\sqrt{2}}. \tag{4.7}$$

Compatibility at joint O furnishes

$$\Delta_1 = \frac{\Delta_2}{\cos \alpha},$$

or, equivalently,

$$\delta_1 \sqrt{C_1} - \frac{\sqrt{C_2} \delta_2}{\sqrt{2} \cos \alpha} = 0. \tag{4.8}$$

The yield condition of the bars are $|X| = \sigma_Y A_1$, $|Y| = \sigma_Y A_2$, or, in non-dimensional form,

$$|x| = \sigma_Y A_1 \sqrt{C_1/2} \equiv x_p,$$

$$|y| = \sigma_Y A_2 \sqrt{C_2} \equiv y_p. \tag{4.9}$$

In a system of cartesian orthogonal axes x and y the yield condition is represented by a rectangle (fig. 4.3). The forces in the bars are the coordinates of the *force point* that, in order to satisfy equilibrium, must, for each value of P, remain on the corresponding straight line with eq. (4.7), called the *equilibrium line*.

Superimpose on the x-, y-axes the ξ-, η-axes, respectively, with

$$\xi = -\delta_{1p},$$

$$\eta = -\delta_{2p}, \tag{4.10}$$

and call the *plastic deformation point* the point with coordinates (ξ, η). The compatibility equation (4.8) can be written

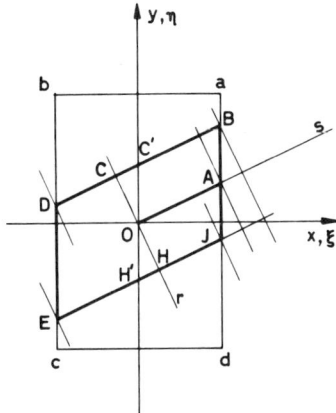

Fig. 4.3.

$$(x-\xi)\sqrt{C_1} - (y-\eta)\frac{\sqrt{C_2}}{\sqrt{2}\cos\alpha} = 0 .\qquad(4.11)$$

It is represented by the *line of compatibility* which is normal to the equilibrium line through the plastic deformation point, as is easily verified from inspection of the coefficients of the variables in eqs. (4.7) and (4.11). The intersection of the line of compatibility with the equilibrium line is the force point $\mathcal{P}$ representing the state of stress of the truss. The position of the force point thus depends on both the value of the load P and the exiting plastic deformation.

Note that the force point has a smaller distance from the plastic deformation point than any other point on the equilibrium line. This minimum distance,

$$\mathcal{E}^* = (x+\delta_{1p})^2 + (y+\delta_{2p})^2 = \delta_1^2 + \delta_2^2 = \Delta_1^2\frac{EA_1}{2l} + 2\Delta_2^2\frac{EA_2}{2l}\cos\alpha ,$$

is the fictitious strain energy computed from the total elongations as if they were purely elastic. This result is a particular form of a more general theorem of Colonnetti [4.1, 4.2].

In a stress-free initial state, the force and permanent deformation points both coincide with the origin, and the equilibrium and compatibility lines are rays Or and Os respectively. When the truss is loaded, the point $\mathcal{P}$ moves on Os whereas the line Or experiences a translation. When $\mathcal{P}$ has the position A (fig. 4.3) we have

$$\xi = \eta = 0,$$

$$x = x_p. \tag{4.12}$$

With use of eqs. (4.12) and (4.9), eq. (4.11) becomes

$$y = y_p \cos^2 \alpha. \tag{4.13}$$

It is seen that the yield limit cannot be reached simultaneously in all three bars for nonvanishing α. The central bar always yields first.

With increasing load intensity P, the force point moves from A towards the corner a where $P = P_l$ (fig. 4.3). Suppose we stop increasing P when we have reached the force point B. The plastic deformation point has moved from O to C' along axis Oy. We then unload and eventually reverse the sign of P. The force point first moves on BC down to D where the central bar yields in compression, and then on bc towards c, whereas the plastic deformation point moves down the Oy axis. If we again reverse the sign of P when we have reached E, the force point climbs on EJ and the plastic deformation point stays at H'. If we cycle the force point along $EHJBD$ (zero load P corresponding to points H and C) the central bar yields alternatively in tension (JB) and compression (DE), though the load is bounded by $P_1 < P_l$ (point B) and $P_2 > -P_l$ (point E). Alternating plasticity of this type is likely to rapidly produce fracture*.

On the other hand, if loading to point B is followed by load variations restricted to values of P corresponding to the points B and D, the system will *shake-down* to purely elastic behavior after the first plastic deformation OC'. This situation occurs in particular for repeated loading (along BC) for all $P < P_l$. As emphasized in Com. V., Section 7.3, shake-down results from a favorable state of selfstress (represented by the coordinates of C in repeated loading).

---

* Loading cycles as above may also produce accumulating plastic deformations of the same sign (see [4.3], p. 28), though this cannot occur in the simple structure considered here.

4.3.3. *Loading depending on several parameters*

For beams, it was shown in Section 7.3 of Com. V. that for loading that depends on several parameters failure can occur from accumulation of plastic deformations even with loads of constant sign. Obviously, failure can also occur from alternating plastic deformation.

The fundamental theorem given in Section 7.3 of Volume 1 can be extended to an elastic perfectly plastic continuum of any shape subjected to arbitrary loads or temperature cycles [4.4–4.9], as follows*: *If a field of self-stress can be found that does not violate the yield condition when superimposed on the (fictitious) purely elastic stress fields produced by the load and temperature cycles, then shake-down will occur.*

Unfortunately we do not know of a single application of this theorem for structures other than beams and frames, except for one-parameter loadings. Moreover, experimental verifications are also practically absent except for beam problems. Hence, the influence on plastic design of the possibility of failure by lack of shake-down is presently very difficult to evaluate.

4.3.4. *Example of one-parameter loading*

Consider [4.7] a thick-wall tube with radii a and b, which is in plane strain under the influence of uniform internal pressure (fig. 4.4). In Chapter 10 we shall prove that, for monotonically increasing pressure, the values p_e and p_l of the pressure for first yielding and complete plastification are respectively given by

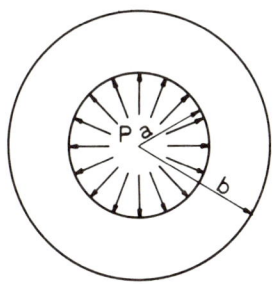

Fig. 4.4.

---

* A kinematical counterpart of the theorem has been given by Koïter [4.10].

$$p_e = \tau_Y \left(1 - \frac{a^2}{b^2}\right), \tag{4.14}$$

and

$$p_l = 2\tau_Y \ln\frac{b}{a}. \tag{4.15}$$

The two formulas above are valid for the yield conditions of both Tresca and von Mises, τ_Y being the yield limit in pure shear (greater by a factor of $2/\sqrt{3}$ in von Mises' condition than in Tresca's).

For repeated loading to p_1, unloading will be elastic as long as the circumferential stress $\sigma_{\theta a}$ at the internal boundary* will satisfy $0 < \sigma_{\theta a} < \sigma_Y$ for $p = p_1$, and $-\sigma_Y < \sigma_{\theta a} < 0$ for $p = 0$. This situation occurs for $p_1 \leq 2p_e$ with the obvious supplementary condition $p_1 < p_l$ to avoid collapse. Hence, the shake-down load p_s for repeated loading is

$$p_s = \min \text{ of } \quad [2p_e, p_l]. \tag{4.16}$$

Eq. (4.16) is represented graphically on fig. 4.5. For $p_1 > p_s$, failure occurs from alternated or cumulative yielding.

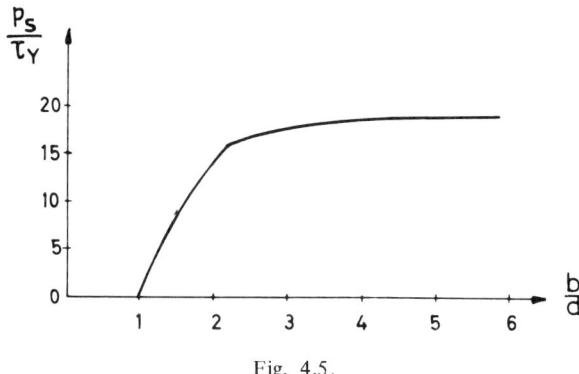

Fig. 4.5.

---

* The radial stress σ_r at the internal boundary is $\sigma_{ra} = p$. It is negligible compared to $\sigma_{\theta a}$. The longitudinal stress is $\sigma_z = \sigma_\theta/2$. Hence, the yield condition is $|\sigma_\theta| = \sigma_Y$.

When r is allowed to vary from p'' to p', the shake-down conditions are

$$-p_l \leqslant p'' \leqslant p' \leqslant p_l \,,$$

$$p' - p'' \leqslant 2p_e \,. \tag{4.17}$$

References

[4.1] G.COLONNETTI, *L'équilibre des corps déformables*, Dunod, Paris, 1955.
[4.2] M.SAVE, "Une interprétation du théorème de Colonnetti", Z.A.M.P., XIII, 5, 1962.
[4.3] W.PRAGER, *An Introduction to Plasticity*, Addison-Wesley Publ. Co., Inc., Reading, Mass., 1959.
[4.4] P.S.SYMONDS, W.PRAGER, "Elastic-Plastic Analysis of Structures Subjected to Loads Varying Arbitrarily Between Prescribed Limits", *J. of Appl. Mech.* **17**: 315, Sept. 1950.
[4.5] Discussion of paper [4.4] by G.WINTER, T.M.CHARLTON, and the authors, *J. Appl. Mech.* **18**: 117, March 1951.
[4.6] P.S.SYMONDS, "Shake-down in Continuous Media", *J. of Appl. Mech.* **18**: 85, March 1951.
[4.7] P.G.HODGE Jr., "Shake-down of Elastic-Plastic Structures", *Residual Stresses in Metals and Metal Constructions*, W.R.OSGOOD, ed., Reinhold Pub. Corp., New York, 1954.
[4.8] W.PRAGER, "Shake-down in Elastic-Plastic Media Subjected to Cycles of Load and Temperature", *Symposium sulla plasticita nella scienza delle custruzioni*, N.ZANICHELLI, Bologna, 1957.
[4.9] W.PRAGER, "Plastic Design and Thermal Stresses", *British Welding J.*, Aug. 1956.
[4.10] W.T.KOÏTER, "General Theorems for Elastic-Plastic Solids", *Progress in Solid Mechanics*, vol. 1, I.N.SNEDDON, R.HILL, eds., North-Holland, Co., Amsterdam, 1960.

5

Generalized Variables

5.1. The concept of generalized variables

5.1.1. *Introduction*

In the following five chapters, we shall deal exclusively with proportional loading.

Limit analysis of a rigid perfectly plastic continuum is based on the three following concepts: (1) yield condition and related flow rule, (2) statically admissible stress field, and (3) kinematically admissible flow mechanism. Great simplification is achieved when these concepts can be applied without the need to discuss three-dimensional stress and displacement fields. This situation arises in linear elasticity when the considered solid is a beam, plate, or shell. Assumptions regarding the deformations of these particular structural elements are accepted as direct consequences of the fact that these elements are "thin" in certain directions (normal to the axis of a beam or to the median surface of a plate or shell).

For beams, the hypothesis of Bernoulli states that plane cross sections remain plane and orthogonal to the deformed material axis. For plates and shells, straight segments normal to the median surface remain straight and normal to the deformed median surface. As long as Hooke's law of linear elasticity is applicable, it is possible to obtain all stress components at every points of a beam, a plate, or a shell when stress resultants and resultant moments are known. To do this, we need only apply the assumption that normals to the median surface are preserved, and use the equilibrium equations and Hooke's law; the latter immediately furnish all strain components [5.1]. Hence, stress resultants and resultant moments are sufficient for a complete description of stresses and strains.

Theory [5.2] and experiments [5.3] show that Bernoulli's hypothesis and its generalization to plates and shells (normals and preserved) are equally valid in the elastic and plastic ranges. Bernoulli's hypothesis will therefore be adopted in the following discussion of rigid-plastic beams, plates, and shells.

5.1.2. Beams without axial force

A generic cross section is subjected to a bending moment M and a shear force V.

From Bernoulli's hypothesis, shear strains are seen to vanish and longitudinal strains ϵ_x are given by*

$$\epsilon_x = y\kappa, \tag{5.1}$$

where y is the distance from the neutral plane and κ is the curvature of the material axis**. (Note that κ is the reciprocal of the radius of curvature.) The strain rate is therefore given by

$$\dot{\epsilon}_x = y\frac{\partial \kappa}{\partial t} = y\dot{\kappa}. \tag{5.2}$$

Because the state of stress is uniaxial***, the power dissipated per unit of length of the beam in a plastic region is

$$D = \int_{-h/2}^{h/2} \sigma_Y |\dot{\epsilon}_x| b(y)\, dy, \tag{5.3}$$

where h is the height of the section of $b(y)$ the width at the level y. With the use of eq. (5.2), eq. (5.3) can be written

---

* Transverse strains are irrelevant.
** For more details on sign conventions, see Com. V., Section 3.1.
*** It is assumed that shear stresses do not influence yielding.

$$D = \int_{-h/2}^{h/2} \sigma_Y |y\dot{\kappa}| b(y)\, dy = |\dot{\kappa}| M_p , \tag{5.4}$$

where M_p is the (ultimate) plastic moment (see Com. V, Section 2.2). The total rate of dissipation D_t is then

$$D_t = \int_{\text{struct}} |\dot{\kappa}| M_p\, ds , \tag{5.5}$$

or, for plastic hinges with rotation rates $\dot{\theta}$,

$$D_t = \Sigma M_{pi} |\dot{\theta}_i| , \tag{5.6}$$

as was established in Com. V., Section 3.4. We see that

1. The yield condition reduces to $|M| = M_p$ and the flow rule to sign $\dot{\theta}_i$ = sign M_i, or $M_i \dot{\theta}_i \geqslant 0$;
2. The stress field reduces to the M diagram;
3. The strain rate field reduces to the distribution of the rate of curvature.

Beam and frame problems have been extensively studied in Com. V.

5.1.3. Arches

In arches, neither the axial strain ϵ_o nor the axial force N can be neglected, even when we are not concerned with instability phenomena (see Com. V., Chapters 6 and 10). The longitudinal strain rate at the level y with respect to the centroid consists of a part due to bending $\dot{\epsilon}_y = y\dot{\kappa}$, and of a part due to axial strain $\dot{\epsilon}_o$.

The total rate of dissipation is

$$D_t = \int_{\text{struct}} (M\dot{\kappa} + N\dot{\epsilon}_o)\, ds , \tag{5.7}$$

where M and N combine to produce complete plastification of the section. Interaction curves M versus N of various sections are given in Com. V., Section 5.4.

Assuming that the shear force V does not influence yielding, the functions M, N, $\dot{\kappa}$ and $\dot{\epsilon}_o$ of the abscissa s are sufficient for the problem at hand.

5.1.4. *Simple plate and shell examples*

In both plates and shells, the thickness t must be small compared to the other dimensions. A plate has a plane median surface and is subjected solely to forces normal to this median plane (when the applied forces are parallel to this plane, the structure is called a disk). A shell has a median surface with at least one finite radius of curvature. A "membrane" is a shell with no bending rigidity.

On the median surface of one of the structures described above, and through a given point P of this surface, draw a line element ds that has P as its center. The normals to the median surface through the points of ds form the "cut based on ds". The stresses transmitted across this cut are statically equivalent to certain forces and couples acting at P, which are proportional to the length of ds. The factors of proportionality are called the "stress resultants" for the considered cut. The "state of stress" at P is specified by the stress resultants for two orthogonal cuts.

1. *Circular plate with constant thickness and rotational symmetry in loading and supports*: With cylindrical coordinates r, θ, z (fig. 5.1), rotational symmetry indicates that the radial and circumferential bending moments, M_r and M_θ, are principal moments (the twisting moment $M_{r\theta}$ vanishes). These bending moments as well as the deflection rate depend solely on the coordinate r.

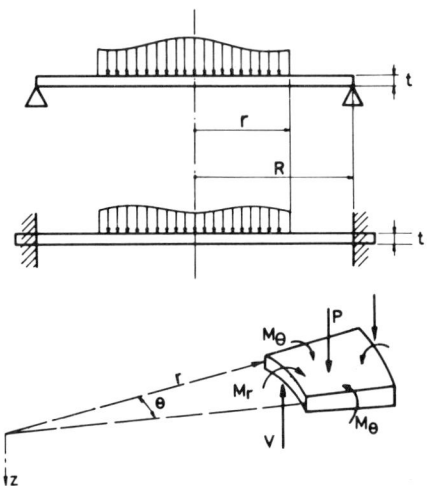

Fig. 5.1.

In accordance with the assumption that material normals remain normal to the deformed median surface, transverse shear strains are neglected. The strain rates are given by

$$\dot{\epsilon}_r = z\dot{k}_r,$$

$$\dot{\epsilon}_\theta = z\dot{k}_\theta,$$

where $\dot{k}_r$ and $\dot{k}_\theta$ are the radial and circumferential (that is the principal) rates of curvature.

In analogy with beams, the dissipation per unit area of the median plane is

$$D = M_r\dot{k}_r + M_\theta\dot{k}_\theta. \tag{5.8}$$

In relation (5.8), M_r and M_θ must combine to completely plastify the volume element $tr\,dr\,d\theta$ at the considered point.

Since the yield condition can be expressed solely in terms of M_r and M_θ, the functions M_r, M_θ, $\dot{k}_r$, $\dot{k}_\theta$ of r are sufficient for the limit analysis of the plate.

2. *Cylindrical shells subjected to rotationally symmetric internal pressure*: Internal resultant forces and moments that the symmetry does not oblige to vanish are shown on fig. 5.2. We immediately note that, because of the rotational symmetry, the circumferential curvature rate $\dot{k}_\theta$ vanishes. Indeed there is no circumferential displacement. Any point of the shell displaces in the meridian plane in which it is contained. Hence, any two neighboring meridian

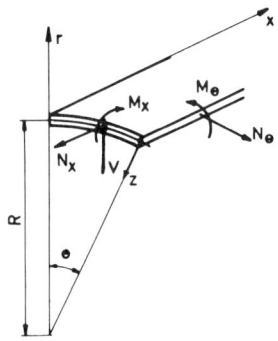

Fig. 5.2.

planes experience no relative rotation, and M_θ does not work. We thus have $\kappa_\theta = \dot{\kappa}_\theta = 0$, because our generalized variables are defined from the expression of the energy dissipated (they are the Lagrange variables). Although the radius of the median surface varies from R to $R + w$, where w is the radial displacement of the median surface, the circumferential strain,

$$\epsilon_\theta = \frac{w}{R - z} \qquad (-t/2 \leqslant z \leqslant t/2),$$

must be regarded as constant (and hence $\kappa_\theta = 0$) because z is negligible with respect to R from the very definition of a shell.

Rates of transversal shear vanish because we assume the material normals to remain normal to the deformed median surface. The dissipation rate is then

$$D = M_x \dot{\kappa}_x + N_x \dot{\epsilon}_{xo} + N_\theta \dot{\epsilon}_{\theta o} . \qquad (5.9)$$

Expressing the yield condition solely in terms of M_x, N_x, and N_θ (see Section 6.4), the functions M_x, N_x, N_θ, $\dot{\kappa}_x$, $\dot{\epsilon}_{xo}$, $\dot{\epsilon}_{\theta o}$ of x and θ will be sufficient for the limit analysis of the shell.

5.2. The general case: choice of the generalized variables

Limit analysis of a structure will use collapse mechanisms of that structure. Denote by $\dot{q}_1, \dot{q}_2, ..., \dot{q}_n$, the *generalized strain rates* suitable for describing these mechanisms. As just seen, the generalized strain rates will be rates of curvature and extension for beams, plates, and shells. For a three-dimensional body, or in the case of plane stress and plane strain, they will be the components of the strain-rate tensor.

The generalized stresses are then, *by definition* [5.4], the stress-type variables $Q_1, ..., Q_n$ that must be associated with the generalized strain rates in order that the specific dissipation be given by

$$D = Q_1 \dot{q}_1 + ... + Q_n \dot{q}_n . \qquad (5.10)$$

The variables Q_i and $\dot{q}_i$ may even be chosen nondimensional, and eq. (5.10) may be rewritten in the slightly more general form

$$D = C(Q_1 \dot{q}_1 + ... + Q_n \dot{q}_n) , \qquad (5.11)$$

where C is a dimensional constant.

We now call "reactions" the generalized stresses that do not *a priori* vanish for reasons of symmetry or equilibrium and that nevertheless do not appear in eq. (5.10) because they correspond to generalized strain rates that, in the considered problem, have been assumed to vanish throughout the structure. For example, in beams, plates, and shells, transversal shear forces are always reactions because normals are assumed to remain normal to the deformed median surface. For the shell of the second example of Section 5.1.4, M_θ is a "reaction" because $\dot\kappa_\theta$ vanishes.

Not only is it always possible to solve problems of limit analysis using only the generalized variables (with no reference to the reactions) but it is also the most efficient way for solving the problems. To that purpose, the reactions must be eliminated from the yield conditions. This will be discussed in Section 5.3.

Let us summarize as follows: The generalized stresses are the only stress-type variables that appear in the expression of the dissipation for the problem at hand. The yield condition is then expressed in terms of these generalized stresses only, by elimination of the reactions.

5.3. Eliminating the reactions

5.3.1. *Introduction*

We now remark that the preceding definitions of generalized stresses and strain rates preserve the validity of formula (2.2) if the stress space Q_i and the strain rate space $\dot q_i$ are superimposed. This fact is sufficient for all fundamental results of Chapter 3 to hold if one substitutes the generalized stresses Q_i for the components of the tensor (σ) and the generalized strain rates $\dot q_i$ for the components of the tensor $(\dot\epsilon)$. Fundamental properties (convexity of yield surface, plastic potential) and fundamental theorems (maximum dissipation, statical and kinematical theorems) are obtained in the very same manner, by mere modification of the terminology.

The only point to clarify is the elimination of the reactions, which, as a rule, initially appear in the most general yield condition.

Consider a structural element and denote by $Q_1, ..., Q_i, ..., Q_n$, the n stress-type variables acting on it. Suppose first that none is a reaction. The yield condition of this element can be written, in a normalized form:

$$F(Q_1,...,Q_i,...,Q_n) = 1 . \qquad (5.12)$$

We assume for the time being that F is a known function. The normality law applies to surface with eq. (5.12) in the superimposed stress space $(Q_1,...,Q_i,...,Q_n)$ and strain-rate space $(\dot{q}_1,...,\dot{q}_i,...,\dot{q}_n)$.

We now suppose that $(n-k)$ relations

$$\dot{q}_{k+1} = 0,$$

$$\dot{q}_{k+2} = 0,$$

$$...$$

$$\dot{q}_n = 0, \qquad (5.13)$$

hold, expressing that, in the particular case under consideration, plastic flow can only occur with $(n-k)$ vanishing generalized strain rates.

According to the normality law, eqs. (5.13) will select a set of points on the surface (5.12) where the projections of a normal vector on the axes $k+1$, ..., n, vanish. This set of points form part of the original yield surface (5.12).

By projecting this part on the $(Q_1,...,Q_k)$ space, one obtains the simplified yield condition:

$$\Phi(Q_1,...,Q_k) = 1, \qquad (5.14)$$

that contains only generalized stresses, and none of the reactions $Q_{k+1}, ..., Q_n$.

5.3.2. *Direct elimination of the reactions through the use of the dissipation function*

Assuming that we know the dissipation function $D(\dot{q}_1,...,\dot{q}_k)$ for a given problem with generalized strain rates $\dot{q}_1, ..., \dot{q}_k$, we can generate the yield surface in the stress space $Q_1, ..., Q_k$ with the technique described in Section 2.4. The normality law obviously applies to that surface. We shall show that the surface obtained in this manner is identical with the surface (5.14) obtained by projection, and that, consequently, the normality law applies to that latter surface.

The basic yield surface (5.12) is, by nature, unique. Hence the manner in which it is obtained is irrelevant. We suppose we construct it from our knowledge of the dissipation function $D(\dot{q}_1,...,\dot{q}_n)$ as described in Section 2.4. We recall that, to every possible mechanism $\dot{\varepsilon}$ with components $\dot{q}_1, ..., \dot{q}_n$ (in the n dimensional space) there corresponds a plane tangent to the yield surface. This plane is normal to $\dot{\varepsilon}$ and distant by $D(\dot{e})$ from the origin (in the direction of $\dot{\varepsilon}$), with $\dot{e}$ the unit vector along $\dot{\varepsilon}$. If we want to select, on the surface

(5.12), the points where eqs. (5.13) are satisfied, we select a subset of tangent planes the normals of which have vanishing projections on axes $k+1, ..., n$.

The wanted simplified surface is the envelope of this subset of planes. Clearly this is identical to constructing the simplified surface directly from the knowledge of $D(\dot{q}_1,...,\dot{q}_k)$ because we so select all mechanisms with $\dot{q}_{k+1} = \dot{q}_{k+2} = ... = \dot{q}_n = 0$ among all possible mechanisms. But this is also identical with the projection procedure of Section 5.3.1 that merely consists of taking the intersection of the subset of planes above in the $(Q_1,...,Q_k)$ space.

To sum up, the same simplified yield condition (5.14) can be obtained either starting from the more general yield condition (5.12) and using conditions (5.13) or directly using the dissipation function $D(\dot{q}_1,...,\dot{q}_k)$.

5.3.3. Remark on the reactions

A distinction must be made between generalized strain rates that vanish because of the very definition of the structure and those that vanish because of special (symmetry) conditions. An example of the first class occurs when a shell or a plate or a beam is defined as a structure in which the direction normal to the median surface is a material direction. Hence, *there never is any transversal rate of shear*. Because of the normality law the yield surface is "cylindrical" with its axis parallel to the shear force axis (fig. 5.3). Consequently, *the shear forces may take any value*. They cannot be determined from mechanism and normality law, but may possibly be obtained from equilibrium conditions.

On the other hand, the stress-type variables that are reactions because of some special (symmetry) conditions are assigned given values by the normality law. For example, in plane stress, $\dot{\epsilon}_2 = 0$ imposes $\sigma_2 = \sigma_1/2 = 1.15\sigma_Y/2$ for the Mises yield condition [fig. 5.4(a)], or $0 \leqslant \sigma_2 \leqslant \sigma_Y$ [fig. 5.4(b)] for

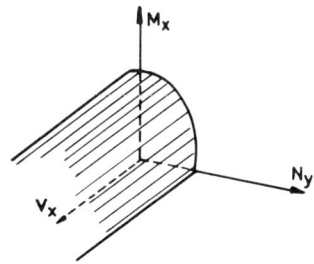

Fig. 5.3.

5.4] OBTAINING YIELD CONDITIONS IN GENERALIZED STRESSES

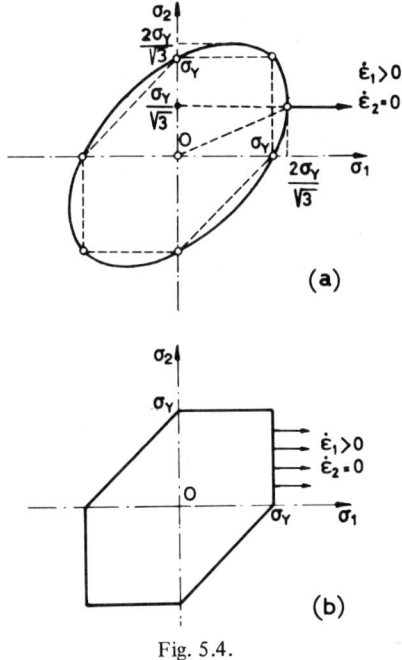

Fig. 5.4.

the Tresca condition. But, at the same time, the equilibrium equations do not contain these reactions because of the special symmetry conditions above.

Hence, when equilibrium and yield conditions are satisfied in terms of generalized stresses only, they can also always be satisfied when reactions are considered. Moreover, note that the equilibrium equations can be obtained from the theorem of virtual work by a variational procedure. No reaction will enter the virtual work equation. This remark proves that it is always possible to eliminate the reactions from the equations of equilibrium.

Obviously, any definition of a structure will have a certain range of validity. Shear forces will have no effect on yielding of shells in most cases, as the thickness-to-span ratio, that ranges from 10 to 30 for beams, goes from 20 to 50 for most plates and even to 500 for some shells [5.5]. Large concentrated forces may change the situation.

5.4. Obtaining yield conditions in generalized stresses

5.4.1. *Method by integration*

Instead of establishing the most general yield condition (5.12) in order to obtain the simplified yield surfaces by section or projection [5.7], it is often desired to obtain the simplified yield condition directly.

To this purpose, the first method is an integration method. On the basis that normals remain normals, generalized strain rates are related to the components of the strain rate tensor at every level in the thickness. The strain rate tensor is related to the stress tensor by the normality law. Integration over the thickness furnishes the yield condition in generalized stresses. We illustrate this method with the example of a *plate*.

Consider a plate of constant thickness, transversally loaded and subjected to arbitrary boundary conditions. Orthogonal cartesian coordinate axes x and y are located in the median plane, and the positive z-axis has the direction of the loads (fig. 5.5). We assume not only that material normals remain normal to the deformed median surface but also that the transversal displacements w are small with respect to the constant thickness t, which, in turn, is small with respect to in-plane dimensions and do not vary with the deformation [5.1].

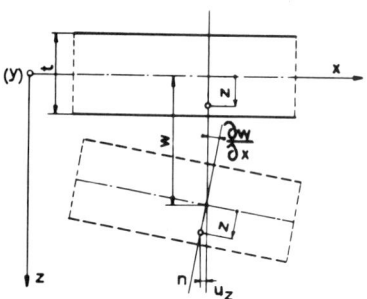

Fig. 5.5.

The deflected shape of the plate is then completely described by the single function $w(x,y)$ because we have (fig. 5.5):

$$u = -z \frac{\partial w}{\partial x}, \qquad v = -z \frac{\partial w}{\partial y}, \qquad w = w(x,y), \qquad (5.15)$$

where u, v, w, are the components of the displacements of the points of the midplane on the x-, y-, and z-axes, respectively.

Using eqs. (1.15) and (1.16), we obtain

5.4] OBTAINING YIELD CONDITIONS IN GENERALIZED STRESSES

$$\epsilon_x = -z \frac{\partial^2 w}{\partial x^2},$$

$$\epsilon_y = -z \frac{\partial^2 w}{\partial y^2},$$

$$\epsilon_z = 0,$$

$$\gamma_{xy} = z \frac{\partial^2 w}{\partial x \partial y},$$

$$\gamma_{xz} = \gamma_{yz} = 0. \tag{5.16}$$

Because the midplane is deformation-free, we consider that the resultant forces parallel to this plane always vanish.

Hence, the remaining resultant forces and moments are shown in fig. 5.6, where all forces and moments are positive. Moments are related to stresses as follows:

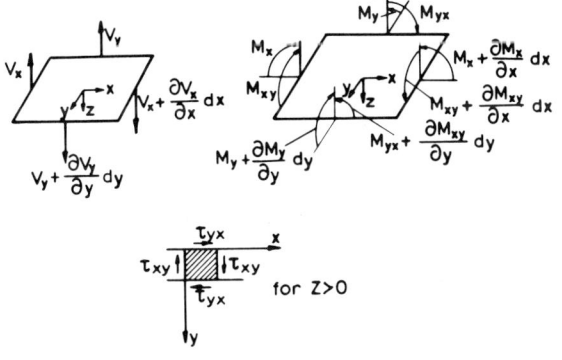

Fig. 5.6.

$$M_x = \int_{-t/2}^{t/2} \sigma_x z \, dz ,$$

$$M_y = \int_{-t/2}^{t/2} \sigma_y z \, dz ,$$

$$M_{xy} = -M_{yx} = \int_{-t/2}^{t/2} -\tau_{xy} z \, dz . \tag{5.17}$$

Note that τ_{xy} is positive as shown in fig. 5.6.

Now, the energy dissipated per unit area of the median plane is

$$\mathcal{E} = \int_{-t/2}^{t/2} (\sigma_x \epsilon_x + \sigma_y \epsilon_y + \tau_{xy} \gamma_{xy}) \, dz . \tag{5.18}$$

Using eqs. (5.16) in eq. (5.18), we obtain

$$\mathcal{E} = \left(-\frac{\partial^2 w}{\partial x^2}\right) \int_{-t/2}^{t/2} \sigma_x z \, dz + \left(-\frac{\partial^2 w}{\partial y^2}\right) \int_{-t/2}^{t/2} \sigma_y z \, dz$$

$$+ \left(2 \frac{\partial^2 w}{\partial x \partial y}\right) \int_{-t/2}^{t/2} \tau_{xy} z \, dz . \tag{5.19}$$

Within the framework of small deflection theory, the factors in parenthesis are the curvatures κ_x, κ_y and twice the torsion κ_{xy} of the deflected surface, respectively:

$$-\frac{\partial^2 w}{\partial x^2} = \frac{1}{\rho_x} \equiv \kappa_x , \quad -\frac{\partial^2 w}{\partial y^2} = \frac{1}{\rho_y} \equiv \kappa_y , \quad -2 \frac{\partial^2 w}{\partial x \partial y} = \frac{2}{\rho_{xy}} \equiv 2\kappa_{xy} . \tag{5.20}$$

With the definitions (5.19) of the moments, relation (5.19) can hence be written $\mathcal{E} = M_x \kappa_x + M_y \kappa_y + 2 M_{xy} \kappa_{xy}$, and the specific power of dissipation is

$$D = M_x \dot{\kappa}_x + M_y \dot{\kappa}_y + 2 M_{xy} \dot{\kappa}_{xy} . \tag{5.21}$$

The generalized stressesses are M_x, M_y, M_{xy} and the corresponding generalized strain rates are $\dot{\kappa}_x$, $\dot{\kappa}_y$ and $2\dot{\kappa}_{xy}$, respectively.

Because we actually consider each layer of thickness dz to be in plane stress, the yield condition is

$$\sigma_R(\sigma_x, \sigma_y, \tau_{xy}) = \sigma_Y .\qquad(5.22)$$

On the other hand, inspection of relations (5.16) reveals that the strain-rate vector has components proportional to z. Hence, the corresponding stress point is the same for all z with same sign. If we now assume that the yield surface with eq. (5.22) is symmetric with respect to the origin (as for the von Mises and Tresca conditions), the stress point goes to a position symmetric with respect to the origin when z changes sign* (fig. 5.7). If the yield state of stress is $(\sigma_x, \sigma_y, \tau_{xy})$ for positive z, it is $(-\sigma_x, -\sigma_y, -\tau_{xy})$ for negative z, and one obtains

$$M_x = \sigma_x \frac{t^2}{4}, \quad M_y = \sigma_y \frac{t^2}{4}, \quad M_{xy} = \tau_{xy} \frac{t^2}{4} .\qquad(5.23)$$

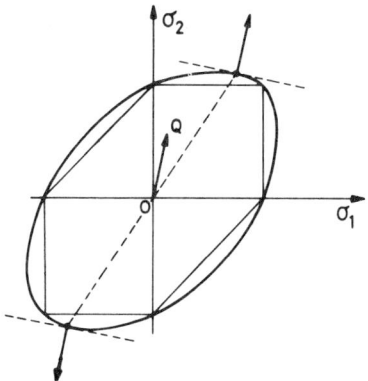

Fig. 5.7.

Because moments are seen to be propositional to the stress components, *the yield surface in the space of moments will have the same form as in the space of stress components.*

It often proves convenient to use nondimensional (also called "reduced")

---

* Flat portions of yield surface must here be regarded as the limits of slightly curved portions.

variables. Stress components are rendered nondimensional by division by σ_Y, and relation (5.22) takes the "canonic" form

$$\Phi\left(\frac{\sigma_x}{\sigma_Y}, \frac{\sigma_y}{\sigma_Y}, \frac{\tau_{xy}}{\sigma_Y}\right) = 1 \ . \tag{5.24}$$

Similarly, we define *reduced moments*

$$m_x = \frac{M_x}{M_p}, \qquad m_y = \frac{M_y}{M_p}, \qquad m_{xy} = \frac{M_{xy}}{M_p} \ , \tag{5.25}$$

where

$$M_p = \sigma_Y \frac{t^2}{4} \tag{5.26}$$

is the yield moment for uniaxial bending. From definitions (5.25) and relations (5.23), we obtain

$$m_x = \frac{\sigma_x}{\sigma_Y}, \qquad m_y = \frac{\sigma_y}{\sigma_Y}, \qquad m_{xy} = \frac{\tau_{xy}}{\sigma_Y} \ . \tag{5.27}$$

With relations (5.27), condition (5.24) becomes

$$\Phi(m_x, m_y, m_{xy}) = 1 \ . \tag{5.28}$$

We see that the yield condition (5.28) in reduced moments is identical to that in reduced stresses.

For example, von Mises' condition (1.34) for plane stress, using reduced stresses, becomes

$$\left(\frac{\sigma_x}{\sigma_Y}\right)^2 + \left(\frac{\sigma_y}{\sigma_Y}\right)^2 - \frac{\sigma_x}{\sigma_Y} \cdot \frac{\sigma_y}{\sigma_Y} + 3\left(\frac{\tau_{xy}}{\sigma_Y}\right)^2 = 1 \ .$$

Hence, the corresponding yield condition for a plate is simply

$$m_x^2 + m_y^2 - m_x m_y + 3 m_{xy}^2 = 1 \ . \tag{5.29}$$

Similarly, Tresca's condition gives

$$\max\ [|m_1|, |m_2|, |m_1 - m_2|] = 1 \ . \tag{5.30}$$

5.4.2. Use of the power of dissipation

Consider the power of dissipation

$$D(\dot{q}_1,...,\dot{q}_k) = Q_1\dot{q}_1 + ... + Q_k\dot{q}_k, \quad (5.31)$$

where $Q_1, ..., Q_k$ and $\dot{q}_1, ..., \dot{q}_k$ are the generalized variables.

Because the function $D(\dot{q}_1,...,\dot{q}_k)$ is homogeneous with the order one, Euler's theorem on homogeneous functions gives

$$D(\dot{q}_1,...,\dot{q}_k) = \frac{\partial D}{\partial \dot{q}_1}\dot{q}_1 + ... + \frac{\partial D}{\partial \dot{q}_k}\dot{q}_k. \quad (5.32)$$

By comparing eqs. (5.31) and (5.32) we find

$$Q_1 = \frac{\partial D(\dot{q}_1,...,\dot{q}_k)}{\partial \dot{q}_1},$$

$$...$$

$$Q_k = \frac{\partial D(\dot{q}_1,...,\dot{q}_k)}{\partial \dot{q}_k}. \quad (5.33)$$

Relations (5.33) are the parametric equations of the yield surface, with parameters $\dot{q}_1, ..., \dot{q}_k$. Actually, because a yield mechanism at a point defines the generalized strain rates except for a common positive factor, there are only $k - 1$ parameters: for example the ratios of the $\dot{q}_i$ to one of them.

We illustrate the method in the example of the yield condition of a shell of revolution with axisymmetric loading [5.8]. Fig. 5.8 shows an element of the shell with the nonvanishing resultant forces and moments (per unit of length) acting on it. Principal directions are ϕ and θ because of the symmetry. We denote by $\dot{\epsilon}_\theta$ and $\dot{\epsilon}_\phi$ the principal rates of strain of the midsurface, and by $\dot{k}_\theta$ and $\dot{k}_\phi$ the rates of curvature of that surface.

Because material normals remain normal to the deformed median surface, the dissipation per unit area of this surface is

$$D = M_\phi \dot{k}_\phi + M_\theta \dot{k}_\theta + N_\phi \dot{\epsilon}_\phi + N_\theta \dot{\epsilon}_\theta, \quad (5.34)$$

so that $M_\phi, M_\theta, N_\phi, N_\theta$ are the generalized stresses.

We use Tresca's condition for plane stress ($\sigma_z \equiv \sigma_3 = 0$; see fig. 5.9). To obtain the direction parameters of the outward pointing normal to the hexagonal cylinder at the various points of the plane hexagonal section of

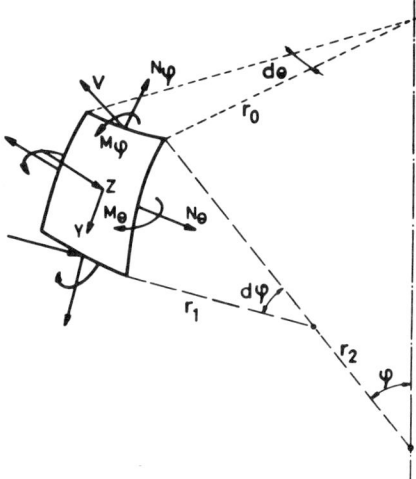

Fig. 5.8.

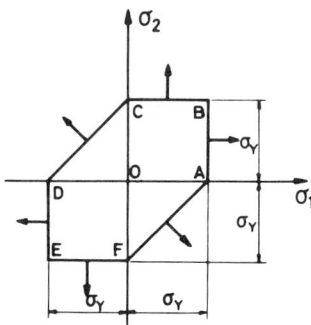

Fig. 5.9.

fig. 5.9, we simply note that the normal to the plane hexagon at one of its point is the projection in the $(O\sigma_1, O\sigma_2)$ plane of the normal to the hexagonal cylinder at the same point. We further recall that the sum of the direction parameters p, q, and r, must vanish (see Section 2.5). We then can write table 5.1, valid for points on the hexagon other than vertices $A, B, ..., F$. From table 5.1 we conclude that, according to eq. (2.8), the dissipation per unit volume is given by

Table 5.1.

Plastic regimes	p	q	r	$\dot{\epsilon}_1$	$\dot{\epsilon}_2$	$\dot{\epsilon}_3$	$|\dot{\varepsilon}|$
AB	1	0	−1	$|\dot{\epsilon}_1|$	0	$-\dot{\epsilon}_1$	$\sqrt{2}|\dot{\epsilon}_1|$
BC	0	1	−1	0	$|\dot{\epsilon}_2|$	$-\dot{\epsilon}_2$	$\sqrt{2}|\dot{\epsilon}_2|$
CD	−1	1	0	$-\dot{\epsilon}_2$	$|\dot{\epsilon}_2|$	0	$\sqrt{2}|\dot{\epsilon}_2|$
DE	−1	0	1	$-\dot{\epsilon}_3$	0	$|\dot{\epsilon}_3|$	$\sqrt{2}|\dot{\epsilon}_3|$
EF	0	−1	1	0	$-\dot{\epsilon}_3$	$|\dot{\epsilon}_3|$	$\sqrt{2}|\dot{\epsilon}_3|$
FA	1	−1	0	$|\dot{\epsilon}_1|$	$-\dot{\epsilon}_1$	0	$\sqrt{2}|\dot{\epsilon}_1|$

$$D_v = \sigma_Y |\dot{\epsilon}_i| . \tag{5.35}$$

When the stress point is at a vertex, we may have a vertex of type A where only one stress component is not zero and hence has the value $\pm \sigma_Y$. At point A, $\sigma_1 = \sigma_Y$, $\sigma_2 = \sigma_3 = 0$. The directions of the vector with components $\dot{\epsilon}_1$, $\dot{\epsilon}_2$, $\dot{\epsilon}_3$ are bounded by those of the vectors associated with the regimes FA and AB. We thus have:

$$\dot{\epsilon}_1 \geqslant |\dot{\epsilon}_2|, \quad \dot{\epsilon}_1 \geqslant |\dot{\epsilon}_3| \quad \text{and} \quad \dot{\epsilon}_1 > 0 .$$

Thus,

$$D_v = \sigma_Y \max |\dot{\epsilon}_i| . \tag{5.36}$$

A second vertex is of type B where $\sigma_1 = \sigma_2 = \sigma_Y$, $\sigma_3 = 0$, $D_v = \sigma_Y(\dot{\epsilon}_1 + \dot{\epsilon}_2)$. We also have $\dot{\epsilon}_1 + \dot{\epsilon}_2 = -\dot{\epsilon}_3$, with $\dot{\epsilon}_1 > 0$ and $\dot{\epsilon}_2 > 0$.

Thus, eq. (5.36) still holds. The same conclusion would be obtained for point E. As eq. (5.35) is a particular case of eq. (5.36) all cases are covered by eq. (5.36). By integration over the thickness t, we find the dissipation D per unit midsurface of the shell. The strain rates $\dot{\epsilon}_i$ vary with z according to

$$\dot{\epsilon}_{1z} = \dot{\epsilon}_1 + \dot{\kappa}_1 z , \quad \dot{\epsilon}_{2z} = \dot{\epsilon}_2 + \dot{\kappa}_2 z , \quad \dot{\epsilon}_{3z} = -(\dot{\epsilon}_{1z} + \dot{\epsilon}_{2z}) . \tag{5.37}$$

The direction 3 is that of the z-axis, and the directions 1 and 2 coincide with the ϕ and θ directions. Note that the directions ϕ and θ are interchangeable as far as the yield condition is concerned.

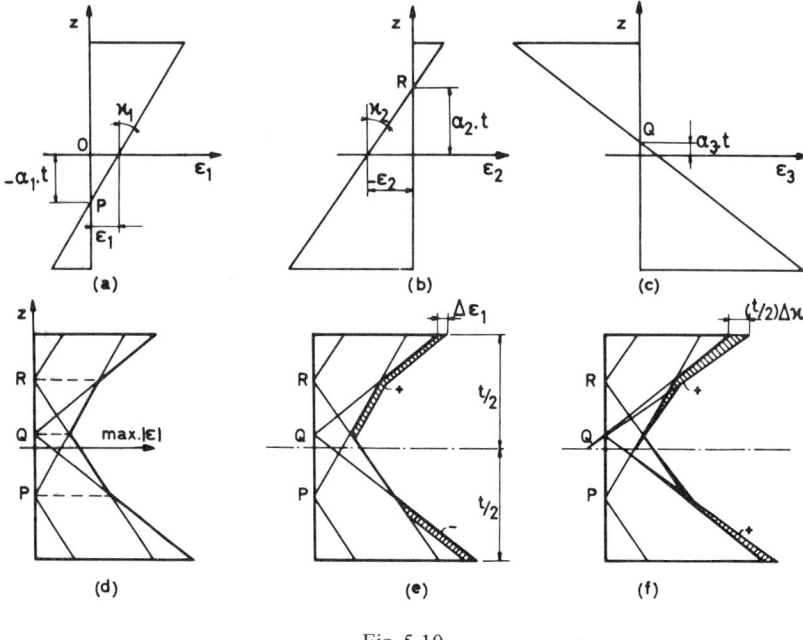

Fig. 5.10.

A typical distribution of $\dot{\epsilon}_{iz}$ is shown on fig. 5.10. The three parameters $\alpha_1, \alpha_2, \alpha_3$, defined by

$$\alpha_1 = -\frac{\dot{\epsilon}_1}{t\dot{\kappa}_1}, \qquad \alpha_2 = -\frac{\dot{\epsilon}_2}{t\dot{\kappa}_2}, \qquad \alpha_3 = \frac{\dot{\epsilon}_1 + \dot{\epsilon}_2}{t(\dot{\kappa}_1 + \dot{\kappa}_2)}, \qquad (5.38)$$

are sufficient to describe this distribution completely (see fig. 5.10 for notations). They locate the points P, Q, R of zero strain rate. The diagram of max $|\dot{\epsilon}_i|$ is then constructed [fig. 5.10 (d)]. The dissipation D is given by the area of that diagram.

Now, relations (5.33) specialize to

$$N_1 = \frac{\partial D}{\partial \dot{\epsilon}_1}, \qquad N_2 = \frac{\partial D}{\partial \dot{\epsilon}_2}, \qquad M_1 = \frac{\partial D}{\partial \dot{\kappa}_1}, \qquad M_2 = \frac{\partial D}{\partial \dot{\kappa}_2}. \qquad (5.39)$$

The derivatives in relations (5.39) are most readily evaluated from the variations of the areas of the diagram in fig. 5.10 (d) for small variations of $\dot{\epsilon}_1, \dot{\epsilon}_2, \dot{\kappa}_1, \dot{\kappa}_2$.

5.4] OBTAINING YIELD CONDITIONS IN GENERALIZED STRESSES

For example, limiting values of the ratios of the dashed areas in fig. 5.10 to the corresponding variations of the parameters give (see [5.8]):

$$N_1 = \sigma_Y t[\tfrac{1}{2}-\alpha_3-(\tfrac{1}{2}+\alpha_1)] = -\sigma_Y t(\alpha_1+\alpha_3),$$

$$M_1 = \sigma_Y \frac{t^2}{4}[\tfrac{1}{4}-\alpha_3^2+(\tfrac{1}{4}-\alpha_1^2)] = \sigma_Y \frac{t^2}{4}[1-2(\alpha_1^2+\alpha_3^2)].$$

Similarly,

$$N_2 = -\sigma_Y t(\alpha_3+\alpha_2),$$

$$M_2 = \sigma_Y \frac{t^2}{4}[1-2(\alpha_2^2+\alpha_3^2)].$$

With the following definitions of the reduced generalized stresses,

$$n_1 = \frac{N_1}{N_p}, \quad n_2 = \frac{N_2}{N_p}, \quad m_1 = \frac{M_1}{M_p}, \quad m_2 = \frac{M_2}{M_p}, \tag{5.40}$$

where $N_p = \sigma_Y t$ and $M_p = \sigma_Y(t^2/4)$, the preceding relations become

$$n_1 = -(\alpha_1+\alpha_3), \quad\quad n_2 = -(\alpha_2+\alpha_3),$$

$$m_1 = 1 - 2(\alpha_1^2+\alpha_3^2), \quad m_2 = 1 - 2(\alpha_2^2+\alpha_3^2). \tag{5.41}$$

Eqs. (5.41) are parametric equations of the desired yield surface, in (n_1, n_2, m_1, m_2) space.

To obtain the complete surface, all relative positions of points P, Q, R of fig. 5.10 must be considered, with corresponding values of α_1, α_2, α_3.

Table 5.2. Points P, Q, R are distinct

Central point	$\pm n_1$	$\pm n_2$	$\pm m_1$	$\pm m_2$
P	$-(\alpha_1+\alpha_3)$	$-(\alpha_3-\alpha_2)$	$1 - 2(\alpha_1^2+\alpha_3^2)$	$2(\alpha_2^2-\alpha_3^2)$
Q	$-(\alpha_1+\alpha_3)$	$-(\alpha_3+\alpha_2)$	$1 - 2(\alpha_1^2+\alpha_3^2)$	$1 - 2(\alpha_3^2+\alpha_2^2)$
R	$-(\alpha_3-\alpha_1)$	$-(\alpha_3+\alpha_2)$	$2(\alpha_1^2-\alpha_3^2)$	$1 - 2(\alpha_3^2+\alpha_2^2)$

Table 5.3. Points P, Q, R are not distinct

Coincidence	Yield surface
$P \equiv Q$	$m_1 = \pm(1-n_1^2)$
$Q \equiv R$	$m_2 = \pm(1-n_2^2)$
$R \equiv P$	$m_1 - m_2 = \pm[1-(n_1-n_2)^2]$

This discussion [5.8] gives the results shown in tables 5.2 and 5.3. Note that points P, Q, R must fall within the shell thickness. Hence α_1, α_2 and α_3 are bounded by $-\frac{1}{2}$ and $+\frac{1}{2}$.

Two important particular situations occur when either one of the axial forces or one of the moments can be eliminated. The yield surface is then an ordinary surface in three-dimensional space (and not a hypersurface in a space of a higher number of dimensions). As the yield surface is symmetric with respect to the origin, representation and specification of one half of it is sufficient.

Consider first $n_2 = 0$ (see [5.8]). The yield surface is shown in fig. 5.11. Part I is a plane with the equation

$$m_2 = 1,\qquad(5.42)$$

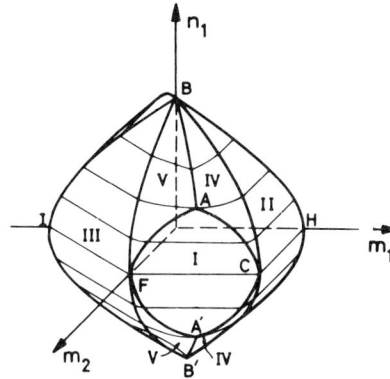

Fig. 5.11.

5.4] OBTAINING YIELD CONDITIONS IN GENERALIZED STRESSES

corresponding to $Q \equiv R$ (table 5.3), + sign, with $n_2 = 0$. Part II corresponds to $P \equiv Q$, + sign. Its equation is

$$m_1 = 1 - n_1^2 \tag{5.43}$$

(cylinder with axis Om_2). Part III corresponds to $P \equiv R$, – sign. Its equation is

$$m_1 - m_2 = n_1^2 - 1 . \tag{5.44}$$

Eqs. of parts IV and V are obtained from table 5.1, taking Q and R for the central points, respectively. Condition $n_2 = 0$ gives $\alpha_3 = -\alpha_2$, and elimination of α_3 and α_2 from the remaining three relations furnishes, using the plus signs,

for part IV, $\quad m_1 = 1 - 2\left[\left(n_1 + \dfrac{\sqrt{1-m_2}}{2}\right)^2 + \dfrac{1-m_2}{4}\right],$ \hfill (5.45)

for part V, $\quad m_1 = 2\left[\left(n_1 + \dfrac{\sqrt{1-m_2}}{2}\right)^2 - \dfrac{1-m_2}{4}\right].$ \hfill (5.46)

The second interesting case occurs when m_2 is a reaction, because $\dot{\kappa}_2 = 0$ (cylindrical shells with axisymmetrical loading, $\dot{\kappa}_2$ denoting circumferential rate of curvature [5.9, 5.10, and 4.3]). Elimination of m_2 results in the yield surface of fig. 5.12. Part I is a plane with equation

$$n_2 = 1 , \tag{5.47}$$

bounded by parabolic arcs

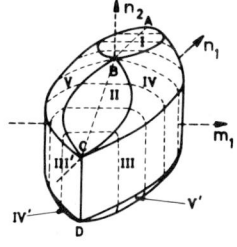

Fig. 5.12.

$$m_1 = \pm 2n_1(1-n_1) \,. \tag{5.47'}$$

Part II is also a plane bounded by parabolic arcs. The relevant equations are

$$n_2 - n_1 = 1 \,, \tag{5.48}$$

$$m_1 = \pm 2n_1(1+n_1) \,. \tag{5.48'}$$

Part III is a portion of the parabolic cylinder with equation

$$m_1 = 1 - n_1^2 \,, \tag{5.49}$$

bounded by its intersection with the planes

$$m_1 = 0 \,,$$

$$2n_2 - n_1 = \pm 1 \,. \tag{5.49'}$$

Parts IV and V belong to the paraboloids with equations

$$m_1 = \pm \tfrac{1}{2}[2-(2n_2-1)^2-(2n_2-2n_1-1)^2] \,. \tag{5.50}$$

Note that the yield surfaces are formed of parts with different analytic expressions, and the normality law can only be formulated explicitly when one knows on what part the stress point is located.

5.4.3. *Purely statical method: adaptation of the reactions*

To begin with, suppose there is no reaction among the stress-type variables, the number of which is, say, three.

A part of the yield surface is shown schematically on fig. 5.13. Choose fixed values Q_1^0 and Q_2^0 such that the point with the coordinates $Q_1^0, Q_2^0, 0$ falls within the yield surface. Now let Q_3 vary from zero to the highest value Q_3 compatible with the yield condition of the material (expressed in terms of the ordinary stress components $\sigma_x, ..., \tau_{xy}, ...$). The coordinates Q_1^0, Q_2^0, Q_3 are that of a point P of the desired yield surface. In a general manner one attributes fixed values to all generalized stresses but one, which will be given the extreme magnitudes compatible with the yield condition of the material. In this way, the yield surface is generated point by point.

Now, if there exist reactions, the only nonfixed generalized stress depends not only on the yield condition of the material but remains a function of the reactions which, as a rule, are not fixed.

5.4] OBTAINING YIELD CONDITIONS IN GENERALIZED STRESSES 81

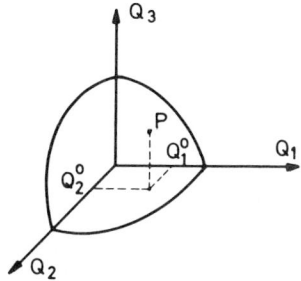

Fig. 5.13.

The following theorem has been proved [5.5]:

Adaptation: If one fixes all generalized stresses but one, the reactions adapt themselves to give the nonfixed generalized stress a maximum positive or minimum negative value (see Section 5.7 for proof).

Thus, the procedure just described still holds when reactions exist, and the reactions may be completely ignored.

We illustrate the results by these two examples:

1. *Bar with square cross section, subjected to two orthogonal bending moments M_x and M_y* (fig. 5.14). We treat this problem by a variational procedure [3.8]. When the cross section is completely plastic, we have at all points $|\sigma| = \sigma_Y$. If $y = \phi(x)$ is the equation of the boundary between the regions of tensions ($\sigma = \sigma_Y$) and compressions ($\sigma = -\sigma_Y$), we have

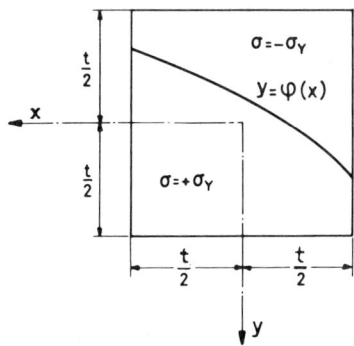

Fig. 5.14.

$$M_x = \int_{-t/2}^{t/2} \left[\int_{-t/2}^{\phi(x)} x(-\sigma_Y) \, dy + \int_{\phi(x)}^{t/2} x\sigma_Y \, dy \right] dx$$

$$= -2\sigma_Y \int_{-t/2}^{t/2} x\phi(x) \, dx \,,$$

and, similarly,
$$M_y = \sigma_Y \int_{-t/2}^{t/2} \left[\frac{t^2}{4} - \phi(x)^2 \right] dx \,. \tag{5.51}$$

Consider a fixed value of M_x and assume M_y to be an analytic maximum. If the stress distribution is then varied by an arbitrary small amount $\delta\sigma$, we have $\delta M_x = 0$ because M_x is fixed; and $\delta M_y = 0$ because M_y is a maximum. Hence, α being a parameter, we can write, using eq. (5.51),

$$\delta M_y + \alpha \delta M_x = -2\sigma_Y \int_{-t/2}^{t/2} [\phi(x) + \alpha x] \, \delta\phi(x) \, dx = 0 \,,$$

for all $\delta\sigma$, that is for all $\delta\phi(x)$. Consequently, $\phi(x) + \alpha x = 0$.

We see from the preceding relation that the boundary between tensions and compressions is a ray emanating from the origin. Hence, we readily obtain

$$M_x = \tfrac{4}{3}\sigma_Y \frac{t^3}{8} \alpha \,, \qquad M_y = \tfrac{2}{3}\sigma_Y \frac{t^3}{8} (3-\alpha^2) \,.$$

Eliminating α from the two equations above, and letting

$$m_x = \frac{M_x}{M_p} \,, \qquad m_y = \frac{M_y}{M_p} \,, \qquad M_p = \sigma_Y \frac{t^3}{4} \,,$$

we finally obtain the desired equation

$$m_y + \tfrac{3}{4} m_x^2 = 1 \tag{5.52}$$

of the yield curve.

Note that eq. (5.52) is valid only for $|\alpha| \leq 1$, that is for $|m_x/m_y| \leq 1$. Because the yield curve is symmetric with respect to the rays $m_x/m_y = \pm 1$, the remaining part is obtained without difficulty.

2. *Circular cylindrical "sandwich" shell without axial force, subjected to axially symmetrical loading.* Because of the absence of axial force and of the symmetry of revolution, the only nonvanishing stress type variables are V_x, M_x, M_θ and N_θ (see fig. 5.15).

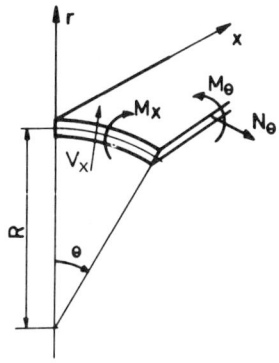

Fig. 5.15.

Shear forces V_x are reactions (see Section 3.3.3).

For a circular cylindrical shell, symmetry of revolution enforces $\dot{k}_\theta = 0$. Hence, M_θ is a reaction and we must simply determine the yield condition in terms of M_x and N_θ.

The "sandwich" shell is formed of a core with thickness H and two face sheets with thickness $t/2$ each (see fig. 5.16). The core carries exclusively shear forces V_x, to which it is always exceedingly resistant. The face sheets carry all other stresses and are assumed in a state of plane stress.

Denote by $\sigma_{\theta e}$ and σ_{xe} the principal normal stresses in the external face sheet, and by $\sigma_{\theta i}$ and σ_{xi} those in the internal sheet. We then have:

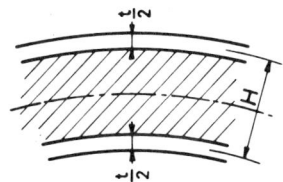

Fig. 5.16.

$$N_\theta = \frac{t}{2}(\sigma_{\theta i}+\sigma_{\theta e}),$$

$$M_\theta = \frac{tH}{4}(\sigma_{\theta i}-\sigma_{\theta e}),$$

$$M_x = \frac{tH}{4}(\sigma_{xi}-\sigma_{xe}). \tag{5.53}$$

Introducing reduced generalized stresses

$$n_\theta = \frac{N_\theta}{N_p}, \quad m_\theta = \frac{M_\theta}{M_p}, \quad m_x = \frac{M_x}{M_p},$$

where $N_p = \sigma_Y t$ and $M_p = \sigma_Y(tH/2)$, we have

$$n_\theta = \frac{\sigma_{\theta i}+\sigma_{\theta e}}{2\sigma_Y},$$

$$m_\theta = \frac{\sigma_{\theta i}-\sigma_{\theta e}}{2\sigma_Y},$$

$$m_x = \frac{\sigma_{xi}-\sigma_{xe}}{2\sigma_Y}. \tag{5.54}$$

Assume the material of the sheets obeys von Mises' yield condition:

$$\sigma_x^2 + \sigma_\theta^2 - \sigma_x\sigma_\theta = \sigma_Y^2, \tag{5.55}$$

represented by the ellipse of fig. 5.17. The state of stress in the shell is represented by points e and i on fig. 5.17, with coordinates $(\sigma_{xe},\sigma_{\theta e})$ and $(\sigma_{xi},\sigma_{\theta i})$, respectively. According to relations (5.54), the coordinates of the midpoint c of segment ei give n_x and n_θ, whereas the projections of the segment ei on the axes give m_x and m_θ (positive factors $1/\sigma_Y$ and $1/2\sigma_Y$ being irrelevant). Because $n_x = 0$, the point c must remain on the σ_θ axis. For a given position of point c (between points A and B) corresponding to some value of n_θ, plastification of the shell element requires that at least one of the two points e and i be on the yield locus.

Now, the adaptation theorem tells us that the slope of segment ei must be such that its projection on the σ_x axis is a maximum. This condition yields

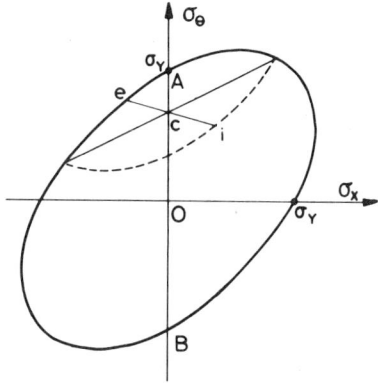

Fig. 5.17.

$$m_\theta = \frac{m_x}{2}, \tag{5.56}$$

and points e and i both lie on the yield locus.

It is easily seen that when point c moves from A to B with condition (5.56) satisfied, the interaction relation is

$$\tfrac{3}{4}m_x^2 + n_\theta^2 = 1. \tag{5.57}$$

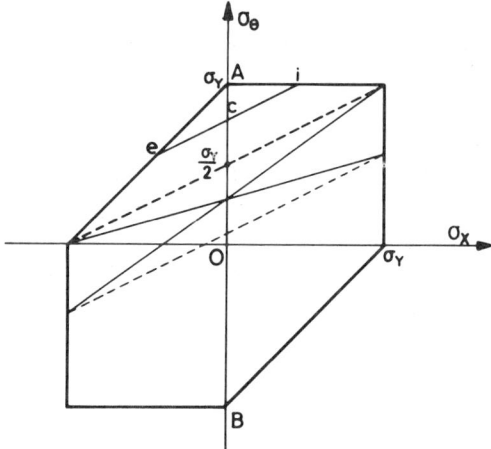

Fig. 5.18.

If the face-sheet material of the shell obeys Tresca's yield condition, represented on fig. 5.18, a similar analysis will furnish (a) for $0 \leq n_\theta \leq \frac{1}{2}$,

$$n_\theta \leq \frac{m_\theta}{m_x} \leq 1 - n_\theta \quad \text{and} \quad m_x = 1, \tag{5.58}$$

and (b) for $\frac{1}{2} \leq n_\theta \leq 1$,

$$m_\theta = \frac{m_x}{2} \quad \text{and} \quad \frac{m_x}{2} + n_\theta = 1. \tag{5.59}$$

The interaction curve is given by eqs. (5.58) and (5.59).

5.4.4. *Method of lower and upper bounds*

Let us imagine an isolated element of a structure. For a beam, such an element may be specified by a line element of length dx along the undeformed axis. This element is bounded by the normal cross sections of the beam through the endpoints of the line element and by part of the lateral surface of the beam. For a plate or shell, an element may be specified by an infinitesimal rectangle of sides ds_1 and ds_2 on the undeformed median surface; it is bounded by the normals of this surface through the points of the rectangle and by parts of the two surfaces of the shell, fig. 5.19.

A structural element of this kind may be regarded as a free body subjected

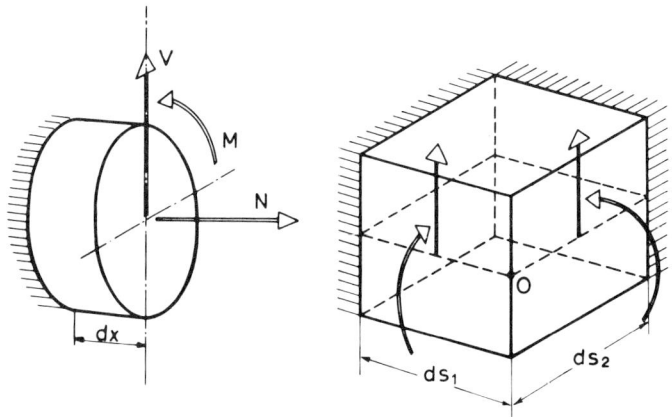

Fig. 5.19.

5.4]   OBTAINING YIELD CONDITIONS IN GENERALIZED STRESSES

to the resultant forces and couples of the stresses transmitted by neighboring elements and such loads as may be directly applied to the considered element. Any combination of stress resultants that causes the element to yield specifies a point of the yield locus.

From this point of view, the fundamental theorems of limit analysis of Sections 3.1 and 3.2 can be used to obtain yield surfaces in generalized stress space:

1. Any statically admissible stress distribution on the element will furnish generalized stresses that will be the coordinates of a point on or within the yield surface;
2. Any kinematically admissible strain rate distribution across the element will be associated, through the normality law, with generalized stresses that will be the coordinates of a point on or outside the yield surface.

The yield surface can thus be bounded from the interior and exterior.

The application of the lower-bound theorem is obviously identical to the statical method of Section 5.4.3. Indeed, reciprocity of shearing stresses is the only condition enforced by local equilibrium, and statical admissibility of a stress distribution reduces to not violating the yield condition (in terms of stress components). Usually, the stress distribution is varied to maximize one generalized stress while fixing the others, in order to move the representing point from the inside onto the yield surface, as explained in Section 5.4.3. Note that the stress distributions will in general correspond to nonvanishing "reactions". For example, suppose that the curvature rate $\dot{\kappa}_x$ vanishes whereas the extension rate $\dot{\epsilon}_x$ does not vanish. Hence, σ_x cannot be eliminated from the yield condition in terms of stress components, and the distribution of σ_x on the cross section will generally correspond to $M_x \neq 0$. However, according to the adaptation theorem, the stress distributions can be chosen without regard to the values of the reactions.

On the other hand, when the internal restraint concerns strain-rate components at every point (for example $\dot{\gamma}_{xy} = 0$ or $\dot{\epsilon}_x = 0$ everywhere), the corresponding stress component (τ_{xy} or σ_x) can be eliminated from the yield condition (see Section 5.3.3 and Chapter 10) as well as the reactions they produce.

Consider now the application of the upper-bound theorem. Suppose that, within the frame work of the basic assumptions, the flow mechanism of a structural element is completely known (as in the shell example treated in Section 5.4.2). Application of the upper-bound theorem will then directly furnish the exact yield surface. Actually, the procedure is identical to using the dissipation power (Section 5.4.2), as we shall show.

Let $\dot{q}_1, ..., \dot{q}_n$ be the generalized strain rates (curvature rates, extension rates of the median surface) used to describe the flow mechanism of an element. The corresponding strain-rate components are given by relations of the type

$$\dot{\epsilon} = \dot{\epsilon}(\dot{q}_1,...,\dot{q}_n,z), \tag{5.60}$$

where z is the distance of the considered point to the midsurface. Now, regard the $\dot{q}_i$ as the parameters of the problem. Through relations (5.60), each set of $\dot{q}_i$ gives a distribution of $\dot{\epsilon}(z)$ to which the yield condition and the normality law relate a distribution $\sigma(z)$. Hence, we can write

$$\sigma = \sigma(\dot{q}_1,...,\dot{q}_n,z). \tag{5.61}$$

Next we obtain the corresponding generalized stress Q_i by integration over the thickness. The type of integration to be done is often obvious: if $\dot{q}_i$ is a curvature rate, Q_i is the corresponding moment, if $\dot{q}_i$ is an extension rate, Q_i is the corresponding axial force, etc. It must however be emphasized that, as a rule, the type of integration to achieve is determined by the *definition* of Q_i. This definition is related to the expression of dissipation power D (see relation 5.10). Integration over z will result in a function of $\dot{q}_1, ..., \dot{q}_n$ as shown by relation (5.61). This function Q_i must be such that $D = \Sigma Q_i \dot{q}_i$. Because D must be homogeneous of the order one in $\dot{q}_i$ (see Section 2.4) we have $Q_i = \partial D / \partial \dot{q}_i$. We conclude that, from the very definition of Q_i, integration over z must yield the same parametric form for Q_i as is obtained from relations (5.33).

It follows that the use of the lower-bound and upper-bound theorems to obtain yield surfaces in generalized stresses will differ from the other methods only if the yield mechanism *of the element* does not correspond by the normality law to the statically admissible state of stress.

We illustrate this discussion by two examples.

1. *Beam with uniform solid cross section subjected to bending and torsion* [3.8]. Let Gx and Gy be the two principal axes of inertia of the cross section of a beam (G being the centroid, fig. 5.20). Tresca's yield condition for pure torsion is $\tau^2 = \tau_{zx}^2 + \tau_{zy}^2 = \sigma_Y^2/4$. Assume a uniform distribution of shear stress τ similar to that corresponding to the limit torque M_{tp} but with $\tau < \sigma_Y/2$. The reduced torque m_t will be

$$m_t = \frac{M_t}{M_{tp}} = \frac{\tau}{\sigma_Y/2}. \tag{5.62}$$

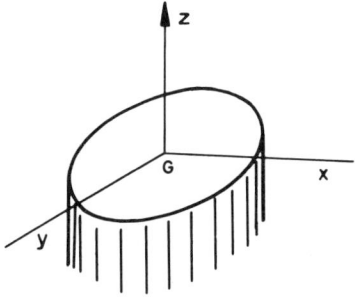

Fig. 5.20.

To this shear stress distribution, we superimpose a distribution of normal stress σ similar to that corresponding to the fully plastic bending moment M_p about axis x, but with $\sigma < \sigma_Y$. The reduced bending moment is

$$m \equiv \frac{M}{M_p} = \frac{\sigma}{\sigma_Y}. \tag{5.63}$$

The two distributions of stresses will best combine to satisfy

$$\sigma^2 + 4\tau^2 = \sigma_Y^2 \tag{5.64}$$

everywhere, to furnish a fully plastified cross section. Substitution of expressions (5.62) and (5.63) for τ and σ, respectively, into eq. (5.64) yields

$$m^2 + m_t^2 = 1. \tag{5.65}$$

Relation (5.65) is a lower bound for the interaction curve because the state of stress is statically admissible but does not correspond to a kinematically admissible strain-rate distribution.

To obtain an upper bound, we arbitrarily *assume* that yielding results in a rate of curvature $\dot{k}_y$ in the G_{zy} plane and a rate of twist $\dot{k}_{xy}$ about G, but no warping of the cross section. The corresponding strain rates are

$$\dot{\epsilon}_z = y\dot{k}_y, \qquad \dot{\gamma}_{zx} = -y\dot{k}_{xy}, \qquad \dot{\gamma}_{zy} = x\dot{k}_{xy}. \tag{5.66}$$

The dissipation is

$$D = \int_A (\sigma_z \dot{\epsilon}_z + \tau_{zx} \dot{\gamma}_{zx} + \tau_{zy} \dot{\gamma}_{zy}) \, dA$$

or

$$D = M \dot{k}_y + M_t \dot{k}_{xy} .$$

Because $\tau^2 = \tau_{zx}^2 + \tau_{zy}^2$, Tresca's condition (5.64) may be written

$$\sigma_z^2 + 4(\tau_{zx}^2 + \tau_{zy}^2) = \sigma_Y^2 . \tag{5.67}$$

The normality law (2.6) applied to condition (5.67) furnishes

$$\dot{\epsilon}_z = 2\lambda \sigma_z , \quad \dot{\gamma}_{zx} = 8\lambda \tau_{zx} , \quad \dot{\gamma}_{zy} = 8\lambda \tau_{zy} . \tag{5.68}$$

From comparison of relations (5.66) and (5.68) we have

$$\sigma_z = \frac{y \dot{k}_y}{2\lambda} , \quad \tau_{zx} = -\frac{y \dot{k}_{xy}}{8\lambda} , \quad \tau_{zy} = \frac{x \dot{k}_{xy}}{8\lambda} . \tag{5.69}$$

Substituting expressions (5.69) for σ_z, τ_{zx}, and $\tilde{\tau}_{zy}$ into relation (5.67), we obtain

$$2\lambda = (\dot{k}_y/\sigma_Y) \sqrt{y^2 + \alpha^2(x^2+y^2)} ,$$

where

$$\alpha = \frac{\dot{k}_{xy}}{2\dot{k}_y} .$$

We now readily obtain

$$m \equiv \frac{1}{M_p} \int_A \sigma_x y \, dA = \frac{\sigma_Y}{M_p} \int_A \frac{y^2}{[y^2 + \alpha^2(x^2+y^2)]^{\frac{1}{2}}} \, dA ,$$

$$m_t \equiv \frac{1}{M_{tp}} \int_A (x\tau_{zy} - y\tau_{zx}) \, dA = \frac{\sigma_Y \alpha}{2M_{tp}} \int_A \frac{(x^2+y^2) \, dA}{[y^2 + \alpha^2(x^2+y^2)]^{\frac{1}{2}}} . \tag{5.70}$$

Relations (5.70) are the parametric equations of the interaction curve m versus m_t, α being used as parameter. Different cross-sections will give rise to different coefficients σ_Y/M_p and σ_Y/M_{tp}. Because the relations (5.70) are obtained from a kinematically admissible strain-rate field [eqs. (5.66)], they furnish an upper bound for the exact interaction curve. Fig. 5.21 shows the two bounds for a circular cross section, curves a; and for a square cross section, curves b.

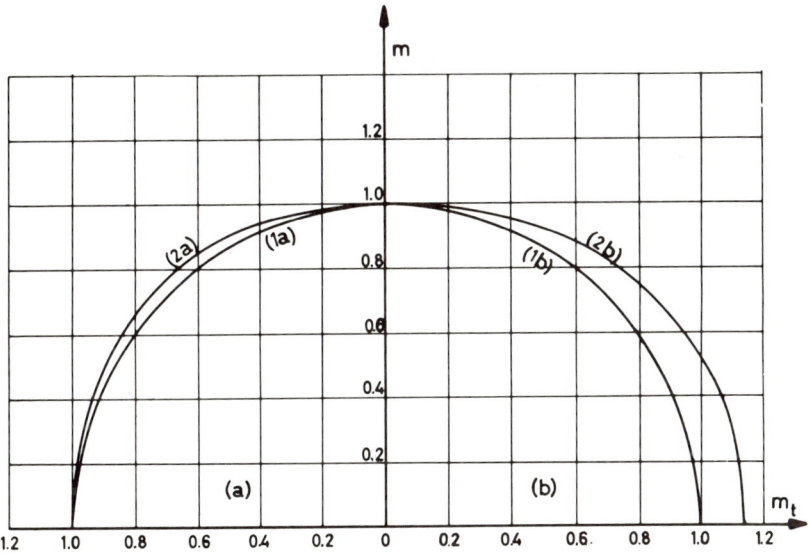

Fig. 5.21. (From *Plastic Analysis of Structures* by P.G. Hodge Jr. Copyright 1959, McGraw-Hill Book Company, Used by permission of McGraw-Hill Book Company, Inc.)

2. *Cylindrical shell without axial force* [3.8]. Consider a circular cylindrical shell as shown in fig. 5.22. It is subjected to an internal pressure that may solely depend on the coordinate x. Because of the symmetry of revolution and the absence of axial load, the only nonvanishing stress resultants and moments are those shown in fig. 5.15. Shear force V_x is a reaction (see Section 5.3.3), as well as bending moment M_θ (see Section 5.1.4, example 2). Tresca's yield condition may be expressed as

$$\max \left[|\sigma_x|, |\sigma_\theta|, |\sigma_x - \sigma_\theta| \right] = \sigma_Y, \qquad (5.71)$$

for all r. Plasticity occurs at a given level r of the shell thickness when one of the six relations (5.71) is satisfied. Obviously, the relation to satisfy may change at some values of r.

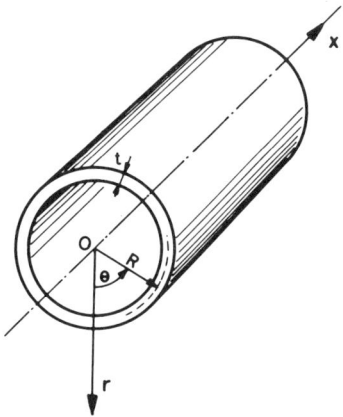

Fig. 5.22.

Let

$$m_x = \frac{M_x}{M_p} = \frac{M_x}{\sigma_Y t^2/4},$$

$$n_\theta = \frac{N_\theta}{N_p} = \frac{N_\theta}{\sigma_Y t}, \tag{5.72}$$

be the reduced bending moment and axial force, respectively. They are the only generalized variables. We now distribute σ_x and σ_θ over the thickness in order to maximize n_θ for a given m_x, while satisfying the yield condition (5.71). We first try

$$m_x = 1. \tag{5.73}$$

It is admissible as long as $n_\theta \leqslant \frac{1}{2}$. Indeed, as shown on fig. 5.23, we have $n_\theta = \frac{1}{2} - \alpha$, and the reaction m_θ is $m_\theta = \frac{1}{2} - 2\alpha^2$, with $0 \leqslant \alpha \leqslant \frac{1}{2}$. For $\frac{1}{2} \leqslant n_\theta \leqslant 1$, we may not have $m_x = 1$ anymore. Fig. 5.23 (b) shows the distribution that gives, for every m_x, the largest value of n_θ compatible with the yield condition (5.71). We obtain

$$m_x = 1 - 4\alpha^2,$$

$$n_\theta = \tfrac{1}{2} + \alpha \tag{5.74}$$

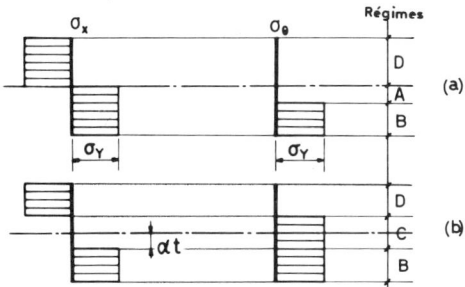

Fig. 5.23.

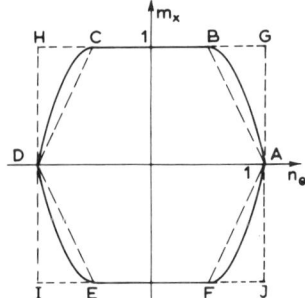

Fig. 5.24. (From *Plastic Analysis of Structures* by P.G. Hodge Jr. Copyright 1959, McGraw-Hill Book Compnay. Used by permission of McGraw-Hill Book Company, Inc.)

(and the reaction $m_\theta = \frac{1}{2} - 2\alpha^2$, with $0 \leq \alpha \leq \frac{1}{2}$). Eqs. (5.73) and (5.74) define the interaction curve shown on fig. 5.24.

To prove that this curve is not only a lower bound but the exact interaction curve, we must associate, to the stress distributions above, corresponding strain-rate distributions. The principal strain rates $\dot{\epsilon}_x$, $\dot{\epsilon}_\theta$ may be represented by the coordinates of a point (fig. 5.25). According to the normality law, all points of the first quadrant with $\dot{\epsilon}_x > 0$ and $\dot{\epsilon}_\theta > 0$ are associated with the vertex B of Tresca's yield hexagon (see the insert in fig. 5.25). Similarly, it is easily seen that the regions bounded by the axes and the diagonals of the second and fourth quadrants correspond to the six vertices, whereas these six rays correspond to the six sides of the hexagon.

We now recall (see Section 5.4.2) that

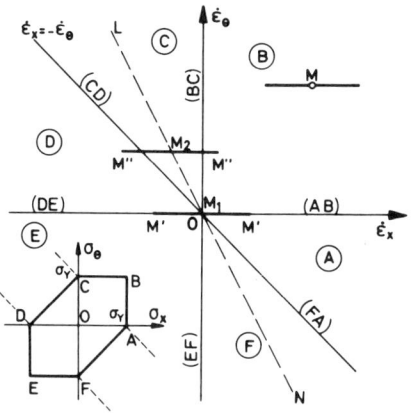

Fig. 5.25.

$$\dot{\epsilon}_x = \dot{\epsilon}_{xo} + z\dot{k}_x ,$$

$$\dot{\epsilon}_\theta = \dot{\epsilon}_{\theta o} + z\dot{k}_\theta = \dot{\epsilon}_{\theta o} , \tag{5.75}$$

where $\dot{\epsilon}_{xo}, \dot{\epsilon}_{\theta o}, \dot{k}_x, \dot{k}_\theta$ are rates of extension and of curvature of the midsurface in the longitudinal and circumferential directions, respectively. Curvature rate $\dot{k}_\theta$ is known to vanish (Section 5.1.4) and z is the radial distance of a layer from the median surface, counted as positive when directed inwards. Hence, if a point M (fig. 5.25) represents the state of strain rate at a point P on the median surface, strain rates of the various points on the normal at P will be represented by the points of a segment parallel to the $\dot{\epsilon}_x$-axis and with center M. But because the axial force N_x must vanish, there must be as many layers in regime D as in regime B (see the insert in fig. 5.25), or as many in regime A as in regime E. Consequently, the point M must fall on a certain dashed ray LON. It is then easily seen that the kinematically admissible strain-rate distributions represented by the points of segments $M'M_1M'$ and $M''M_2M''$ in fig. 5.25 correspond to the stress distributions in fig. 5.23. The interaction curve shown in fig. 5.24 is therefore exact. Finally, we note that the parametric eqs. (5.73) and (5.74) could have been deduced from the results of Section 5.4.2 where the dissipation function was used. Indeed, if we relabel x and θ as 1 and 2, respectively, eqs. (5.75) show that the central point in fig. 5.10 is either P or Q because R goes to infinity. In both cases $\alpha_1 = -\alpha_3$ because $n_1 = 0$, and $m_1 = 1 - 4\alpha^2$, whereas $n_2 = \frac{1}{2} + \alpha_1$, because $\alpha_2 = \frac{1}{2}$ or $-\frac{1}{2}$, respectively (see [5.9]).

5.5. Simplified yield surface

5.5.1. *Convenience of a simplified yield surface*

A linear yield condition is very attractive from the mathematical point of view. Indeed, if the yield surface consists of plane facets, as long as the stress point remains on a given plane the yield vector retains the same direction and the yield "mechanism" does not change. All possible yield mechanisms can thus be classified into a finite number of plastic "regimes", each regime corresponding to the contact of the stress point with one plane, one edge, or one vertice (fig. 5.26). Limit analysis proves much easier in these circumstances than when the yield surface is curved, especially when the principal directions are known beforehand. These reasons explain the preference given to Tresca's yield condition over the condition of von Mises. But when generalized stresses are used, even Tresca's linear yield condition need not result in a piecewise linear yield surface, as shown in the second example of Section 5.4.4. The exact yield surface must then be replaced by a polyhedron, either inscribed (dashed lines in fig. 5.24) or circumscribed (dotted lines in fig. 5.24).

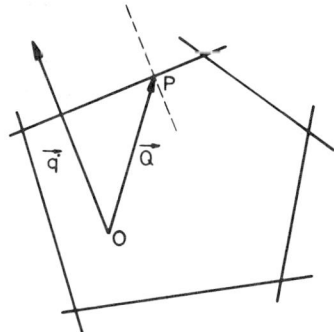

Fig. 5.26.

5.5.2. *Influence on the limit load of linearizing the yield condition*

For the sake of simplification, consider a yield condition that involves only two generalized stresses Q_1 and Q_2. Let the exact yield curve be represented by the heavy line in fig. 5.27, and let P_l be the corresponding exact limit load for a given structure and a given loading scheme. If, instead of the exact yield curve e, the inscribed polygon i (dashed lines) is used, the state of stress at collapse for polygon i is statically admissible for curve e. If the corresponding limit load is P_i, the static theorem furnishes the inequality

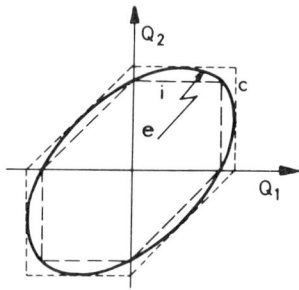

Fig. 5.27.

$$P_i \leqslant P_l. \tag{5.76}$$

If P_i^- is a statically admissible load for i, we have

$$P_i^- \leqslant P_i. \tag{5.77}$$

On the other hand, using a circumscribed polygon c would result in

$$P_l \leqslant P_c. \tag{5.78}$$

If P_c^+ is a kinematically admissible load for c, the kinematic theorem asserts that

$$P_c \leqslant P_c^+. \tag{5.79}$$

Inequalities (5.76) to (5.79) may be combined into one continued inequality,

$$P_i^- \leqslant P_i \leqslant P_l \leqslant P_c \leqslant P_c^+, \tag{5.80}$$

that enables us to bound the error introduced by the linearization process as follows.

The limit load of a structure is directly proportional to the yield stress σ_Y of the material. If the yield stress is multiplied by a factor that is greater or smaller than unity, the yield surface is similarly expanded or contracted with respect to the origin. Hence, if polygons i and c are homothetical with factor k, we have

$$P_c = kP_i .\tag{5.81}$$

If we want to bound P_l from above and below, it is sufficient to know either P_i or P_c, say P_i. We then determine the lowest expansion factor k that makes polygon i become circumscribed to e. Note also that, if the stress point, for all plastic regions, remains on a certain part of curve e, as AB in fig. 5.28, the yield polygon obtained by expansion of polygon i must be external to curve e in that part only. This remark enables us to use the smallest possible expansion factor k.

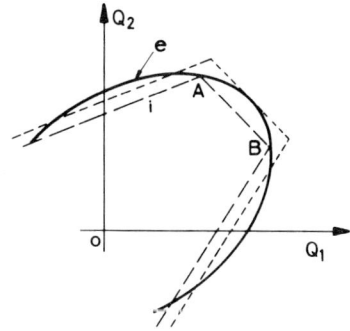

Fig. 5.28.

5.5.3. Linearization process

A first method consists of finding the exact yield surface and then inscribe (or circumscribe) more or less arbitrarily a polyhedron in order to simplify the subsequent analysis.

A second point of view consists of using an approximation not to the yield surface but to the structure itself [5.11].

Consider (fig. 5.29) a structure (plate, shell) of the "ideal sandwich" type defined in Section 5.4.3, example 2. On a cross section, the axial force and bending moment per unit of length are [see eq. (5.53)]

$$N = \frac{t^*}{2}(\sigma_i + \sigma_e) ,\tag{5.82}$$

$$M = \frac{t^* H}{4}(\sigma_i - \sigma_e) ,\tag{5.83}$$

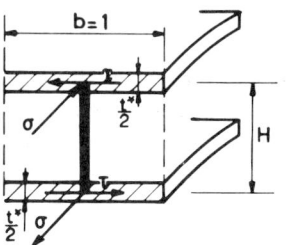

Fig. 5.29.

where σ_i and σ_e are the normal stresses in the internal and external sheets, respectively, $t^*/2$ the thickness of each sheet and H the core thickness. A twisting moment M_t (per unit of length) will produce shear stresses τ such that

$$M_t = \tau \frac{t^*}{2} H . \tag{5.84}$$

If we denote by τ_Y^* the yield shearing stress of the sandwich sheets and σ_Y^* its yield stress in tension, the full plastic axial force, bending moment and twisting moment are

$$N_p^* = \sigma_Y^* t^* , \tag{5.85}$$

$$M_p^* = \sigma_Y^* H \frac{t^*}{2} , \tag{5.86}$$

$$M_{tp}^* = \tau_Y^* H \frac{t^*}{2} . \tag{5.87}$$

For the sandwich structure to be substituted for a structure with uniform cross section of thickness t and yield stresses σ_Y and τ_Y, H and t^* must be so chosen as to satisfy

$$\sigma_Y t = \sigma_Y^* t^* \equiv N_p ,$$

$$\sigma_Y^* H \frac{t^*}{2} = \sigma_Y \frac{t^2}{4} \equiv M_p ,$$

$$\tau_Y^* H \frac{t^*}{2} = \tau_Y \frac{t^2}{4} \equiv M_{tp} .$$

Relations (5.75) to (5.87) then give

$$\sigma_Y^* t^* = \sigma_Y t,$$

$$H = \frac{t}{2},$$

$$\frac{\tau_Y^*}{\sigma_Y^*} = \frac{\tau_Y}{\sigma_Y}. \tag{5.88}$$

The last relation of eqs. (5.88) just means that the same physical yield condition must hold for both structures.

Because the generalized stresses N, M, M_t are *linear* functions of the stress components (relations (5.82) to (5.84)), any linear yield condition in terms of the stress components will thus generate a linear yield condition in terms of generalized stresses. Hence, the preceding procedure directly furnishes an approximate yield polyhedron without recourse to the exact yield surface. This polyhedron is *inscribed* in the exact yield surface because the latter is convex and, according to relations (5.88), both have in common the points on the axes. Fig. 5.30 shows, for example, the yield polyhedron of a sandwich cylindrical shell axisymmetrically loaded and made of a Tresca material. Equations of the various planes are:

I: $\quad n_\theta = 1,$

II: $\quad n_\theta - n_x = 1,$

III: $\quad n_x - m_x = -1,$

IV: $\quad 2n_\theta - n_x + m_x = 2,$

V: $\quad 2n_\theta - n_x - m_x = 2. \tag{5.89}$

Coordinates x and θ are as indicated in fig. 5.22. This yield polyhedron is a linearization of the yield surface in fig. 5.12, with which it should be compared (subscripts 1 and 2 becoming x and θ, respectively).

5.5.4. *Example of application*

Following Hodge and Sawczuk [5.12], consider an infinitely long cylindrical shell without axial force, loaded in a cross section by a radial uniform line load of total magnitude $2F_o$ (fig. 5.31).

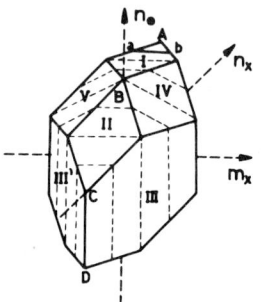

Fig. 5.30.

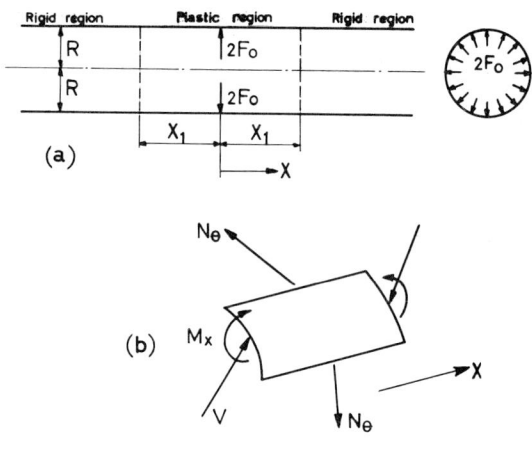

Fig. 5.31.

Because of the symmetry, the plastic region at collapse will extend over an (unknown) length x, on each side of the loaded cross section where we locate the origin of the abscissae. The only generalized stresses are M_x and N_θ (fig. 5.31).

We assume the real shell to exhibit uniform thickness and satisfy von Mises' yield criterion. We compare it to these shells regarded as approximations to the former: (a) sandwich shell made of von Mises material; (b) uniform shell made of Tresca material; (c) sandwich shell made of Tresca material; (d) shell with the (arbitrarily simple) "limited interaction" yield curve.

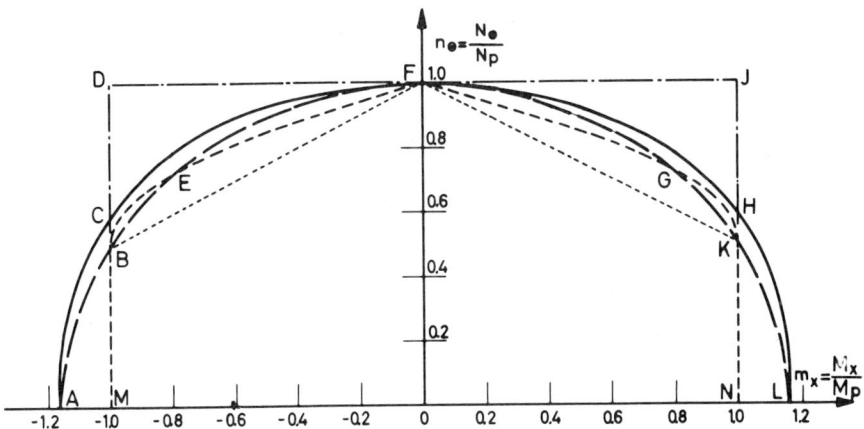

Fig. 5.32. Interaction curves.

The various yield curves are shown in fig. 5.32. In the absence of twisting moment, conditions (5.88) reduce to

$$\sigma_Y^* l - \upsilon_Y t,$$

$$H = \frac{t}{2}.$$

The lines 2, 3, and 4 in fig. 5.32 were obtained in Section 5.4. Line 1 was given by Hodge [5.13] as a special case of the yield surface of a shell of revolution loaded with rotational symmetry. Line 5 was arbitrarily chosen for the sake of simplicity.

Line 4 is a linear approximation to lines 1, 2, and 3. Line 5 is an arbitrary linear approximation to lines 1 to 4.

The load was nondimensionalized as follows: $f_o = R^{1/2}(M_p N_p)^{-1/2} F_o$.

The various exact limit loads obtained by Hodge and Sawczuk are given in column 3 of table 5.4.

In column 4 we find the deviations of the preceding limit loads from those of the real shell (line 1 in fig. 5.32). Column 5 contains the relative deviations. The figures given in columns 6 and 7 were obtained as follows: each yield curve was similarly enlarged or reduced by a factor k (see Section 5.5.2) to become: (a) external to curve 1 for curves 2, 3, and 4 or (b) internal to curve 1 for curve 5 (see fig. 5.33). The exact yield curve is thus bounded,

Table 5.4. Comparison of reduced limit loads $[f_o = \sqrt{(R/M_pN_p)F_o}]$

1	2	3	4	5	6	7
Type of shell	Yield condition	f_o	Deviation from uniform von Mises	Actual	Deviation Lower bound	Upper bound
Uniform	von Mises	1.949	0	0	0	0
Sandwich	von Mises	1.905	−0.044	− 2.3	− 2.3	2.6
Uniform	Tresca	1.826	−0.123	− 6.3	− 6.3	7.9
Sandwich	Tresca	1.732	−0.217	−11.1	−11.1	5.7
Limited interaction curve		2.000	+0.051	2.6	−20.0	18.2

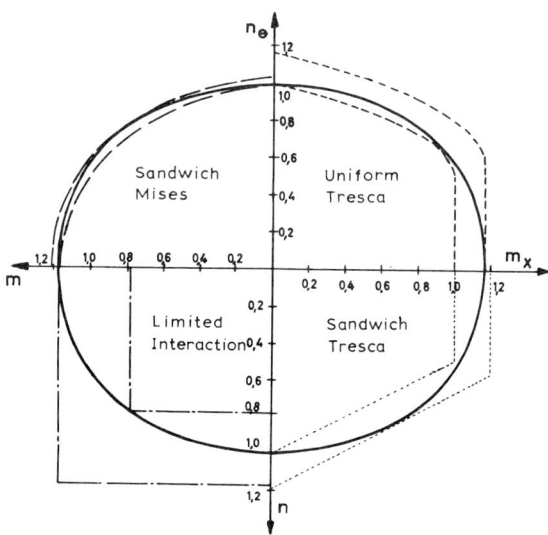

Fig. 5.33. Approximate interaction curves.

for each "approximation" shell, by two approximate curves that furnish lower and upper bounds for the limit load. The relative deviations of these bounds are indicated in columns 6 and 7.

Note that the original "limited interaction" curve (line 5 in fig. 5.32) gives neither a lower bound nor an upper bound as it arbitrarily cuts across the exact yield curve. After geometrically similar expansion or reduction, it does furnish bounds to the limit load, which differ appreciably from the other bounds. Nevertheless, the original curve itself furnishes a fairly good approximate limit load.

Diagrams of the reduced bending moment $m_x = M_x/M_p$ and axial force $n_\theta = N_\theta/N_p$ are shown in figs. 5.34 and 5.35, where elastic diagrams corresponding to the maximum load in elastic range are also given. It is easily seen that an important redistribution of both n_θ and m_x takes place prior to collapse.

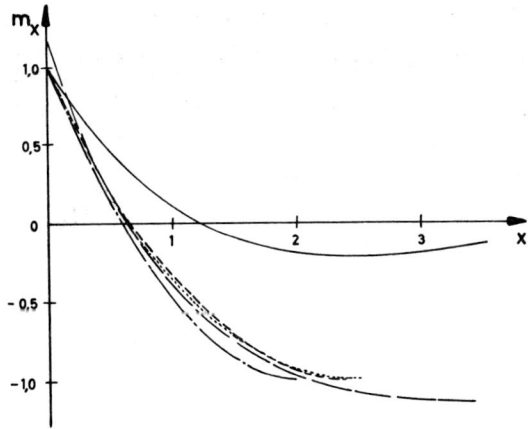

Fig. 5.34. Moment distribution.

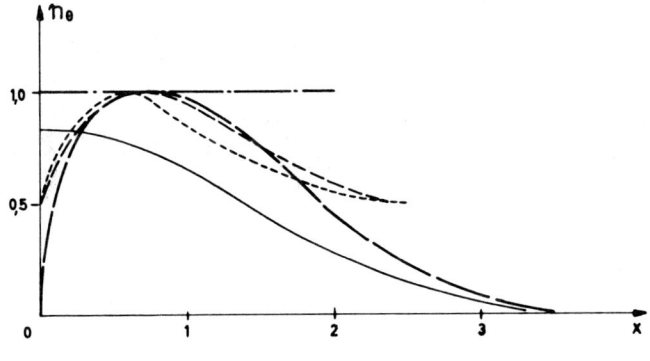

Fig. 5.35.

5.6. Discontinuities

5.6.1. *Introduction*

When searching for fields of generalized stresses Q_i and strain rates $\dot{q}_i$, it often proves useful to introduce discontinuities. We already know some of these discontinuities from Com. V.: at a plastic hinge in a beam the slope jumps by a finite amount and in a plastically bent segment of the beam the stress σ jumps from σ_Y to $-\sigma_Y$ when the neutral layer is crossed. For a given theory, that is, for a given degree of idealization, these discontinuities are permissible because perfect plasticity relaxes some of the restraints of geometrical compatibility.

As a rule, displacements normal to the midsurface of a shell will be kept continuous if the shell is not to break, but slopes might be discontinuous, as well as displacements in the tangent plane. Obviously, these discontinuities must be regarded as limiting cases (that is, idealizations consistent with the level of the theory used) of very rapid variations over very narrow regions, in which compatibility remains fully satisfied. The stress fields will also be allowed to exhibit some discontinuities because stresses are no longer related in a unique manner to a continuous strain field.

It is important to know what the admissible discontinuities are and what relations they obey. We refer the reader to Prager [5.14] and Hill [5.15]. We restrict ourselves hereafter to the minimum amount of indications necessary for the applications in the coming chapters.

5.6.2. *Stress discontinuities*

Consider (fig. 5.36) a point O on a surface of discontinuity S, and a elementary parallelepiped with center O. The general rule is as follows: *elementary forces and moments that balance each other across the discontinuity surface must be continuous.*

On the other hand, elementary forces and moments that balance each other along the discontinuity surface may experience discontinuities across that surface.

If we examine, for example, a plate subjected to bending, the discontinuity surface reduces to a discontinuity line DD as shown in fig. 5.37. The action at O across the line is a moment vector with components M_x and M_{xy}. This vector must be continuous across the line, and hence there is no discontinuity admissible on M_x and M_{xy}. We express this result symbolically as

$$M_x] = 0, \qquad M_{xy}] = 0. \tag{5.90}$$

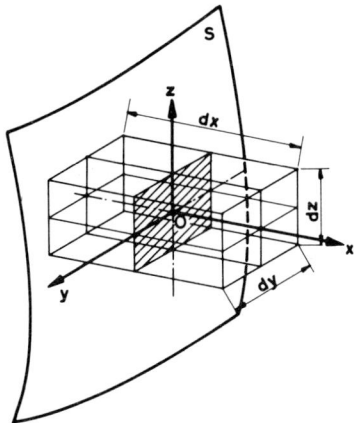

Fig. 5.36.

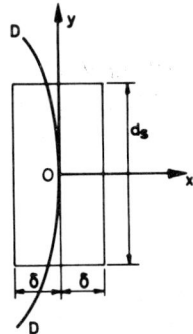

Fig. 5.37.

The remaining component M_y of the moment tensor at O may be discontinuous:

$$M_y] \neq 0 . \tag{5.91}$$

The equilibrium of an element with center O furnishes the relations:

$$\frac{\partial M_x}{\partial x} - \frac{\partial M_{xy}}{\partial y} = V_x , \tag{5.92}$$

$$\frac{\partial M_y}{\partial y} - \frac{\partial M_{xy}}{\partial x} = V_y, \qquad (5.93)$$

$$\frac{\partial V_x}{\partial x} + \frac{\partial V_y}{\partial y} = -p. \qquad (5.94)$$

Eq. (5.92) shows that

$$V_x] = 0 \qquad (5.95)$$

whereas eq. (5.93) does not exclude

$$V_y] \neq 0 \qquad (5.96)$$

because $\partial M_y/\partial y$ may be discontinuous as well as M_y.

The preceding considerations will be used in Chapter 9 in dealing with reinforced concrete shells, as well as in Chapter 10 where plane stress and strain are studied.

5.6.3. Strain-rate discontinuities

As noted in Section 5.6.1, *the field of transversal displacement rates $\dot{w}$ must be continuous*. If we examine the consequences of this statement for the first derivatives $\partial \dot{w}/\partial x$ and $\partial \dot{w}/\partial y$, we see that we must have

$$\frac{\partial \dot{w}}{\partial y}\Big] = 0 \qquad (5.97)$$

(to exclude different displacement rates for points with coordinates $(-\delta, ds/2)$ and $(+\delta, ds/2)$ with vanishingly small δ, fig. 5.37). But we may have

$$\frac{\partial \dot{w}}{\partial x}\Big] \neq 0, \qquad (5.98)$$

a situation where the discontinuity line *DD* (fig. 5.37) is a "yield line" or "hinge line" that generalizes the plastic hinge of beams.

Obviously, a certain quantity of energy is dissipated in the strain-rate discontinuities, and due account must be taken of this fact. Indeed, some mechanisms will be made solely of discontinuities, as we shall see in the coming chapters.

5.7. Appendix

Proof of the adaptation theorem

For the sake of brevity, we restrict the discussion to strictly convex yield surfaces. Extension to yield surfaces with flats and vertices is straightforward (see [5.5]).

Consider the yield condition

$$F(Q_1,...,Q_n) = 1, \tag{a}$$

where all stress-type variables Q_1 to Q_n are included. Give fixed values K_i to $n - k$ of these Q_i:

$$Q_i = K_i \qquad (i=k+1,...,n). \tag{b}$$

Points with coordinates $Q_1, ..., Q_n$ that satisfy both relations (a) and (b) form a degenerate surface

$$\Phi(Q_1,...,Q_k) \equiv F(Q_1,...,Q_k,K_{k+1},...,K_n). \tag{c}$$

The normality rule applied to surface (a) gives

$$\dot{q}_i = \frac{\partial}{\partial Q_i} F(Q_1,...,Q_n), \tag{d}$$

for $i = 1$ to n, and in particular for $i = 1$ to k. But, because the K_i are *constants*, we obviously have

$$\frac{\partial}{\partial Q_i} F(Q_1,...,Q_n) = \frac{\partial}{\partial Q_i} F(Q_1,...,Q_k,K_{k+1},...,K_n) \qquad (i=1,...,k), \tag{e}$$

Comparison of relations (e), (d), and (c) then furnishes

$$\dot{q}_i = \frac{\partial \Phi}{\partial Q_i} \qquad (i=1,...,k), \tag{f}$$

and we have the following: *the normality rule remains applicable to a simplified yield surface obtained giving fixed values to some stress-type variables.*

We now suppose that $Q_1, ..., Q_k$ are generalized stresses, whereas $Q_{k+1}, ..., Q_n$ are reactions. This means that

$\dot{q}_i = 0 \qquad (k+1,...,n)$. \hfill (g)

Assign to all generalized stresses but one, say Q_k, fixed values. We have

$$Q_i = K_i \qquad (i=1,...,k-1).$$ \hfill (h)

With conditions (h), yield surface (a) becomes

$$F(K_1,...,K_{k-1},Q_k,Q_{k+1},...,Q_n) = 1.$$ \hfill (j)

Relation (j) expresses Q_k as a function of $Q_{k+1}, ..., Q_n$. Apply the normality rule to (j) and take account of (g). Then,

$$\dot{q}_i = \frac{\partial}{\partial Q_i} F(K,...,K_{k-1},Q_k,...,Q_n) = 0 \quad \text{for} \quad i = k+1, ..., n.$$ \hfill (h)

Relations (h) are a set of equations expressing that, to satisfy both (j) and (g), the function Q_k of $Q_{k+1}, ..., Q_n$ must exhibit an extremum.

We thus have the *adaptation theorem: If all generalized stresses but one are assigned fixed values, reactions adapt themselves to give to the remaining generalized stress a maximum positive value or a minimum negative value.*

5.8. Problems

5.8.1. Show that the yield condition for a sandwich shell of revolution axisymmetrically loaded and made of a von Mises material is:

$$(n_\theta + m_\theta)^2 - (n_\theta + m_\theta)(n_\phi + m_\phi) + (n_\phi + m_\phi)^2 = 1,$$

$$(n_\theta - m_\theta)^2 - (n_\theta - m_\theta)(n_\phi - m_\phi) + (n_\phi - m_\phi)^2 = 1,$$

with $n = N/N_p$ and $m = M/M_p$.

5.8.2. Determine analytically the minimum amplification coefficient k to apply to the yield curve of the sandwich structure to have it circumscribe the exact yield curve (corresponding to uniform cross section of a Tresca material).

(a) For a rectangular beam subjected to bending and axial force the exact interaction curve being [see Com. V., relation (5.4)], $m = 1 - n^2$. *Answer:* $k = 1.25$.

(b) For a cylindrical shell without axial force and axisymmetrically loaded. *Answer*: $k = 1.225$.

5.8.3. Determine analytically the minimum reduction coefficient k to apply to the "square" yield condition (of the type of curve 5, fig. 5.32) to have it inscribed in the exact Tresca yield curve:

(a) When the yield curve m_1 versus n_1 is considered. *Answer*: $k = 0.50$.
(b) When the yield curve m_1 versus n_2 is considered. *Answer*: $k = 0.75$.

5.8.4. Determine the interaction curve for biaxial bending (Section 5.4.3, example 1) in the presence of a given nonvanishing twisting moment. *Hint*: use statical approach. *Answer*:

$$m_y \sqrt{1 - m_t^2} + \tfrac{3}{4}m_x^2 + m_t^2 = 1 \qquad (|m_x| \leqslant m_y),$$

with $m_t = M_t/M_{tp}$.

References

[5.1] K.GIRKMAN, *Flächentragwerke*, Springer, 1959.
[5.2] C.MASSONNET, "Faut-il introduire l'hypothèse de Bernouilli en résistance des matériaux?", *Bull. Soc. Roy. des Sci.*, **12**: 301, Liège, 1947.
[5.3] A.R.RIANITSYN, *Calcul à la rupture et plasticité des constructions*, p. 34, Eyrolles, Paris, 1959.
[5.4] W.PRAGER, "The General Theory of Limit Design", *Proc. 8th Int. Congr. Appl. Mech.*, Istambul, 1952, **2**: 65, 1956.
[5.5] M.SAVE, "On Yield Conditions in Generalized Stresses", *Quart. of Applied Math.*, **XIX**: 3, October 1961.
[5.6] "Structures", *L'architecture d'aujourd'hui*, March 1956.
[5.7] A.SAWCZUK, and J.RYCHLEWSKI, "On Yield Surfaces for Plastic Shells", *Archiwus Mechaniki Stosowanej*, **1**: 12, 1960.
[5.8] E.T.ONAT and W.PRAGER, "Limit Analysis of Shells of Revolution", *Koninkl. Nederl. Akademie van Wetenschappen*, Amsterdam, **57**: 5, 1954.
[5.9] P.G.HODGE Jr., "The Rigid-Plastic Analysis of Asymmetrically Loaded Cylindrical Shells", *J. of Appl. Mech.*, **21**: 336, 1954.
[5.10] E.T.ONAT, "The Plastic Collapse of Cylindrical Shells under Axially Symmetrical Loading", *Quart. Appl. Math.* **13**: 63, 1955.
[5.11] P.G.HODGE Jr., "The Linearization of Plasticity Problems by Means of Nonhomogeneous Materials", *Proc. I.U.T.A.M. Symp.*, 1958. *Nonhomogeneity in Elasticity and Plasticity*, W.OLSZAK, ed., pp. 147-156, Pergamon Press, 1959. See also W.PRAGER, "On the Plastic Analysis of Sandwich Structures", in *Problems in Continuum Mechanics*, pp. 342-349, Soc. for Ind. and Appl. Math., Philadelphia, 1961.

[5.12] A.SAWCZUK, P.G.HODGE Jr., "Comparison of Yield Conditions for Circular Cylindrical Shells", *J. of the Franklin Inst.*, **269**:, no. 5, May 1960.

[5.13] P.G.HODGE Jr., "The Mises Yield Condition for Rotationally Symmetric Shells", *Quart. of Appl. Math.*, **18**: 305, 1961.

[5.14] W.PRAGER, "Discontinuous Fields of Plastic Stress and Flow", *Proc. 2nd U.S. Nat. Congr. Appl. Mech.*, 21-32, A.S.M.E., Ann Arbor, 1954.

[5.15] R.HILL, "Discontinuity Relations in Mechanics of Solids", *Progress in Solid Mechanics*, Vol. II, Sneddon, Hill, ed., North-Holland Publ. Co., Amsterdam, 1961.

Part Two

Applications to Plates
Shells and Disks

6

Metal Plates

6.1. Introduction

Consider a plane horizontal region A, which is bounded by one or several curves. At a generic point P of this region, erect a straight vertical segment of length t that has P as its midpoint.

Imagine that the point P occupies all possible positions in A and assume: (1) that t is a continuous and "sufficiently smooth" function of the position of P; and (2) that t always remains small with respect to the dimensions of A. The body generated in this manner is a *plate* or a *disk* (fig. 6.1).

It is called a plate when loaded transversally to its median plane. It then carries the applied loads essentially by bending stresses; its median plane will bend but not stretch or contract*.

In many important cases, the region A will be a circle, a circular ring, or a polygon, and the thickness will be constant. When assumptions 1 and 2 are satisfied, material normals to the median plane will transform into material normals of the curved surface into which the median plane deforms, both in the plastic and the elastic ranges. When the thickness t does not vary smoothly enough in some regions, elastic stress concentrations will occur there, which

---

* See Sections 6.2 and 6.7 for further details.

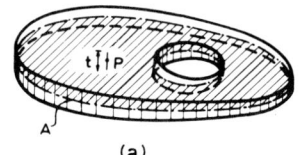

Fig. 6.1 (a).

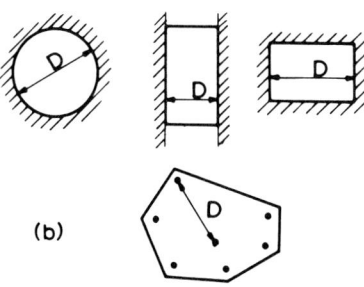

Fig. 6.1 (b).

later on may be leveled out by plastic flow. Alternatively, very abrupt thickness variation may result in a local elastic state of stress with three principal tensile stresses and this is likely to cause brittle failure (see Com. V., Section 5.2). Hence, condition 1 must be imposed when the emphasis is on plastic deformation. Condition 2 is the basis for material normals to the median plane to remain normal to the median surface in the deformed state. To make this condition more precise, we may introduce the "slenderness" of the plate, that is a geometrical nondimensional parameter defined by

$$\mu = \frac{D}{t}, \tag{6.1}$$

where D is some "span" of the plate with thickness t. This span will be taken to be [see fig. 6.1 (b)]:

1. for circular or annular plates: the outside diameter;
2. for rectangular plates: (a) simply supported or built-in along two parallel edges: the span; (b) simply supported or built-in along four edges: the largest span;

3. for plates supported on columns: the largest span between two neighboring supports.

The parameter μ is the extension to plates of the span-to-height ratio used in beam theory. For a plate of a given material to carry transverse loads essentially by bending stresses, μ must be bounded both from below and from above. Indeed, if μ is unduly small, we no longer have a thin plate but a body with comparable horizontal and vertical dimensions.

On the other hand, if μ is unduly large, we have a membrane that carries transverse loads by direct stresses after undergoing deflections that are comparable to its thickness. This situation is more often encountered with metallic plates that tend to be very slender, than with reinforced concrete plates.

The limit load obtained from simple plastic theory based on the rigid perfectly plastic scheme will thus prove to have a real physical meaning only for a limited domain of values of μ — that we shall define as precisely as possible for the various problems treated. Even in this domain of μ, the limit load will not correspond to large plastic deformations under constant load, such as usually occur in frame structures. In most cases, favorable geometry changes due to unrestricted plastic flow will cause membrane action that eventually enables the plate to carry a load in excess of the limit load (see Sections 3.5 and 6.2).

Since the treatment of reinforced concrete plates justifies a complete chapter, in this chapter we will only consider metal plates, either isotropic or (structurally) orthotropic. We will also devote a long paragraph to minimum-weight design (Section 6.7).

6.2. Experimental information on metal plates

Tests on metal plates in the plastic range are unfortunately still few, and all of them [6.1–6.7] deal with circular plates except one work on rectangular plates that are built-in at the small edges, free at the others, and carry a central concentrated load [6.8].

The essential results of the experiments on circular plates are given by the load versus deflection diagrams of figs. 6.2 – 6.8. The theoretical limit loads have been computed as indicated in the following sections, assuming a rigid perfectly plastic material. The theoretical load versus deflection diagram coincides with the load axis up to the limit value of the load and, for this value, becomes a parallel to the deflection axis when changes in geometry

114   METAL PLATES   [Ch. 6

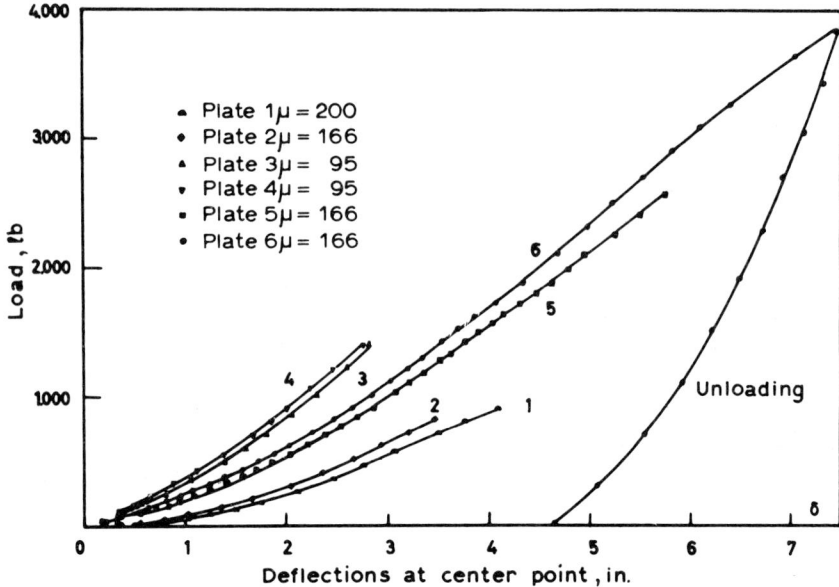

Fig. 6.2.

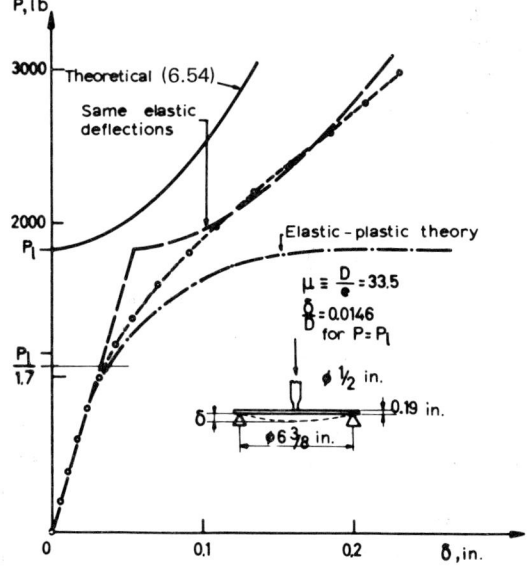

Fig. 6.3.

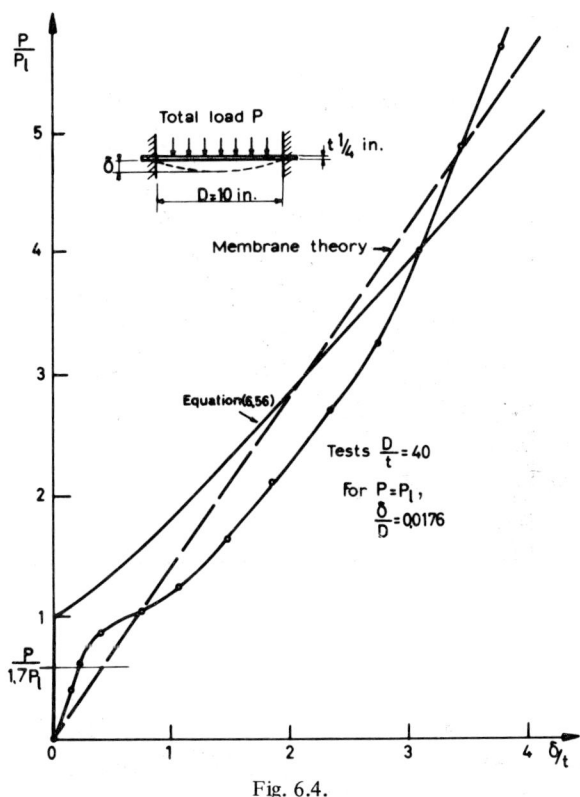

Fig. 6.4.

caused by unrestricted plastic flow are neglected. On the other hand, when these changes are taken into account, the parallel to the deflection axis at the level of the limit load must be replaced by a raising curve of parabolic shape, as drawn in figs. 6.3 to 6.5.

Simple inspection of these figures show that the limit load cannot be regarded as a physical failure load or carrying capacity.

Indeed, the main difficulty is to define what will be considered as failure of the plate. In the case of beams and frames, experimental diagrams exhibit a sharp bend near the limit load P_l, preceded by very small deflections and followed by very large permanent deflections for very small further load increments (see Com. V., fig. 2.26). In experiments on plates, the sharp bent in the load versus deflection diagram tends to disappear (see fig. 6.6, $D/t = 40$). Both the slenderness D/t of the plate and the boundary condi-

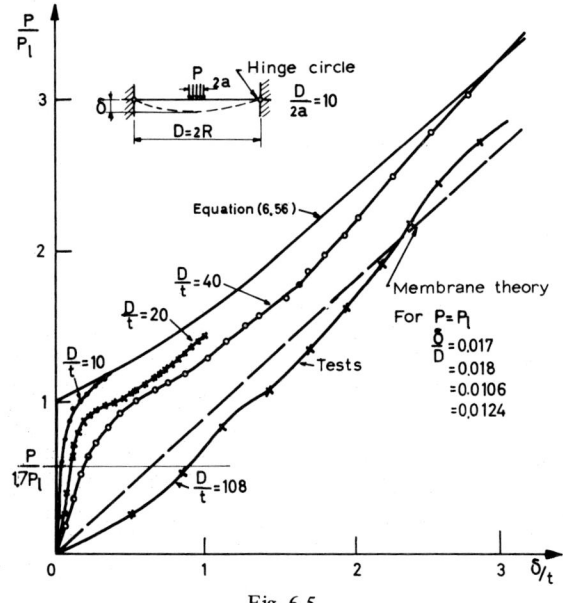

Fig. 6.5.

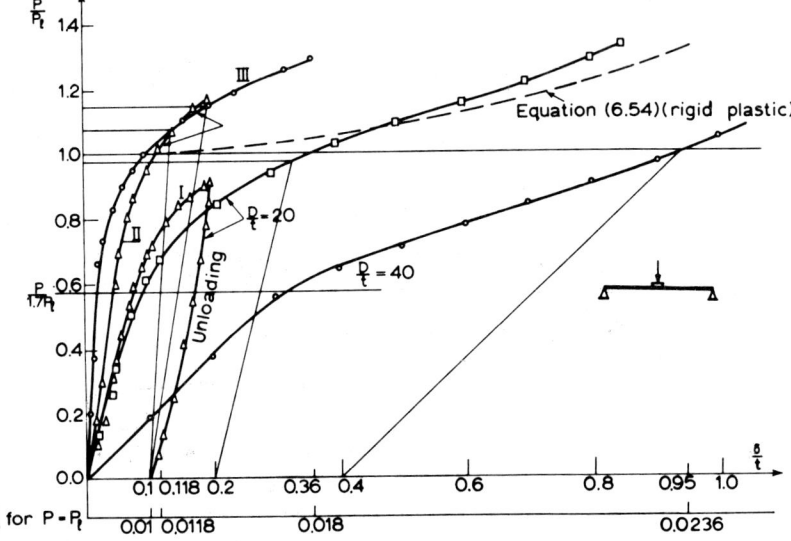

Fig. 6.6.

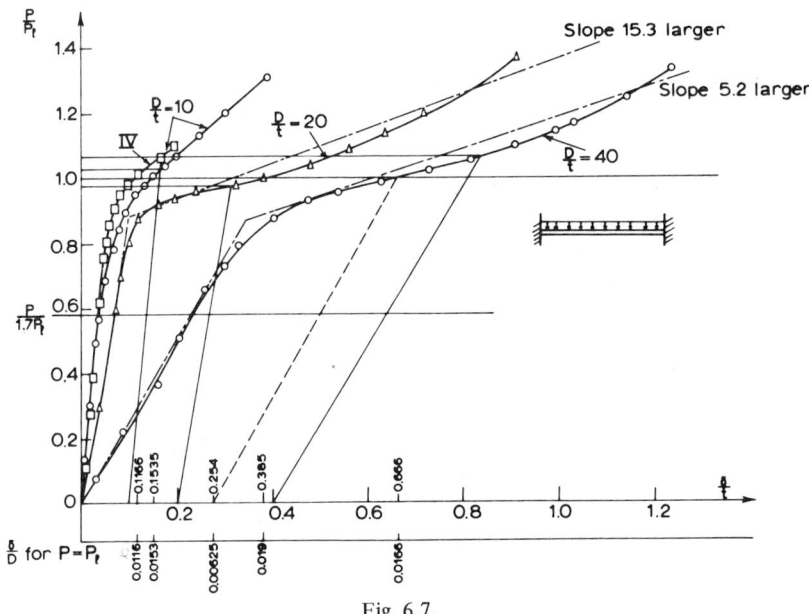

Fig. 6.7.

tions are responsible for this smoothing of the curve. For extremely slender plates as those studied by Cooper and Shifrin [6.6] (see fig. 6.2), with slenderness μ varying from 95 to 200, membrane effects are prominent from the very beginning of loading and bending analysis is meaningless even in the elastic range. Plates of this kind must be treated as membranes, as will be done in Section 6.6. The load versus deflection curves then obtained are rays (represented by dashed lines in figs. 6.4 and 6.5). From the experiments summarized in figs. 6.3 to 6.7, we see that bending is the relevant phenomenon for isotropic plates with $\mu \equiv D/t \leqslant 40$.

When fatigue or buckling failures are not to be considered, failure of the plate may reasonably be defined as exceeding a given limit set to the maximum deflection. To choose this limit in the light of currently available experiments, we note that:

1. For $P = P_l$, the total deflection δ remains a small fraction of the diameter, always smaller than 2% (except for the simply supported plate with $\mu = 40$, fig. 6.6, where $\delta/D = 2.36\%$);
2. For $P = P_l/1.7$ the plate is in the elastic range and its maximum deflection does not exceed 1% of the diameter;

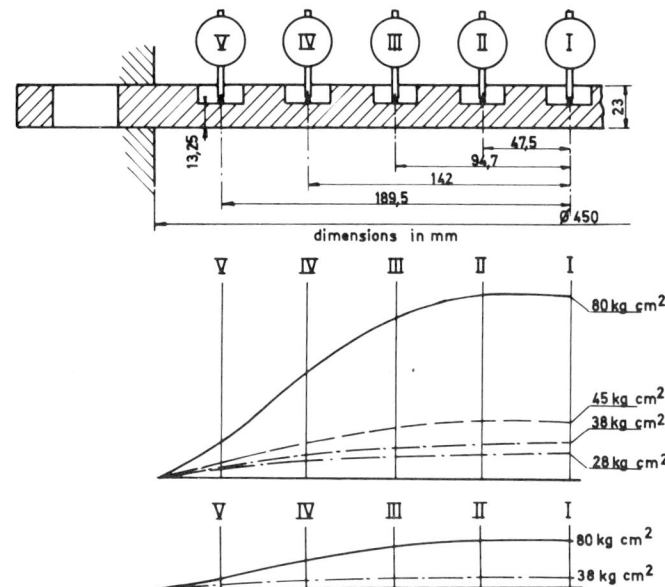

Fig. 6.8 (a).

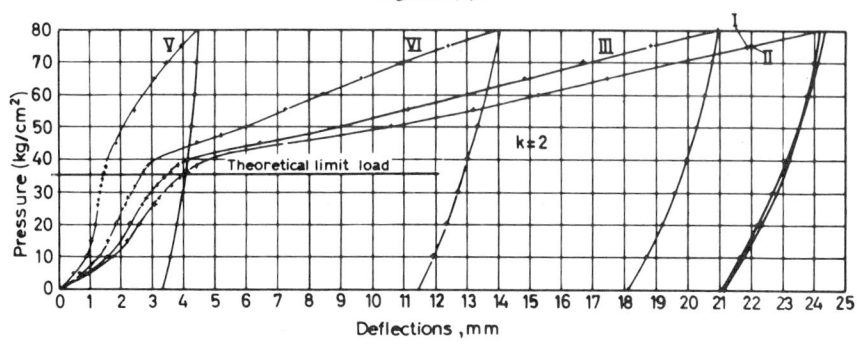

Fig. 6.8 (b).

3. For the built-in plates of fig. 6.7, the average ratio δ/P increases rapidly as P increases from $0.8P_l$ to P_l.

The two-segment approximations to the diagrams give the following results:

for $D/t = 10$, δ/P becomes 11 times larger ;

   $D/t = 20$, δ/P becomes 15 times larger ;

   $D/t = 40$, δ/P becomes 5 times larger .

Hence, the built-in plate with $D/t = 20$ is such that its limit load may be regarded as the failure load. In this particular case we have, for $P = P_l$, $\delta/D = 1.9\%$. Consequently, *we choose the failure deflection as 2% of the diameter**.

It is readily verified that the corresponding experimental load is always larger than P_l, except for the simply supported plate with $D/t = 40$, where it is approximately $0.9 P_l$. It is interesting to remark that a *permanent* deflection of 1% of the span corresponds to loads between $0.975 P_l$ and $1.15 P_l$ for the simply supported plate, between $0.975 P_l$ and $1.055 P_l$ for the built-in plate. when

$$10 \leqslant D/t \leqslant 40 . \tag{6.2}$$

Recent tests by Ohashi and Murakami [6.9, 6.10] and by Ohashi, Murakami and Endo [6.11] give further support to the preceding conclusions. Though their main purpose was to verify the validity of an elastic plastic analysis, they show that, for a circular plate with $D/t = 25$, subjected to uniformly distributed load, and built-in or simply supported [6.9, 6.10], relative deflections δ/D at limit load are 1.4% and 2% respectively, with a sharp bent in the load versus deflection diagram near the limit load. Tests on annular plates [6.11] with $D/t = 40$ show that the bent in the experimental load versus deflection diagram is less and less noticeable with increasing central hole diameter. From the preceding discussion it may be concluded that, in the slenderness range defined by eq. (6.2), the limit load may be regarded as the actual failure load and its determination is therefore useful. Moreover, not only has the limit load a physical significance, but the predicted collapse mechanism is fairly well supported by the experiments [6.1]. Design for safety with respect to the theoretical limit load will result in safety with respect to failure deflection, without necessitating the more difficult analysis that would have furnished the load versus deflection curve.

Experiments on built-in, circular structurally orthotropic, uniformly loaded plates [6.7] give very similar results. Typical diagrams (as well as dial locations) are shown in fig. 6.8. When the breadth of the reinforcing rings is close to the plate thickness, the limit load can be given the same physical interpretation as above provided that

$$22.9 \leqslant \frac{D}{t_{av}} \leqslant 38 , \tag{6.3}$$

when $t_{av} = (t_{max} + t_{min})/2$.

---

* This rule is also accepted in the Belgian Code on Plastic Design of Beams and Frames.

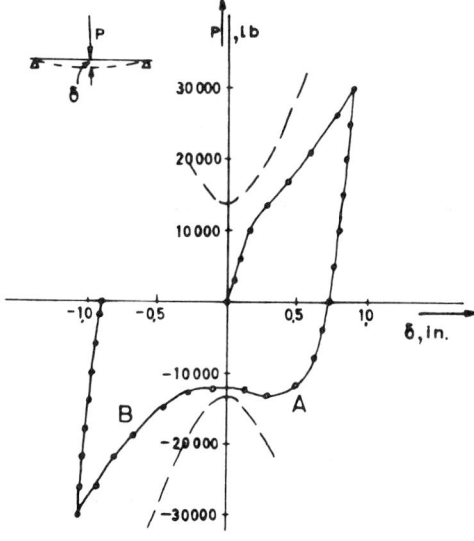

Fig. 6.9.

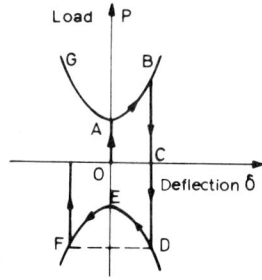

Fig. 6.10.

When the load can be reversed, as for the simply supported plate with the central concentrated load of fig. 6.9, the theoretical load versus deflection diagram is shown in fig. 6.10, based on the rigid perfectly plastic scheme [6.2]. The experimental curves differ from the theoretical ones in that, upon reversal of the load, unrestrained plastic flow occurs in the neighborhood of the (negative) limit load (range AB, fig. 6.9). Hence, this later limit load is here a real carrying capacity. A similar situation is seen on the diagrams of fig. 6.11, dealing with reversed loading of six rectangular built-in beams of 10-in. span, 1 in. thick, subjected to a concentrated load at their central point, and with 1, 3, 5, 7, 8.88, and 18 in. breadths, respectively [6.8]. With the

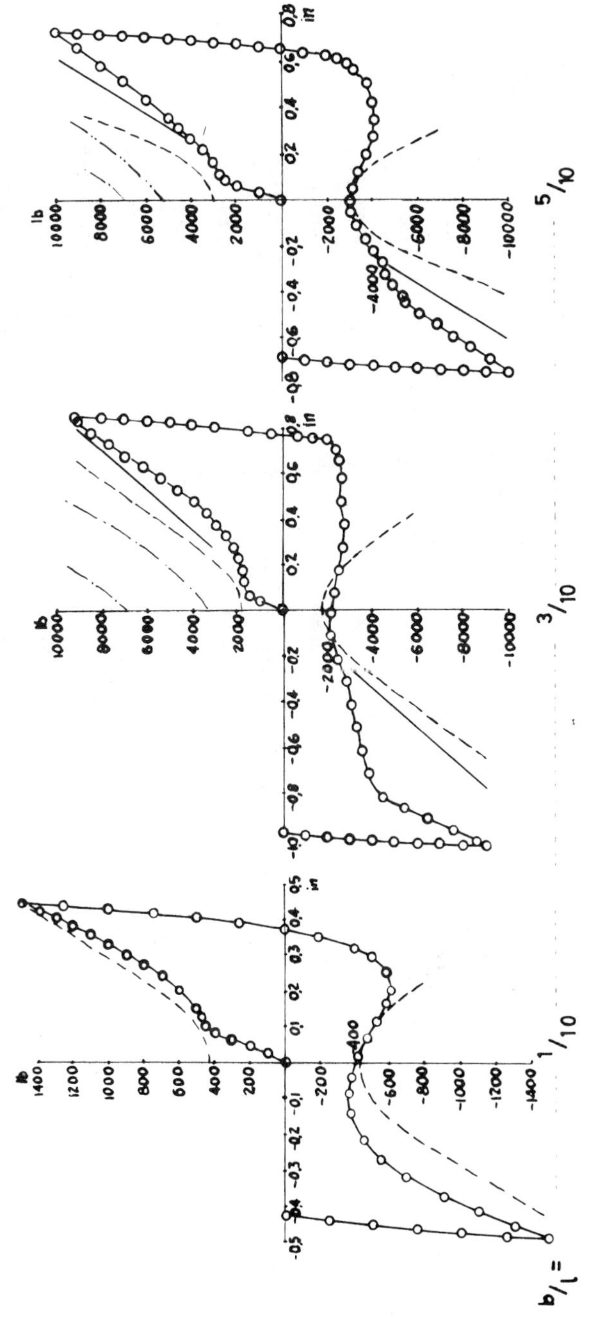

Fig. 6.11.

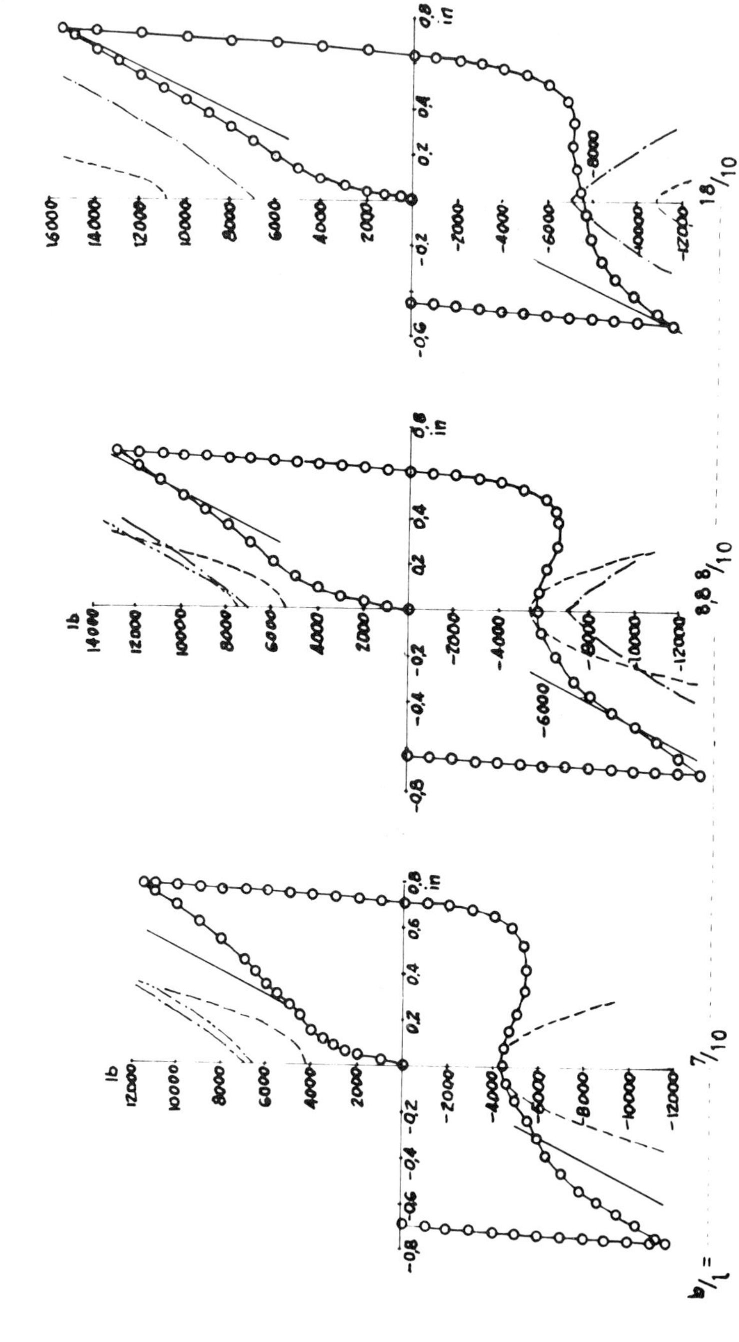

Fig. 6.11.

exception of the two first, these beams are actually rectangular plates built-in along the two short edges and free along the others. Plastic collapse can occur either by a mechanism appropriate for the built-in beam or by a local plate mechanism around the central load, or by a combination of both.

The slenderness ratio is $\mu = 38.6$ and it is seen that, as for circular plates, when $P = P_l$ the total maximum deflection is about 2% of the span and the permanent maximum deflection is about 1% of the span.

Finally it should be emphasized that much more experimental information is needed, especially concerning noncircular plates and plates made of other metals than steel, before the limit load can be attributed a physical meaning of wider applicability.

6.3. Circular isotropic plates

6.3.1. *General relations*

We refer to fig. 6.12 for notations. The distributed load p is assumed to depend on r exclusively and, hence, the problem is axially symmetric.

Equilibrium of the transverse forces acting on a circular central part with radius r requires that

$$2\pi r V = - \int_0^r p 2\pi r \, dr , \qquad (6.4)$$

where p is the load per unit surface. Moment equilibrium of an annular plate element (fig. 6.12) requires that

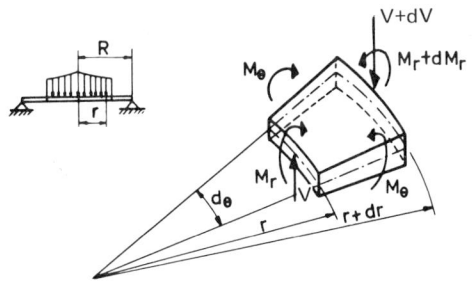

Fig. 6.12.

$$\frac{d}{dr}(rM_r) = M_\theta + rV. \tag{6.5}$$

Elimination of V from eqs. (6.4) and (6.5) furnishes the following fundamental equation:

$$\frac{d}{dr}(rM_r) = M_\theta - \int_0^r pr\, dr. \tag{6.6}$$

A velocity field will be described by a function $\dot{w}$ of r only. The corresponding curvature field is

$$\dot{\kappa}_r = -\frac{d^2\dot{w}}{dr^2}, \qquad \dot{\kappa}_\theta = -\frac{1}{r}\frac{d\dot{w}}{dr}. \tag{6.7}$$

6.3.2. Simply supported plate, circular loading

The load is uniformly distributed over a central circular area of radius a, and the plate is simply supported along its outer edge, which has radius R [6.12] (fig. 6.13).

We first assume Tresca's yield condition to hold. The state of stress at the various points of the plate will be represented by stress points located on or inside the yield curve. The locus of these stress points will be called the "stress profile". The stress profile must start from point A for $r = 0$, because the axial symmetry requires that $M_r = M_\theta$ at the center. The stress profile must end at point B for $r = R$ where $M_r = 0$. Regime AF must be excluded because, according to normality law, eq. (6.7), and boundary conditions, it is not compatible with nonvanishing $\dot{w}$.

Hence, assuming the plate to be entirely plastic at collapse, we have

$$M_\theta = M_p \quad \text{for} \quad 0 \leqslant r \leqslant R. \tag{6.8}$$

Substituting M_p for M_θ in eq. (6.6) and integrating we obtain:

$$\text{for } 0 \leqslant r \leqslant a, \quad \text{where} \quad \int_0^r pr\, dr = p\frac{r^2}{2},$$

$$M_r = M_p - \frac{pr^2}{6} + \frac{C_1}{r}. \tag{6.9}$$

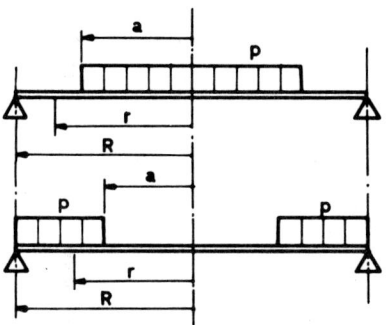

Fig. 6.13(a).

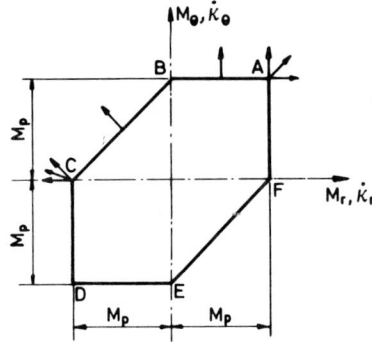

Fig. 6.13(b).

for $a \leqslant r \leqslant R$, where $\int_0^r pr\, dr = p\dfrac{a^2}{2}$,

$$M_r = M_p - \frac{pa^2}{2} + \frac{C_2}{r}. \tag{6.10}$$

Because M_r remains finite (and equal to M_p) at $r = 0$, we have $C_1 = 0$.
Continuity of M_r for $r = a$ then yields $C_2 = pa^3/3$. Finally, $M_r = 0$ at $r = R$ yields the load

$$p = \frac{6M_p R}{a^2(3R-2a)}. \tag{6.11}$$

The corresponding moment field is obtained upon substitution of expression (6.11) for p in eqs. (6.9) and (6.10), with the values of C_1 and C_2 found above.

We obtain

$$M_r = M_p \left[1 - \frac{r^2}{a^2(3-2a/R)} \right] \qquad 0 \leqslant r \leqslant a. \qquad (6.12)$$

$$M_\theta = M_p$$

$$M_r = M_p \left[1 - \frac{3-2a/r}{3-2a/R} \right] \qquad a \leqslant r \leqslant R. \qquad (6.13)$$

$$M_\theta = M_p$$

To verify that eq. (6.11) actually gives the exact limit load, we must find a kinematically admissible velocity field corresponding to the moment field (6.12) and (6.13). For the stress profile AB, the normality law requires that

$$\dot{\kappa}_r = 0, \qquad \dot{\kappa}_\theta \geqslant 0.$$

Using the first of these conditions, eq. (6.7), and the boundary condition $\dot{w}(R) = 0$, we obtain the conical collapse mechanism

$$\dot{w} = \dot{w}_o \left(1 - \frac{r}{R} \right), \qquad (6.14)$$

where $\dot{w}_o$ denotes the transversal velocity of the central point.

From eqs. (6.7) and (6.14) it is readily verified that $\dot{\kappa}_\theta > 0$. We thus have found a complete solution and eq. (6.11) furnishes the exact limit load p_l.

Let $P_l = \pi a^2 p_l$ denote the total load. Eq. (6.11) may also be written as

$$P_l = \frac{6\pi M_p}{3 - (2a/R)}. \qquad (6.15)$$

When a tends to zero or to R, we obtain *the limiting value of a concentrated central load*,

$$P_l = 2\pi M_p, \qquad (6.16)$$

and *the limiting value of a uniformly distributed load*,

$$P_l = 6\pi M_p ,\qquad (6.17)$$

respectively. Note that, for a concentrated load, the moment field is $M_r = 0$, $M_\theta = M_p$ (given by eq. (6.13) with $a = 0$) with a singularity for $r = 0$. It should however be kept in mind that bending solutions neglect the influence of shear forces and, consequently, cannot be expected to be valid for concentrated loads. For $D/t \geqslant 20$, it has been found [6.13] that bending theory neglecting shear forces effects apply when $a/t > 1.5$.

The preceding problem has been treated by Hopkins and Wang using the von Mises yield condition [6.14]. The curved stress profile AB on fig. 6.14 is used. From the very nature of von Mises' condition, the resulting differential equilibrium equation is nonlinear and must be integrated numerically. The limit load versus the ratio a/R is given in fig. 6.15.

Velocity fields that correspond to the moment fields obtained in this manner can be found. The limit load is therefore exact. The results of Hopkins and Wang practically coincide with those found by Sokolovsky [6.15] by the use of a deformation theory. This is in accordance with our remark at the end of Section 3.6. For $a/R = 1$, that is for a *uniformly loaded plate*, we have

$$p_l = \frac{6.51 M_p}{R^2} \qquad \text{from [6.14] ,}$$

$$p_l = \frac{6.46 M_p}{R^2} \qquad \text{from [6.15] ,} \qquad (6.18)$$

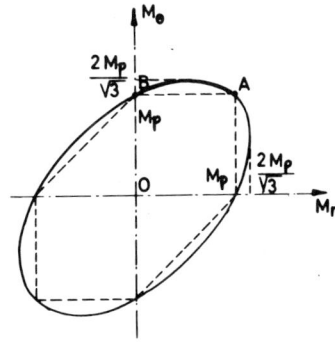

Fig. 6.14.

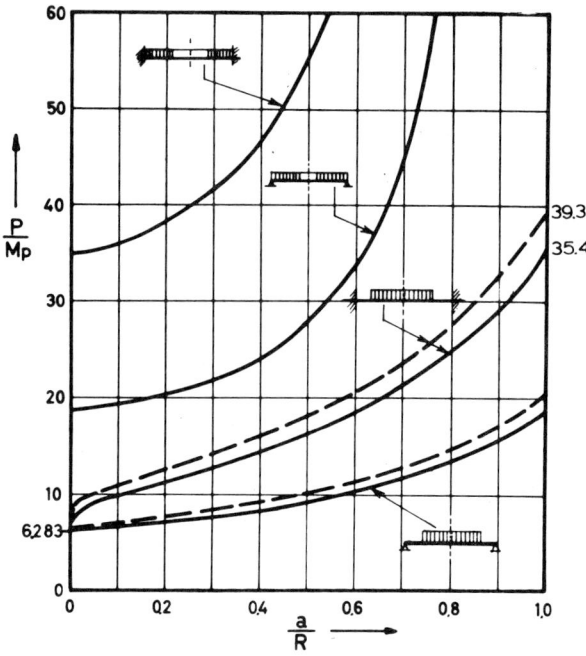

Fig. 6.15.

the discrepancy being smaller than 1%.

For a concentrated load P, the limit value is found to be

$$P_l = 2\pi M_p, \tag{6.19}$$

as for the Tresca criterion. This is easily understood. Indeed, in both cases the stress profile reduces to two points, A (at the center) and B (everywhere else) that are on both yield curves.

6.3.3. *Simply supported circular plate, annular loading*

The load is uniformly distributed over an external annulus [6.12]. The analysis is similar to that of Section 6.3.2, and the limit load is shown in fig. 6.15. No corresponding solution for the von Mises condition is known at present.

6.3.4. Built-in circular plate, circular loading

For Tresca's yield condition, the stress profile is ABC, fig. 6.13. The side CD is excluded because the requirements of the normality law cannot be reconciled with eqs. (6.7) for nonvanishing $\dot{w}$. Complete solutions are obtained as in Section 6.3.2, but the limit load depends on a/R through the radius ρ at which the stress regime changes from AB to BC, and this radius is given by a transcendental equation that must be solved numerically. Limit loads are given in fig. 6.15 and velocity distributions in fig. 6.16. For $a/R = 1$, the limit load is $P_l \cong 35.4M_p$, approximately 88% higher than for the simply supported plate. Limit loads with the von Mises condition [6.14] are also given in fig. 6.15.

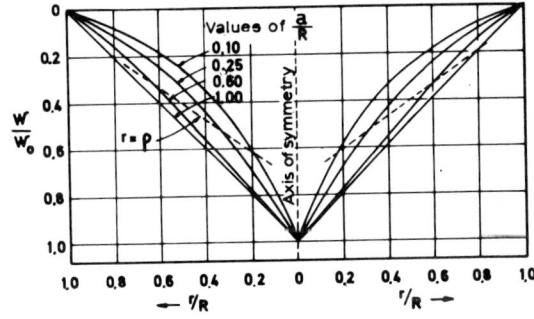

Fig. 6.16.

With the von Mises condition the limiting intensity of a uniform load acting over the entire plate is ($a/R=1$)

$$p_l \cong \frac{12.5M_p}{R^2}, \tag{6.20}$$

or,

$$P_l \equiv p_l \pi R^2 = 39.3M_p. \tag{6.21}$$

The limiting value of a concentrated central load is

$$P_l = \frac{2}{\sqrt{3}} 2\pi M_p = 7.26M_p. \tag{6.22}$$

This compares with $2\pi M_p$ for the condition of Tresca.

6.3.5. Built-in plate, annular loading
The limit load is given in fig. 6.15, for the yield condition of Tresca [6.12].

6.3.6. Circular line load with radius a and total magnitude Q
From Sawczuk and Jaeger [6.16] we have:
for the simply supported plate

$$Q_l = 2\pi M_p \frac{1}{1 - a/R}, \qquad (6.23)$$

for the built-in plate

$$Q_l = 2\pi M_p \left(1 - \frac{1}{\ln \rho}\right), \qquad (6.24)$$

where ρ is given by

$$\rho - \frac{a}{R}(1 - \ln \rho) = 0. \qquad (6.25)$$

6.3.7. Annular plates subjected to uniform load
Limit loads for various boundary conditions are given in figs. 6.17 and 6.18 from Sawczuk and Jaeger [6.16].

6.3.8. Various plates with line loads
Various plates with line loads are shown in fig. 6.19. The relations in this figure are taken from Iliouchine ([1.2] pp. 237-244)*. They were obtained by a purely statical approach with the Tresca yield condition. Hence they are lower bounds Q_-, and are represented in fig. 6.20. Relation (6.25) was found independently by Hodge [3.8] as an exact limit load.

6.3.9. Plate with overhang
Drucker and Hopkins [6.17] have discussed a plate of radius $\alpha R (\alpha \geq 1)$,

---

* Limit loads for circular loading obtained by Hopkins and Prager [6.12] can also be found in ref. [1.2].

6.3] CIRCULAR ISOTROPIC PLATES 131

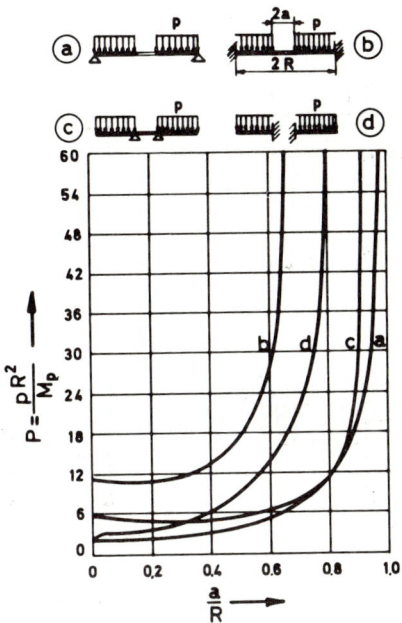

Fig. 6.17.

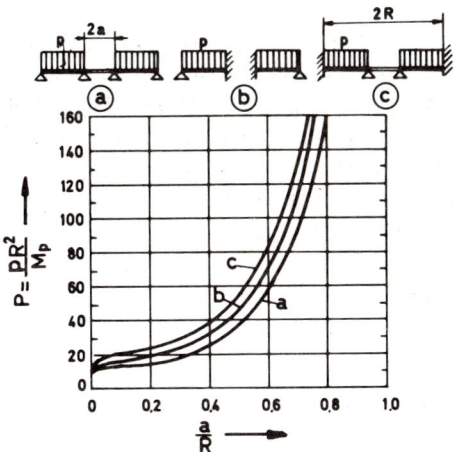

Fig. 6.18.

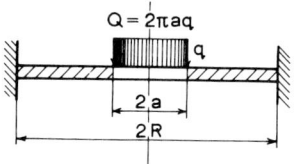

Fig. 6.19 (a) $\dfrac{Q}{M_p} = 2\pi \left(1 + \dfrac{1}{\ln a/R}\right)$

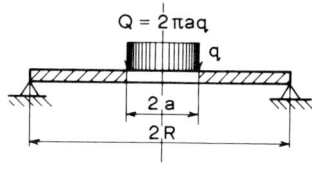

Fig. 6.19 (b) $\dfrac{Q}{M_p} = 2\pi$

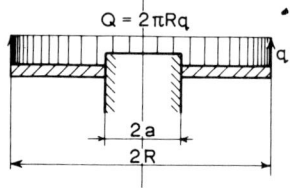

Fig. 6.19 (c) $\dfrac{Q}{M_p} = \dfrac{1}{1 - a/R}$

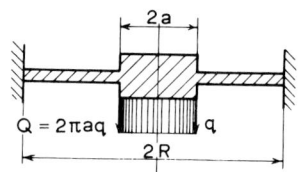

Fig. 6.19 (d) $\dfrac{Q}{M_p} = 2\pi k$

$$\dfrac{k}{k-1} e^{1/k-1} = \dfrac{R}{a}$$

simply supported on a circle of radius R, and subjected simultaneously to a uniform pressure p over the area interior to the circle of support and to a central concentrated load P (fig. 6.21). This problem is of particular interest because it contains the simply supported plate ($\alpha=1$) and the built-in plate (α sufficiently large) as limiting cases, and is likely to represent more accurately real support conditions intended to provide simple support. The yield condition of Tresca is used. The results of the integration of eq. (6.6) in the various plastic regimes are given on fig. 6.22. In the present case, the relevant regimes are AB and BC.

For small values of r, we first have

$$M_r = M_p - \dfrac{P}{2\pi} - p\dfrac{r^2}{6}, \qquad 0 \leqslant r \leqslant r_1, \tag{6.26}$$

where r_1 corresponds to $M_r = 0$, that is,

6.3] CIRCULAR ISOTROPIC PLATES 133

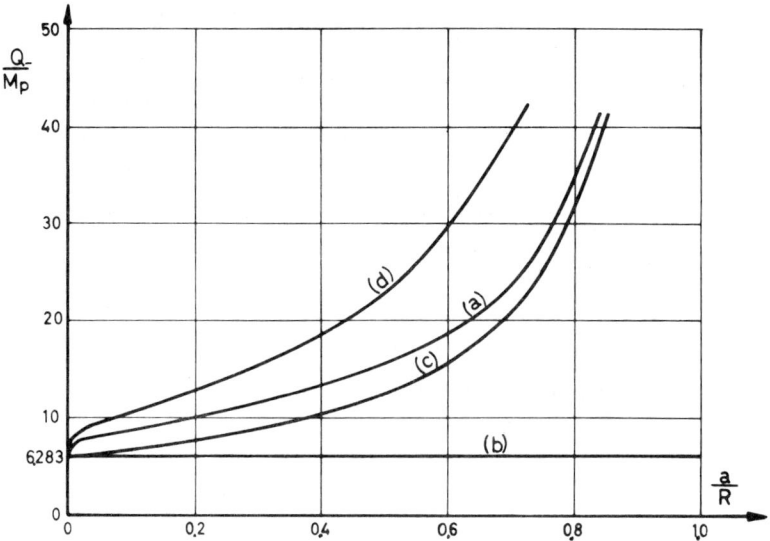

Fig. 6.20.

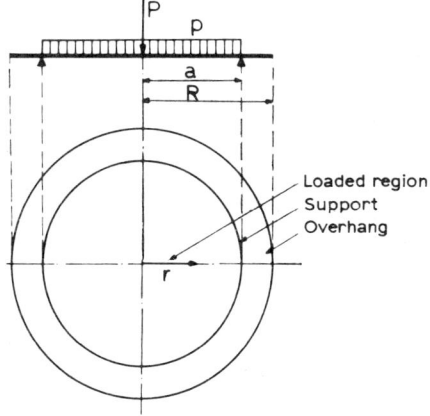

Fig. 6.21.

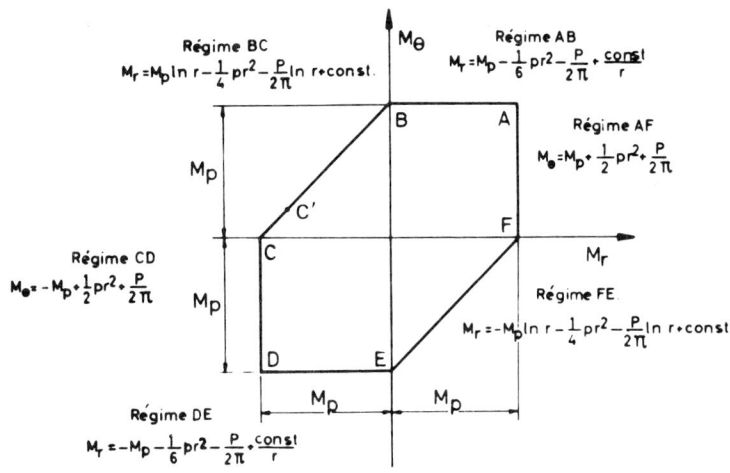

Fig. 6.22.

$$r_1 = \frac{6}{p}\left(M_p - \frac{P}{2\pi}\right). \tag{6.27}$$

Note that there is a singularity at $r = 0$ because M jumps from M_p to $M_p - P/2\pi$. This is due to the use of a bending theory in the presence of the concentrated load P.

If $\alpha = 1$ (simply supported plate), $r_1 = R$ and we get from eq. (6.27)

$$\frac{pR^2}{6M_p} + \frac{P}{2\pi M_p} = 1. \tag{6.28}$$

Hence, we see that the uniformly distributed load p and the central concentrated load P for collapse are linearly related. Relation (6.28) reduces to eq. (6.16) or eq. (6.17) when p or P vanish, respectively. If $\alpha > 1$, higher loads are possible. A detailed analysis [6.17] shows that

1. If $0 < \ln \alpha < 1$, that is, if $1 < \alpha < 2.718$, the plastic regime is $ABC'B$, points C' and B corresponding to $r = R$ and $r = \alpha R$, respectively. The abscissa of point C' is $M_r(R) = -M_p \ln \alpha > -M_p$. Eq. (6.28) is replaced by

$$p\pi R^2 \cdot \left(\frac{1 - r_1^2/R^2}{4}\right) + \frac{P}{4\pi} \ln \frac{R^2}{r_1^2} = M_p \left(\ln \frac{R}{r_1} + \ln \alpha\right), \tag{6.29}$$

where r_1 is given by eq. (6.27).

2. If $\ln \alpha = 1$, that is, if $\alpha = 2.718$, point C' coincides with point C. We have $M_r(R) = -M_p$ and relation (6.29) holds, with $\ln \alpha = 1$.

3. If $\ln \alpha > 1$, that is, if $\alpha > 2.718$, the overhang remains rigid and horizontal, and eq. (6.29) must be used with $\ln \alpha = 1$ as in case 2.

Results are summarized in figs. 6.23 and 6.24, where dimensionless variables have been used.

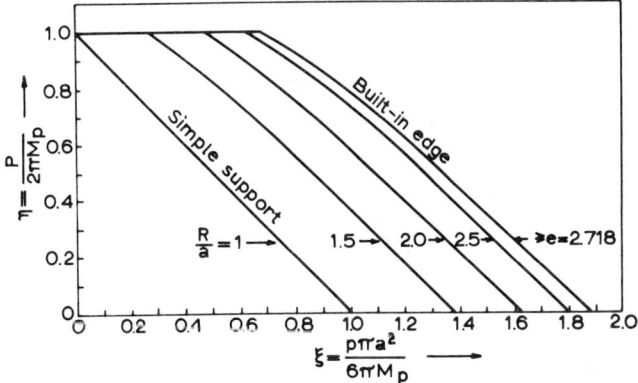

Fig. 6.23.

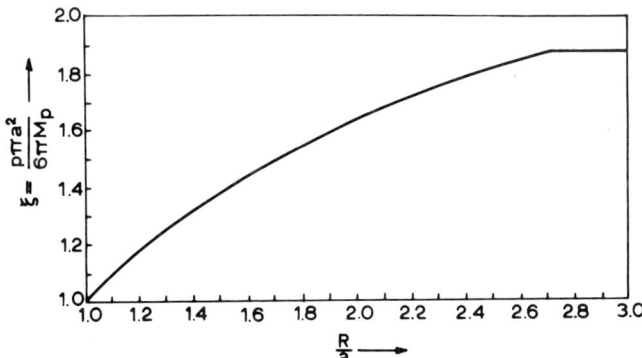

Fig. 6.24.

6.4. Circular orthotropic plates

6.4.1. *Introduction*

Many plates used in practice are strengthened by stiffeners, to achieve high strength with small structural weight. Stiffeners are most often placed along the lines of an orthogonal net and the plate so constructed exhibits structural plane orthotropy, that is, structural plane anisotropy with two orthogonal axes of symmetry.

Material orthotropy can also arise from the cold forming process: yield stresses in various directions are then different. Whatever the source, the

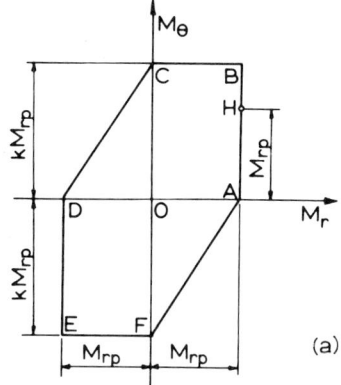

Fig. 6.25 (a).

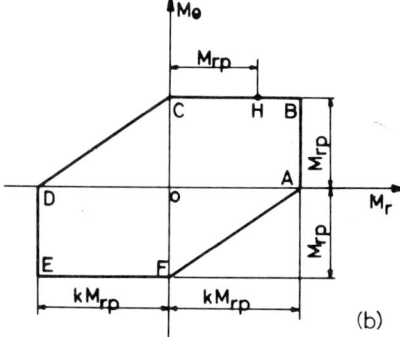

Fig. 6.25 (b).

orthotropy will be described by a proper yield condition, which must ultimately be justified by experimental evidence. In the following, two types of orthotropy will be considered, generalizing the yield condition of Tresca. The normality law is assumed to apply to the yield conditions used, and the principal directions for the stresses and the orthotropy are supposed to be the radial and circumferential directions.

6.4.2. *Orthotropy of the first kind*

The assumed yield condition is shown in fig. 6.25 [6.18,6.19]. The full plastic moments $M_{\theta p}$ and M_{rp} in the circumferential and radial directions, respectively, differ by a given factor $k = M_{\theta p}/M_{rp}$, called the orthotropy coefficient.

Restricting our considerations to axisymmetric loading, we immediately note that the central point of the plate is special. Indeed, any two orthogonal directions at this point are principal directions for stresses and orthotropy. Hence the plate is locally isotropic at the center. Any stress profile must start, at $r = 0$, from point H in fig. 6.25, corresponding to $M_r = M_\theta$.

Suppose first that $k \geq 1$, $M_{\theta p} \geq M_{rp}$, and consider a simply supported uniformly loaded plate (fig. 6.26). The stress profile is HBC in fig. 6.25 (a).

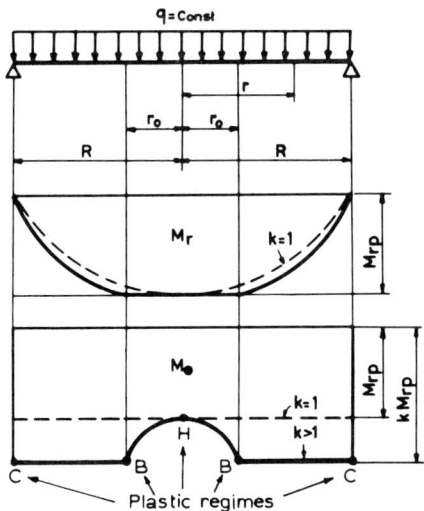

Fig. 6.26.

Integration of eq. (6.6) with this stress profile and adequate continuity conditions at the radius r_o corresponding to point B gives the moment field shown in fig. 6.26.

The value of r_o is given by

$$2\left(\frac{r_o}{R}\right)^3 (k-1) - 3\frac{r_o k}{R} + k - 1 = 0 . \qquad (6.30)$$

The limit load is

$$P_l \equiv p_l \pi R^2 = 6\pi \frac{k-1}{(r_o/R)^2} M_{rp} , \qquad (6.31)$$

and the velocity field is represented in fig. 6.27. The limit load (6.31) is shown graphically in fig. 6.28, with the limit load for the built-in plate obtained in a similar manner. An approximate formula for this latter case is

$$P_l = \pi M_{rp} [11.26+4.84(k-1)] . \qquad (6.32)$$

When $k \leqslant 1$, the stress profile for simple supports is BC in fig. 6.25 (b), as for an isotropic plate with plastic moment M_p equal to $M_{\theta p}$. Hence the limit load is

$$P_l = 6\pi M_{\theta p} < 6\pi M_{rp} ,$$

and it is concluded that purely radial strengthening is completely useless.

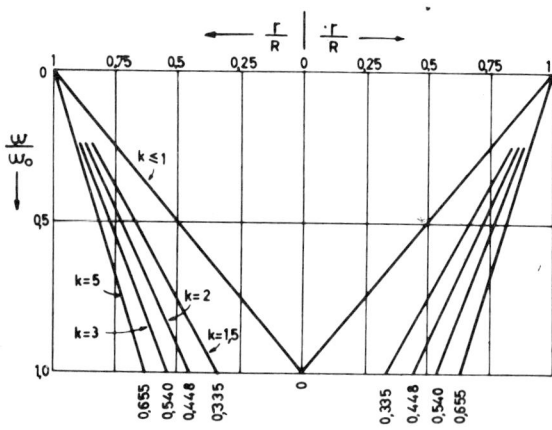

Fig. 6.27.

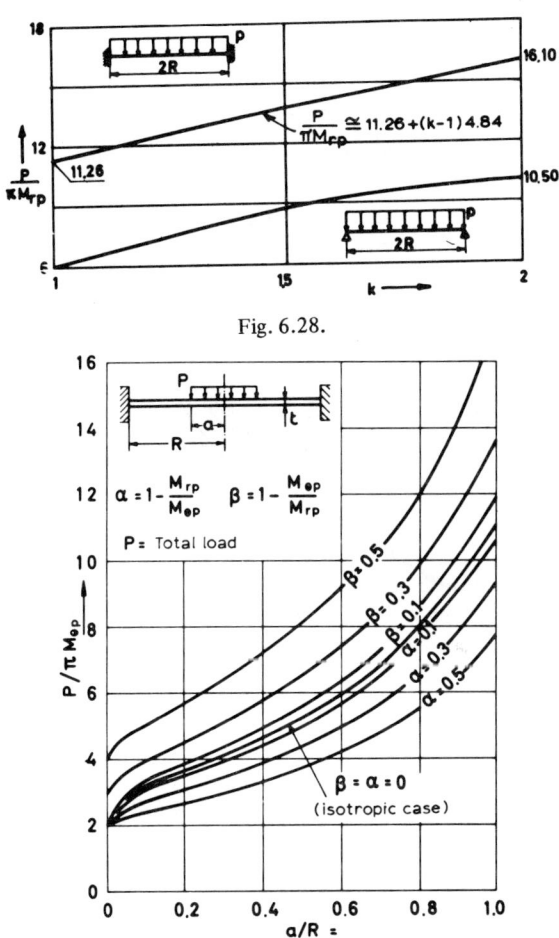

Fig. 6.28.

Fig. 6.29. Collapse load: built-in, uniformly loaded circular plate. (Reprinted by permission of American Society of Civil Engineers from J.Markowitz and L.W.Hu, "Plastic Analysis of Orthotropic Circular Plates", *Proc. A.S.C.E., J. Eng. Mech. Div.*, **90**: 251, 1964.)

The preceding results, taken from Olszak and Sawczuk [6.18,6.19,6.20], have been extended to various loading cases by Markowitz and Hu [6.21] from whose paper we borrow the next ten figures (figs. 6.29 to 6.38). The reader is referred to the original publication for details. Orthotropy is described either by α or β, with

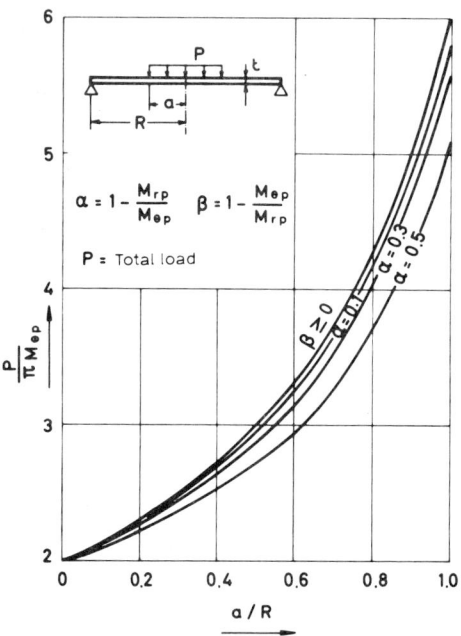

Fig. 6.30. Collapse load: simply supported, uniformly loaded circular plate. (Reprinted by permission of American Society of Civil Engineers from J.Markowitz and L.W.Hu, "Plastic Analysis of Orthotropic Circular Plates", *Proc. A.S.C.E., J. Eng. Mech. Div.*, **90**: 251, 1964.)

$$\beta = 1 - k, \qquad \alpha = 1 - \frac{1}{k}, \qquad \alpha = \frac{\beta}{\beta - 1}, \qquad \beta = \frac{\alpha}{\alpha - 1}.$$

Limit loads in figs. 6.29 to 6.38 are exact limit loads obtained from complete solutions. It is seen from the graphs that the orthotropy may: (a) influence the limit load for all degrees of orthotropy (fig. 6.29); (b) influence the limit load only if $M_{\theta p} > M_{rp}$ (fig. 6.30); or (c) not influence the limit load at all (fig. 6.34).

We finally remark that, for large central bosses or large central holes, shear action may become important and the limit load from bending theory might lose its physical significance (see Section 6.7.2, fig. 6.51). As already noted

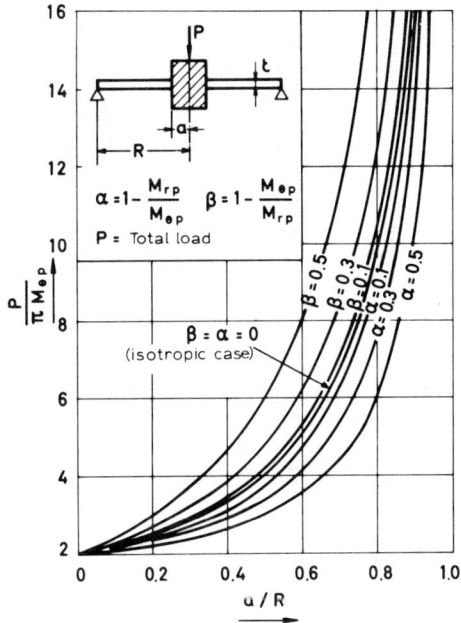

Fig. 6.31. Collapse load: simply supported circular plate with shear force applied to inner edge; rotation prevented at inner edge. (Reprinted by permission of American Society of Civil Engineers from J.Markowitz and L.W.Hu, "Plastic Analysis of Orthotropic Circular Plates", *Proc. A.S.C.E., J. Eng. Mech. Div.*, **90**: 251, 1964.)

in Section 6.2, experiments [6.7] support formula (6.32) for a structural orthotropy with numerous reinforcing rings.

6.4.3. *Orthotropy of the second kind*

An alternative yield condition has been proposed by Sawczuk [6.20]. It is shown in fig. 6.39.

For a simply supported, uniformly loaded plate, the stress profile is BC as r varies from 0 to R; it furnishes the statically admissible moment field (fig. 6.40).

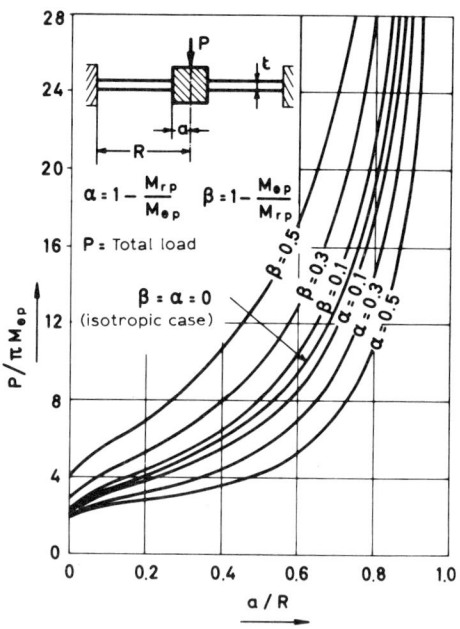

Fig. 6.32. Collapse load: built-in circular plate with shear force applied to inner edge, rotation prevented at inner edge. (Reprinted by permission of American Society of Civil Engineers from J.Markowitz and L.W.Hu, "Plastic Analysis of Orthotropic Circular Plates", *Proc. A.S.C.E., J. Eng. Mech. Div.*, **90**: 251, 1964.)

$$M_r = \left[1 - \left(\frac{r}{R}\right)^2\right] M_p,$$

$$M_\theta = \left[1 + \left(\frac{r}{R}\right)^2 (k-1)\right] M_p, \qquad (6.33)$$

where $k = M_{\theta p}/M_p$.

Associated kinematically admissible velocity fields are shown in fig. 6.41, and the exact limit load is

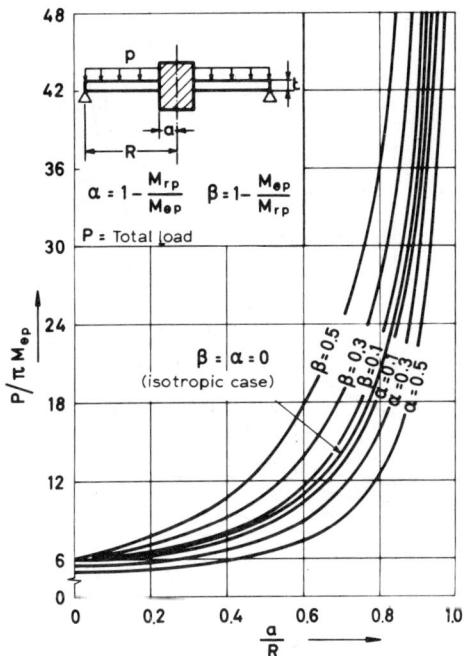

Fig. 6.33. Collapse load: simply supported, uniformly loaded circular plate; rotation prevented at inner edge. (Reprinted by permission of American Society of Civil Engineers from J.Markowitz and L.W.Hu, "Plastic Analysis of Orthotropic Circular Plates", *Proc. A.S.C.E., J. Eng. Mech. Div.*, **90**: 251, 1964.)

$$P_l \equiv p_l \pi R^2 = 6\pi M_p \frac{2+k}{3}. \tag{6.34}$$

We note that the limit load does not depend on M_{rp}/M_p as it was to be expected from the stress profile used. Hence, radial strengthening is useless as far as the limit load is concerned.

For the yield curve of fig. 6.39 to remain convex, the condition

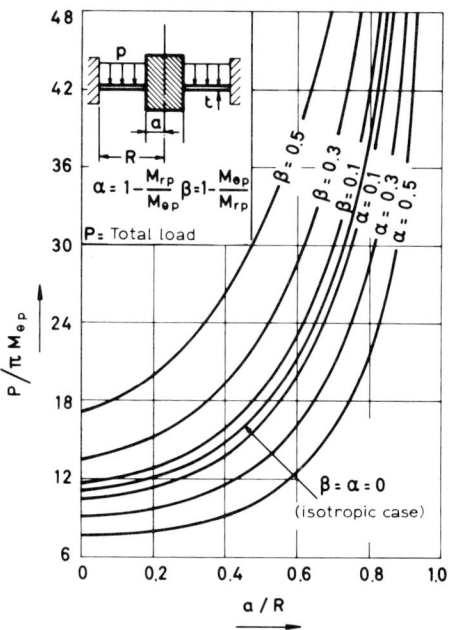

Fig. 6.34. Collapse load: built-in uniformly loaded circular plate; rotation prevented at inner edge. (Reprinted by permission of American Society of Civil Engineers from J.Markowitz and L.W.Hu, "Plastic Analysis of Orthotropic Circular Plates", *Proc. A.S.C.E., J. Eng. Mech. Div.*, **90**: 251, 1964.)

$$k \geqslant \frac{M_{rp}}{M_p + M_{rp}}$$

must be satisfied [6.20].

6.5. Isotropic rectangular plates

6.5.1. *Introduction*

Limit analysis of rectangular metal plates is still in its infancy. Contrary to the case of circular plates, the principal directions of bending moment and

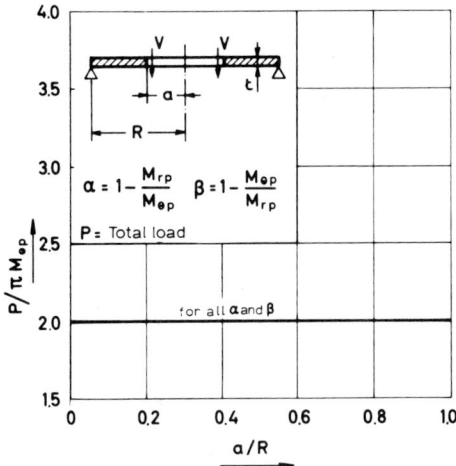

Fig. 6.35. Collapse load: simply supported circular plate with shear force applied to inner edge. (Reprinted by permission of American Society of Civil Engineers from J.Markowitz and L.W.Hu, "Plastic Analysis of Orthotropic Circular Plates", *Proc. A.S.C.E., J. Eng. Mech. Div.*, **90**: 251, 1964.)

curvature are not known beforehand. The mathematical difficulties therefore are considerably greater. On the other hand, experiments are completely lacking, despite the obvious practical interest of the problem. Hence, we restrict ourselves to presently available theoretical solutions.

6.5.2. *Tresca yield condition*

Consider a simply supported rectangular plate, with side lengths l and L, subjected to uniformly distributed load. In an elementary approach, Drucker [6.22] imagines a mechanism formed of straight yield lines (hinge lines) (fig. 6.42), as is customary in reinforced concrete (see Chapter 7). The hinge lines must be regarded as the limiting case of narrow strips exhibiting cylindrical curvature. Hence, in the considered mechanism, the curvature rates vanish everywhere except in the hinge lines, where the principal directions of curvature rate are that of the hinge line and the normal to it. The curvature rate along the hinge line always vanishes. Because of isotropy and

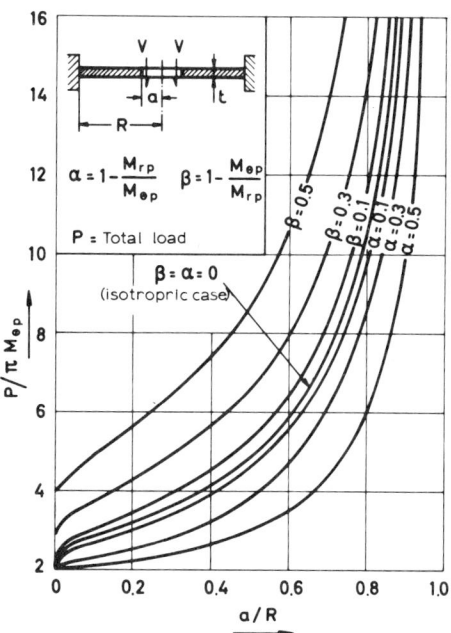

Fig. 6.36. Collapse load: built-in circular plate with shear force applied to inner edge. (Reprinted by permission of American Society of Civil Engineers from J.Markowitz and L.W.Hu, "Plastic Analysis of Orthotropic Circular Plates", *Proc. A.S.C.E., J. Eng. Mech. Div.*, **90**: 251, 1964.)

normality law, principal directions of the moment tensor coincide with those of the curvature rate tensor.

From these considerations, we see that the plastic regime in a (straight) hinge line must be either *AB* or *DE* (fig. 6.43) and the dissipation per unit length of hinge line is simply the product $M_p|\dot{\theta}|$, where $\dot{\theta}$ denotes the relative rotation rate of adjacent parts.

Inspection of fig. 6.42 shows that the relative rotation rates are $4\delta/l$ for segment *EF*; and $2\sqrt{2}\delta/l$ for segments *AE, EB, FD,* and *FC*, where δ is the transversal velocity of segment *EF*. The power of the applied load is p times the volume bounded by the initial midplane of the plate and its deformed situation at collapse.

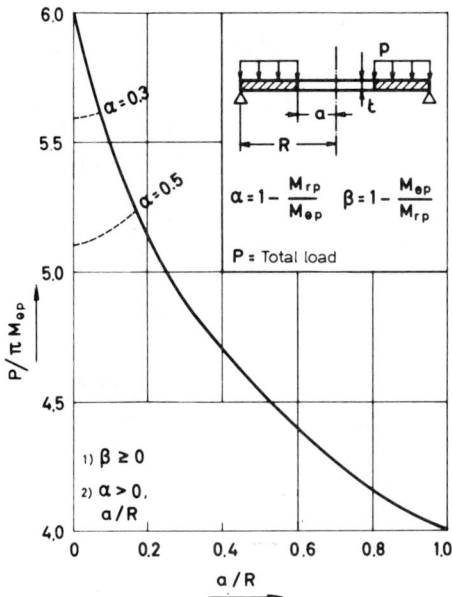

Fig. 6.37. Collapse load: simply supported, uniformly loaded circular plate with center cutout. Solid curve valid for (1) $\beta > 0$, and (2) $\alpha > 0$, except for small a/R as indicated. (Reprinted by permission of American Society of Civil Engineers from J.Markowitz and L.W.Hu, "Plastic Analysis of Orthotropic Circular Plates", *Proc. A.S.C.E., J. Eng. Mech. Div.*, **90**: 251, 1964.)

From the kinematic theorem, we then obtain

$$p_+ [\tfrac{1}{3}l^2\delta + l(L-l)\delta] = M_p \left[4\frac{\delta}{l}(L-l) + 4(2\sqrt{2})\frac{\delta}{l}\frac{l}{\sqrt{2}} \right],$$

or

$$p_+ \frac{l^2}{M_p} = 24\,\frac{1 + L/l}{3L/l - 1}. \tag{6.35}$$

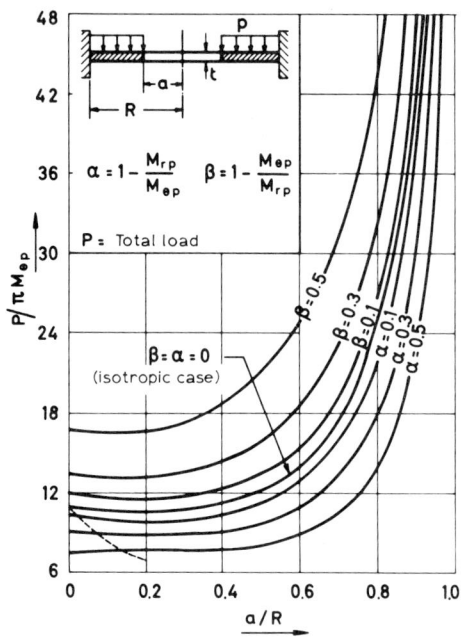

Fig. 6.38. Collapse load: built-in, uniformly loaded circular plate with center cutout. (Reprinted by permission of American Society of Civil Engineers from J.Markowitz and L.W.Hu, "Plastic Analysis of Orthotropic Circular Plates", *Proc. A.S.C.E., J. Eng. Mech. Div.*, **90**: 251, 1964.)

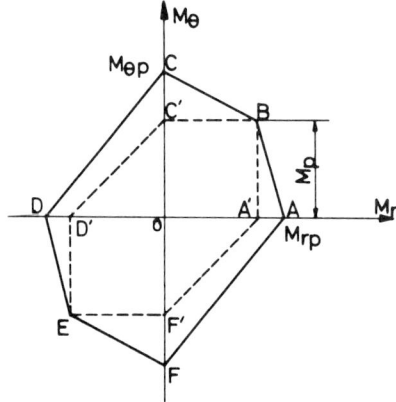

Fig. 6.39.

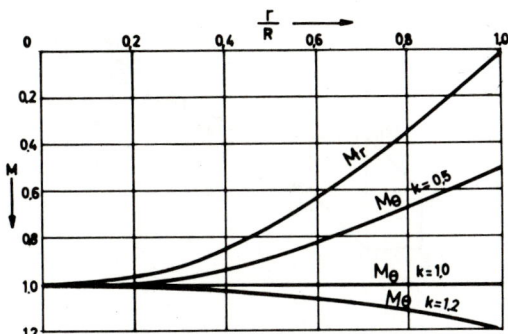

Fig. 6.40.

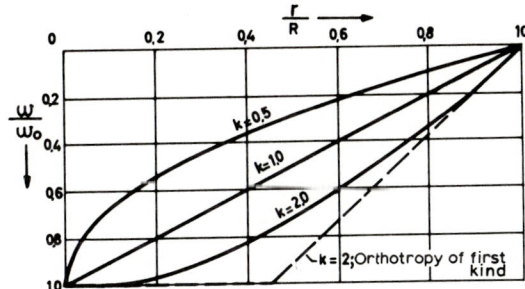

Fig. 6.41.

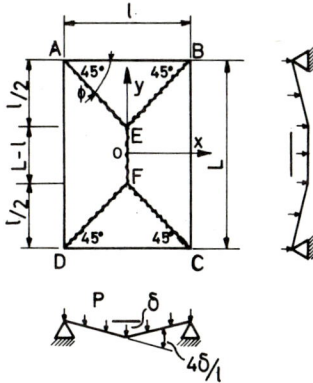

Fig. 6.42.

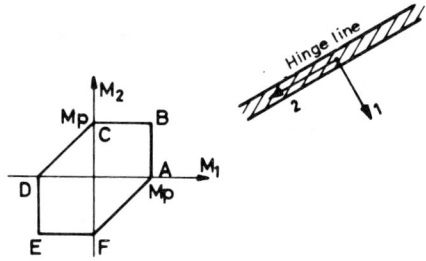

Fig. 6.43.

To obtain a lower bound p_-, let us imagine a moment field with no twist, inspired from the analogous beam problem:

$$M_x = C_1\left(\frac{l^2}{4} - x^2\right),$$

$$M_y = C_2\left(\frac{L^2}{4} - y^2\right),$$

$$M_{xy} = 0. \tag{6.36}$$

To satisfy the equilibrium equation*

$$\frac{\partial^2 M_x}{\partial x^2} + \frac{\partial^2 M_y}{\partial y^2} - 2\frac{\partial^2 M_{xy}}{\partial x\, \partial y} = -p, \tag{6.37}$$

we must have

$$C_1 + C_2 = \frac{p}{2}. \tag{6.38}$$

Maximum moments occur for $x = 0$ and $y = 0$. Highest load is obtained with regime B (fig. 6.43) at central point, because both maximum moments attain the plastic value simultaneously. Substitution of $x = y = 0$ and $M_x = M_y = M_p$ in eqs. (6.36) results in

---

* Shear forces are given by $V_x = \partial M_x/\partial x - \partial M_{xy}/\partial y$, $V_y = \partial M_y/\partial y - \partial M_{xy}/\partial x$.

$$C_1 = 4\frac{M_p}{l^2},$$

$$C_2 = 4\frac{M_p}{L^2}. \tag{6.39}$$

From eqs. (6.38) and (6.39) we obtain the lower bound

$$p_- = 8M_p\left(\frac{1}{l^2} + \frac{1}{L^2}\right). \tag{6.40}$$

For a square plate with side l, eqs. (6.35) and (6.40) are combined into

$$16 \leqslant \frac{p_l l^2}{M_p} \leqslant 24. \tag{6.41}$$

The bounds given by eq. (6.41) are far apart because the lower bound is much too crude.

A refined approach to the problem of the rectangular plate has been given by Shull and Hu [6.23]. An improvement of the moment field is obtained by setting

$$M_x = M_y = C\left(1 - \frac{8x^3}{l^2}\right)\left(1 - \frac{8y^3}{L^2}\right). \tag{6.42}$$

Substitution of expressions (6.42) for M_x and M_y in eq. (6.37) furnishes M_{xy} as a function of p, C, x, and y. Now, the yield condition of Tresca in the moment space M_x, M_y, M_{xy} is needed. We know that relations (1.12) and (1.13) are also valid for the (plane) moment tensor, by simply replacing σ_1, σ_2, σ_x, σ_y, τ_{xy} by M_1, M_2, M_x, M_y, $-M_{xy}$, respectively. Hence, condition (5.30) represented in fig. 6.43 can be written

(a) $\quad \dfrac{m_x + m_y}{2} + \left[\left(\dfrac{m_x - m_y}{2}\right)^2 + m_{xy}^2\right]^{1/2} = \pm 1,$

(b) $\quad \dfrac{m_x + m_y}{2} - \left[\left(\dfrac{m_x - m_y}{2}\right)^2 + m_{xy}^2\right]^{1/2} = \pm 1,$

(c) $\quad 2\left[\left(\dfrac{m_x - m_y}{2}\right)^2 + m_{xy}^2\right]^{1/2} = \pm 1,$ \hfill (6.43)

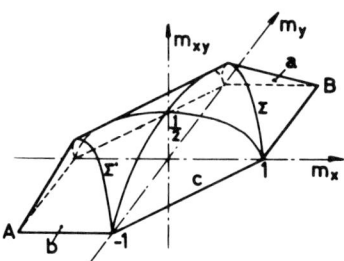

Fig. 6.44.

using dimensionless moments as defined by eqs. (5.25).

Condition (6.43) is represented in fig. 6.44. The surface consists of a cylindrical central part c with eq. (6.43c) and two conical caps a and b with eqs. (6.43a) and (6.43b) respectively. The intersections Σ and Σ' of the surfaces above satisfy the condition

$$m_x m_y = m_{xy}^2 . \tag{6.44}$$

Because we have chosen a particular moment field (6.42) in which $M_x = M_y$, the yield surface reduces to its intersection polygon with the plane $m_x = m_y$. The equations of the sides of this polygon are

$$m_x + |m_{xy}| = 1 ,$$

$$m_x - |m_{xy}| = -1 ,$$

$$2m_{xy} = \pm 1 . \tag{6.45}$$

For the stress point to be on or within the polygon with eq. (6.45) for $0 \leqslant x \leqslant l$ and $0 \leqslant y \leqslant L$, p must not be larger than some values that depend on C and $\eta = l/L$. Best lower bounds were obtained for each η, adjusting C to maximize the limiting values of p, with the aid of an IBM-650 computer.

The upper bound [eq. (6.35)] is improved by considering the length λ of segment EF as a parameter in the mechanism (or equivalently, the angle ϕ, which was previously taken as 45°). The value of λ is chosen to minimize the load, according to the kinematic theorem. As will be shown in details in Section 7.4.8, the resulting load is

$$p_+ = \frac{24M_p}{l^2[\sqrt{3+(l/L)^2}-l/L]^2}.\tag{6.46}$$

Results are summarized in table 6.1 and fig. 6.45.

Table 6.1. Load-carrying capacities of rectangular plates

Ratio of sides $\eta = l/L$	Lower bound p_*^-	Upper bound p_*^+	p_*	Max. possible percent error
1.00	0.826	1.000	0.913 ± 0.087	9.6
0.95	0.870	1.055	0.963 ± 0.092	9.6
0.90	0.921	1.117	1.019 ± 0.098	9.6
0.85	0.979	1.186	1.083 ± 0.103	9.6
0.80	1.049	1.274	1.162 ± 0.112	9.6
0.75	1.108	1.372	1.240 ± 0.132	10.6
0.70	1.193	1.494	1.344 ± 0.150	11.2
0.65	1.293	1.639	1.466 ± 0.173	11.8
0.60	–	1.828	–	–
0.55	1.561	2.053	1.087 ± 0.246	13.6
0.50	1.744	2.358	2.050 ± 0.308	15.0
0.45	1.979	2.747	2.363 ± 0.384	16.3
0.40	2.289	3.289	2.789 ± 0.500	17.9
0.35	2.708	4.049	3.379 ± 0.670	19.8
0.30	3.309	5.236	4.273 ± 0.963	22.5
0.25	4.222	7.042	5.632 ± 1.410	25.0
0.20	5.471	10.417	8.079 ± 2.338	28.9
0.15	8.615	17.241	12.928 ± 4.303	33.3

$$p_* = pL^2/24M_p$$

The yield condition of von Mises has been applied to the problem of a square, uniformly loaded, simply supported plate, with the corners prevented to lift. Hodge [3.8] has found

$$p_- = 20.6 \frac{M_p}{l^2} \leqslant p_l \leqslant 27.7 \frac{M_p}{l^2} = p_+,\tag{6.47}$$

and Iliouchine [1.2] has obtained

$$p_l \leqslant p_+ = 26.4 \frac{M_p}{l^2}.\tag{6.48}$$

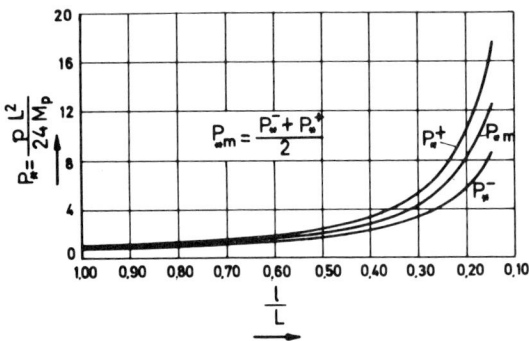

Fig. 6.45. Upper and lower bounds on load-carrying capacities of rectangular plates; $p_* = pL^2/24M_p$.

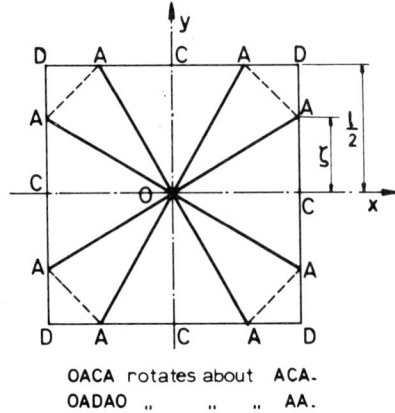

OACA rotates about ACA.
OADAO "   "   " AA.

Fig. 6.46. *OACA* rotates about *ACA*. *OADAO* rotates about *AA*.

When corners are permitted to lift, Hodge uses the mechanism of fig. 6.46 to find

$$p_+ = 25.65 \frac{M_p}{l^2}.$$

An extension of Hodge's results to some class of orthotropic plates was given by Kao, Mura, and Lee [6.24].

6.6. Review of other work

Complete limit analysis solutions for plates resting on an ideally plastic foundation have been given by Sawczuk and Kaliszky ([6.25]; see also [6.16], p. 186). The plate satisfies the Tresca yield condition and the subgrade is considered as yielding in uniaxial compression, with a yield stress that may vary with position (nonhomogeneous foundation). An infinite plate uniformly loaded over a finite circular area, and a circular plate centrally loaded by a concentrated force have been studied.

The influence of shear forces on the limit loads of circular plates has been theoretically investigated by Sawczuk and Duszek [6.26].

A first step towards accounting precisely for structural orthotropy due to stiffening has been made by Nemirovski [6.27], but did not furnish many practical results.

Circular plates with either continuous or discontinuous thickness variations (plastically nonhomogeneous) have been widely studied, mainly by Polish authors [6.18, 6.19, 6.20, 6.28]. We shall consider this problem in some detail in Section 6.8.

With a suitably linearized yield condition, limit analysis (and minimum-weight design) of plates may be formulated as problems of linear programming, as was advocated by Koopman and Lance [6.29]. Indeed, this approach is quite general [6.30, 6.31, 6.32], and some applications to reinforced concrete plates will be considered in Section 7.5.9. In their paper, Koopman and Lance [6.29] give the lower bound

$$p_- \frac{L^2}{24 M_p} = 0.964,$$

to compare with

$$p_- \frac{L^2}{24 M_p} = 0.826$$

found by Shull and Hu [6.23]. Caution should be exercised, however, in using the former value because the mesh used in the finite difference approach was rather coarse and, hence, the yield condition might be violated at intermediate points.

Simply supported plates of arbitrary shape have been considered by Villaggio [6.33], using a complex variable technique and the yield condition of von Mises. Presently available numerical applications of the method compare poorly with previous solutions.

Complete solutions for metal plates with polygonal (or more complicated) boundaries seem otherwise to be absent. Schumann has shown [6.34] that the limit load p of a uniformly loaded plate with arbitrary simply supported edge satisfies

$$p_l \geqslant \frac{6\pi M_p}{S}, \qquad (6.49)$$

where S is the plate area. In the case of a regular polygonal plate we have $S = \pi R^2 \tan(\pi/n)$ where n is the number of sides and R the radius of the inscribed circle. An upper bound can be obtained with a pyramidal mechanism, as will be seen in Section 7.4. It gives

$$p_l \leqslant \frac{6\pi M_p}{R^2}.$$

Hence,

$$\frac{6\pi R^2}{n \tan(n/\pi)} \leqslant p_l \leqslant \frac{6\pi M_p}{R^2}.$$

We finally note that the limit value of a concentrated load acting on a plate obeying the Tresca condition is

$$P_l = 2\pi M_p \qquad (6.50)$$

for any boundary condition (built-in, simply supported, partially restrained) and any load location [6.35, 6.36, 6.37].

More detailed information on plate problems can be found in the book by Sawczuk and Jaeger [6.16].

6.7. Deformations of metal plates

6.7.1. *Introduction*

In the case of beams and frames studied in Com. V., the (theoretical) limit load P_l can most often be regarded as a real failure load because changes in geometry, either in elastic-plastic range or due to the collapse mechanism, very seldom have an stabilizing effect (except, for example, for a built-in beam [3.7]). Hence, the limit load can as a rule be used as a good approximation to the carrying capacity of the structure. Calculation of the limit load must be supplemented by a sufficiently accurate estimate of maximum deflections at impending collapse, to ensure that the assumption of negligible changes in geometry is satisfied up to the onset of unrestricted plastic flow*. Fortunately, there exist general methods for computing these deflections (see Com. V. and [6.29]).

For metal plates, the situation is quite different. Deformations, either prior to attaining the limit load or subsequent, always have a stabilizing effect and the load can be considerably increased beyond the (theoretical) limit value. Hence, *the failure load will be determined by a limit assigned to the deflection.* As was seen in Section 6.2, for circular plates with a slenderness μ ranging from 10 to 40, a maximum deflection of 2% of the diameter gives a failure load very near to the limit load P_l. Unfortunately, we do not know of a general method to obtain the load versus deflection diagram of a plate up to large deflections. In the following, we discuss currently available particular solutions that are useful in estimating whether, in every particular situation, the deflections under the limit load are acceptable. It must be emphasized, however, that both theory and experiments on the subject of noncircular plates are almost completely lacking.

6.7.2. *Elastic-plastic range*

In the elastic-plastic range, displacements remain of the order of magnitude of elastic displacements. Accordingly, the analysis refers to the undeformed state, as usual in elasticity.

Haythornthwaite [6.3] considers a circular plate, simply supported and loaded on a rigid central boss of radius a by a load P (fig. 6.47). He uses Tresca's yield condition, the associated flow law, and the sandwich-type moment versus curvature diagram of fig. 6.48. He obtains the load versus deflection diagrams shown in fig. 6.47. Deflections corresponding to the limit load P_l are given with good accuracy by the relation

$$\delta_{pl} = \frac{12(1-a/R)R^2}{Et^3} M_p , \qquad (6.51)$$

where E is Young's modulus. They correspond to the small circles in fig. 6.47. We remark that for $a/R = 0$, we obtain the central concentrated load.

---

* The Belgian Code N.B.N.I. restricts the maximum deflection at impending collapse to 2% of the corresponding span.

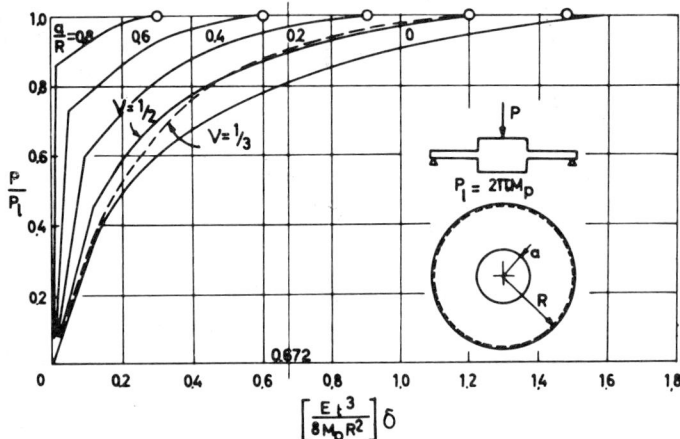

Fig. 6.47.

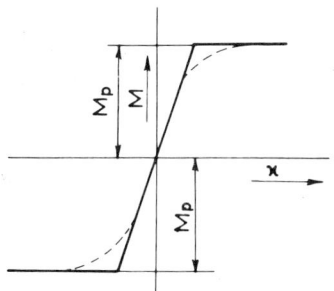

Fig. 6.48.

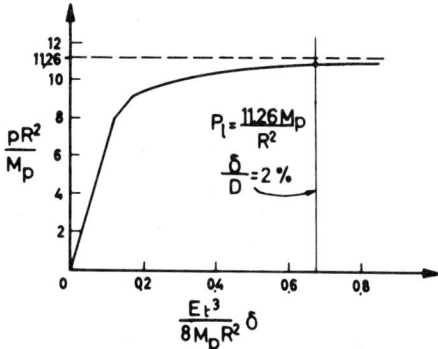

Fig. 6.49.

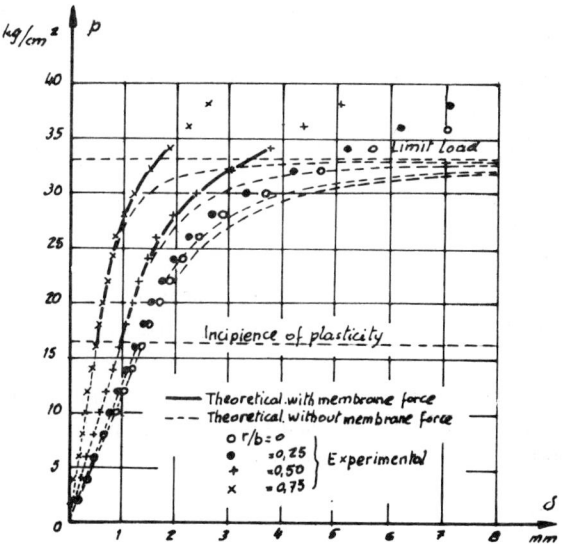

Fig. 6.50 (a). Built-in plate, uniformly loaded; dials at indicated r/R.

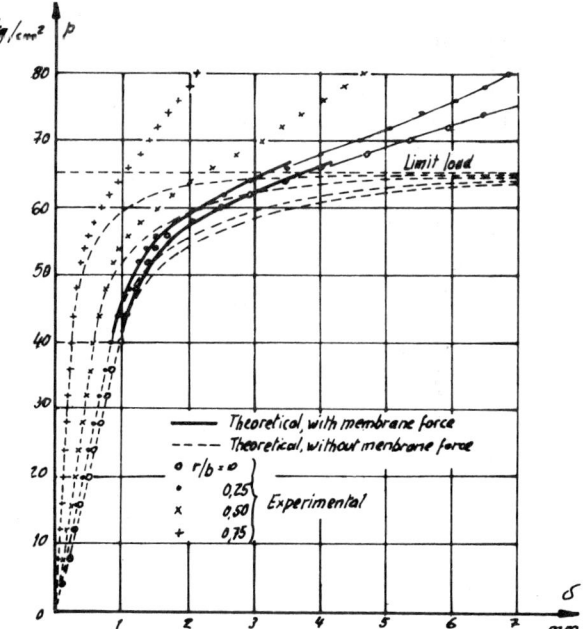

Fig. 6.50 (b). Simply supported plate, uniformly loaded; dials at indicated r/R.

On the same basis, Tekinalp [6.38] has studied the built-in plate, uniformly loaded.

The load versus deflection diagram is given in fig. 6.49. In both cases, we have indicated points corresponding to $\delta/D = 2\%$ with $\mu = D/t = 50$, $E = 21000$ kg/mm^2 = 29.6 × 10^3 ksi, $\sigma_Y = 25$ kg/mm^2 = 35.6 ksi, that is to

$$\frac{Et^3\delta}{8M_p R^2} = 0.672 .$$

It is seen that the corresponding load is very near to P_l for the built-in plate and remains larger than $0.8P_l$ for the simply supported plate.

The actual situation is better, because favorable membrane action begins already in the leastic-plastic range* as found by Oshaki and Murakami [6.9,6.10], from whose paper we take fig. 6.50. For plates with large holes [6.11] the limit load has less practical meaning (fig. 6.51).

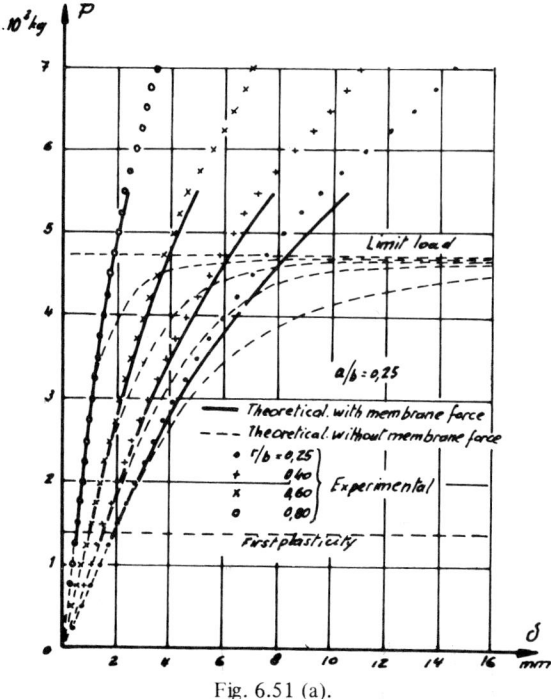

Fig. 6.51 (a).

---

* Work hardening may also contribute to some strengthening of the plate.

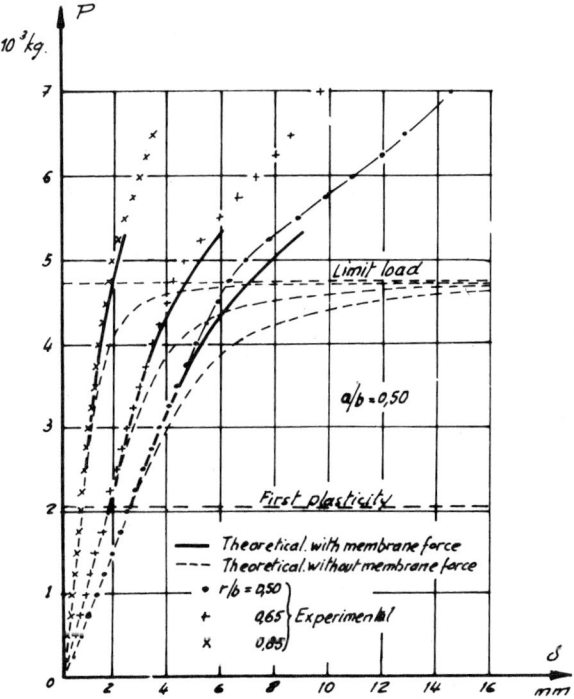

Fig. 6.51 (b).

6.7.3. Post-limit behavior

The behavior of the plate at loads larger than the limit load P_l will be called "post-limit" behavior. It is essentially dependent on the changes of geometry of the structure during the plastic flow. Hence, the analysis can be made assuming rigid perfectly plastic material. Approximation will be poor in the neighborhood of P_l but will become very good for large deflections. The basic difficulty of the problem is that the collapse mechanism may not remain the same as deflections become larger (see [6.39], for example). Fortunately, for very flat simply supported conical shells considered by Onat [3.10], fig. 6.52, the collapse mechanism is also conical and the problem is greatly simplified. Using the Tresca yield condition, and with the notations $\alpha = a/R$ and $\beta = \delta/t$, the following relation is obtained:

$$P = 2\pi M_p \left[\frac{1}{1-\alpha} + \frac{\beta^2}{3}\left(1+\alpha-5\alpha^2+3\alpha^3 - \frac{3t^2}{R^2(1-\alpha)}\right) \right]. \quad (6.52)$$

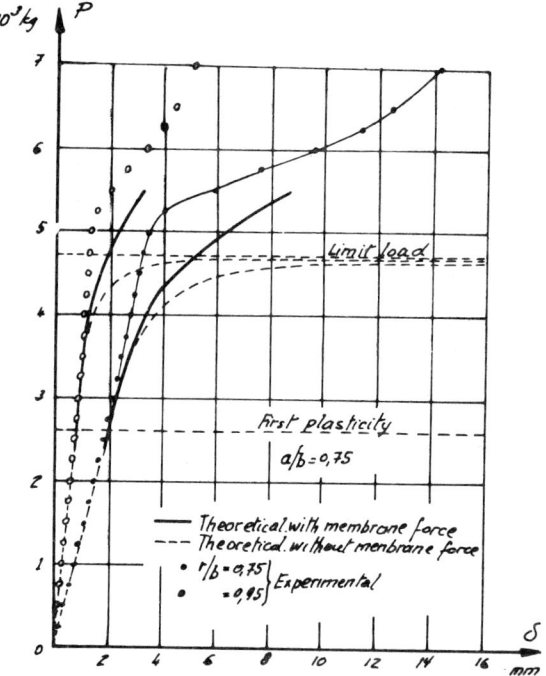

Fig. 6.51 (c). Simply supported plates, central hole with radius a. Total line load P along hole P.

If we let $\beta = 0$ in eq. (6.52) we first obtain the limit load

$$P_l = \frac{2\pi M_p}{1 - \alpha} \tag{6.53}$$

of the flat plate. Next, assuming small boss sizes so that $\alpha \ll 1$, and taking account that $t/R(1-\alpha) \ll 1$ we have at moderate δ

$$\frac{P}{P_l} = 1 + \tfrac{1}{3}\left(\frac{\delta}{t}\right)^2, \tag{6.54}$$

where P_l is given by eq. (6.53).

Eq. (6.54) is theoretically valid only for

$$\beta^2 \leq \frac{3}{2}\frac{\alpha}{1-\alpha}\frac{1}{1-(\alpha/2)(1+\alpha)+\tfrac{3}{2}(\alpha^4-\alpha^3)}, \tag{6.55}$$

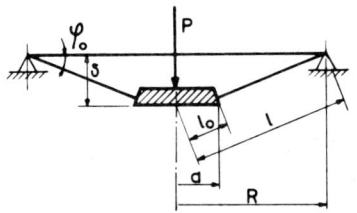

Fig. 6.52.

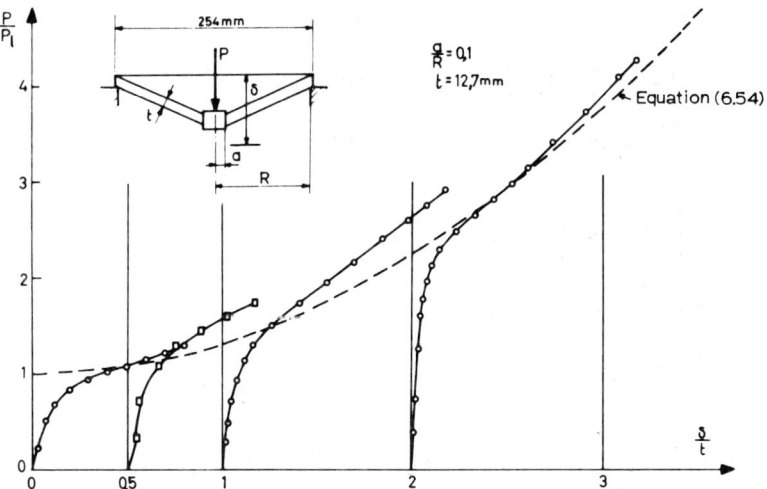

Fig. 6.53.

but comparison with experimental data in figs. 3.6 and 6.53 shows that it is usable with good accuracy even for larger β.

The built-in circular plate under "circular loading" (see Section 6.3.2) had been considered earlier by Onat and Haythornthwaite [6.5], who made additional assumptions concerning the deformation. They obtained

$$\frac{P}{P_l} = \begin{cases} 1 + \alpha_1 \left(\frac{\delta}{t}\right) + \alpha_2 \left(\frac{\delta}{t}\right)^2 & \text{for} \quad \frac{\delta}{t} \leq \frac{1}{2} + \frac{1}{2}\ln\frac{R}{\rho} \\ \beta_1 + \beta_2 \left(\frac{\delta}{t}\right) + \beta_3 \left(\frac{t}{\delta}\right) & \text{for} \quad \frac{\delta}{t} \geq \frac{1}{2} + \frac{1}{2}\ln\frac{R}{\rho} \end{cases} \quad (6.56)$$

where

$$\alpha_1 = \frac{1 + 2\ln R/\rho}{(2+\ln R/\rho)(1+\ln R/\rho)},$$

$$\alpha_2 = \frac{2(1+3\ln R/\rho)}{3(2+\ln R/\rho)(1+\ln R/\rho)^2},$$

$$\beta_1 = \frac{3 + \ln R/\rho}{2(2+\ln R/\rho)}, \qquad \beta_2 = \frac{2(1+2\ln R/\rho)}{(2+\ln R/\rho)(1+\ln R/\rho)},$$

$$\beta_3 = \frac{1 + \ln R/\rho}{12(2+\ln R/\rho)}.$$

Here, ρ is given by the following equations:

$$1 - \frac{2}{3}\frac{a}{\rho}\left(1+\ln\frac{R}{\rho}\right) = 0 \qquad \text{if} \qquad \frac{a}{R} \leqslant 0.606,$$

$$1 - \frac{a^2}{\rho^2}\left(1+2\ln\frac{R}{\rho}\right) + \frac{2}{3}\left(1+\ln\frac{R}{\rho}\right) = 0 \qquad \text{if} \qquad \frac{a}{R} \geqslant 0.606.$$

In the formulas above, we have

$$P_l = 2\pi M_p \frac{A}{B},$$

where

$$A = 2 + \ln\frac{R}{\rho},$$

$$B = \begin{cases} 1 + \ln\dfrac{R}{\rho} - \dfrac{2a}{3\rho} & \text{if} \quad \rho \geqslant a, \\ \dfrac{1}{2} + \ln\dfrac{R}{a} - \dfrac{\rho^2}{6a^2} & \text{if} \quad \rho \leqslant a. \end{cases}$$

Curves of P/P_l versus δ/t for $a/R = 1$ and $a/R = \frac{1}{10}$ are given in figs. 6.4 and 6.5, respectively, together with experimental results.

6.7.4. Membrane analysis

For large D/t ratios (for example, $D/t = 108$ in fig. 6.5) there is no difference in the plate behavior for $P < P_l$ and for $P > P_l$. Membrane forces are of primary importance from the very beginning. Hence, we must consider plastification by stretching without bending of a very thin circular plate hinged at the boundary and loaded on a central circular area with radius a. The plate deforms into a surface of revolution. We assume that the slope ϕ of the meridian curve of that surface remains small. Consequently, the meridian curvature is given by d^2y/dr^2 and $\sin\phi \cong \tan\phi \cong dy/dr$. Equilibrium equations (see, for example, [6.40]) are accordingly simplified. They contain the unknown functions $y(r), dy/dr, d^2y/dr^2$, the membrane forces N_ϕ and N_θ, and the load p. It turns out that the plastic regime, $N_\phi = N_\theta = N_p$, that must exist at the central point by symmetry, is valid in the whole structure. Integration of equilibrium equations then furnishes the deflected shape $y(r)$ and the corresponding total load

$$P = 2\pi N_p \frac{\delta}{\frac{1}{2} + \ln R/a}, \qquad (6.57)$$

where $P = \pi a^2 p$ and $\delta = y(0)$. We see that δ is directly proportional to P. Eq. (6.57) was used in the cases $D/t = 40$ and $D/t = 108$ in figs. 6.4 and 6.5, respectively.

6.7.5. Summary of practical results for circular plates

For circular plates made of mild steel, the following three cases will be considered:

1. For $D/t < 10$, the theory does not apply because of the influence of work hardening (and shear forces). The theoretical limit load P_l remains useful, however, as a conventional failure load corresponding to a relative deflection δ/D smaller than 2%.

2. For D/t ranging from 10 to approximately 30 for simply supported plates and approximately 50 for built-in plates, the load versus deflection relations are eqs. (6.54) and (6.56), respectively, applicable for $P > P_l$. In these relations, δ is to be regarded as purely plastic. For every value of P, the corresponding elastic deflection δ_E should be added to δ. It is given by

$$\delta_E = \frac{P}{16\pi F}\left[\frac{3+\nu}{1+\nu}R^2 + a^2 \ln\frac{a}{R} - \frac{7+3\nu}{4(1+\nu)}a^2\right] \qquad (6.58)$$

for the simply supported plate (see [6.40]), and by [6.41]

$$\delta_E = \frac{P}{4}\frac{1}{16\pi F}\left[4R^2 - 4a^2 \ln\frac{R}{a} - 3a^2\right], \qquad a \neq 0,$$

$$\delta_E = \frac{P}{4}\frac{R^2}{4\pi F}, \qquad a = 0, \qquad \text{concentrated load}, \qquad (6.59)$$

for the built-in plate. In the two preceding formulas, E is Young's modulus, ν is Poisson's ratio, $F = Et^2/12(1-\nu^2)$, and $P = \pi a^2 p$.

This procedure is not valid in the vicinity of P_l (approximately, for $0.9 P_l < P < 1.1 P_l$) where there is a sharp bent in the P, δ curve due to elastic-plastic behavior. In the range $0.9 P_l < P < 1.1 P_l$, we shall estimate the deflection to be approximately $2\delta_E$;

3. For $D/t > 30$, plates hinged at the boundary, $D/t > 50$, built-in plates, eq. (6.57) will be used.

Simple supports with no lateral restraints at all will not be considered.

6.7.6. *Strip subjected to concentrated load*

We consider, after Haythornthwaite and Boyce [6.8], a rectangular plate built-in along two parallel supports, free along the two others and loaded at the center by a uniformly distributed load over a small circular area of radius a.

Hence, the total load is $P = \pi a^2 p$.

1. *The breadth b of the plate between the free edges is small* [3.7]. We actually deal with a built-in beam with rectangular cross section, loaded at mid span. The limit load is (see Com. V., Section 3.3)

$$P_l = \frac{8 M_p b}{l}, \qquad (6.60)$$

where l is the span and M_p the plastic moment per unit of breadth. The beam has three sections of plastic hinges A, B, C (fig. 6.54) where the rotation axes must be at the common level of the centroids of the sections in the undeformed state under P_l (impending collapse). When the deflection at C is δ,

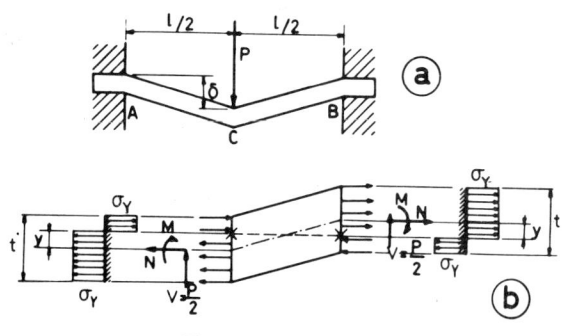

Fig. 6.54.

an axial force N appears, which displaces the rotation axes of sections A, B, C. Because of the symmetry, these axes remain at a common level in sections A and B. The beam being rigid except in the sections of hinges, the deflection at C is possible only if the rotation axis in section C is at the same level as in A and B. In the conditions above, an infinitely small increase of deflection $\Delta\delta$ is possible for every position of the level of the rotation axes. This level is located at the same distance y beneath the centroids of sections A and B and above the centroid of section C, for equilibrium of half a beam in the longitudinal direction (fig. 6.54) to be satisfied. Therefore,

$$y = \frac{\delta}{2}. \qquad (6.61)$$

On the other hand, the stress distributions shown in fig. 6.53 indicate that

$$\frac{N}{N_p} = \frac{2y}{t} \qquad (6.62)$$

(N and N_p being axial forces per unit breadth). From rotational equilibrium of one half of the beam [fig. 6.54 (b)], we have

$$P - (2M + \delta N)\frac{4b}{l} \qquad (6.63)$$

Dividing the two sides of eq. (6.63) by the corresponding sides of eq. (6.60), we obtain

$$\frac{P}{P_l} = \frac{M}{M_p} + \delta\frac{N}{2M_p}. \qquad (6.64)$$

We set $M/M_p = m$ and $N/N_p = n$. Because $N_p/M_p = t/(t^2/4) = 4/t$, eq. (6.64) becomes

$$\frac{P}{P_l} = m + 2\delta\frac{n}{t}. \qquad (6.65)$$

Taking account of eq. (6.61), eq. (6.62) can be rewritten as

$$n = \frac{\delta}{t}. \qquad (6.66)$$

We also know (Com. V., Section 5.4, eq. (5.4)) that in a section with a plastic hinge,

168   METAL PLATES   [Ch. 6

$$|m| + n^2 = 1 .\tag{6.67}$$

Using eqs. (6.66) and (6.67) in eq. (6.65) to eliminate m and n, we obtain

$$\frac{P}{P_l} = 1 + \left(\frac{\delta}{t}\right)^2 ,$$

valid for $\delta \leqslant t$. When $\delta \geqslant t$, the neutral axes fall outside the cross sections according to eq. (6.61). Bending moments vanish and condition (6.67) is replaced by $n = 1$, $m = 0$. From relation (6.65) we then find $P/P_l = 2(\delta/t)$. Hence, the "post-limit" behavior of a rigid-plastic built-in beam with rectangular cross section, loaded at midspan by a concentrated force, is described by

$$\frac{P}{P_l} = \begin{cases} 1 + \left(\frac{\delta}{t}\right)^2 & \text{for} \quad \frac{\delta}{t} \leqslant 1 , \\ 2\frac{\delta}{t} & \text{for} \quad \frac{\delta}{t} \geqslant 1 . \end{cases}\tag{6.68}$$

The preceding analysis was refined by Haythornthwaite to account for elastic deformability of the beam [6.42]. The load versus deflection relation cannot be obtained in closed form anymore, but is determined numerically.

Theoretical and experimental curves are given in figs. 3.4 and 6.55.

2. *The breadth b of the plate between free edges is very large* (strip problem). The plate will exhibit a local collapse mechanism with axial symmetry about the line of action of the concentrated load. As noted in Section 6.6, the limit value is $P = 2\pi M_p$.

3. *The breadth b of the plate between free edges is "intermediate"* (0.5l, for example). The mechanism is most likely a combination of the preceding mechanisms. It has been found [6.8] that the range of that combined mechanism is confined to $b/l \leqslant 1$. The purely local mechanism already is a good approximation for $b = l$.

Experiments on six mild steel plates with b/l ratios of 0.1, 0.3, 0.5, 0.888, and 1.8 have resulted in the diagrams in fig. 6.11. The conclusions are:

1. For the first loading sequence, the simple beam mechanism is most adequate, up to $b/l = 0.888$, as regards both the limit load ($P_l=8M_p b/l$) and the post-limit load versus deflection curve, provided that elastic deflections are added. For $b/l = 1.8$, the local mechanism should be considered;

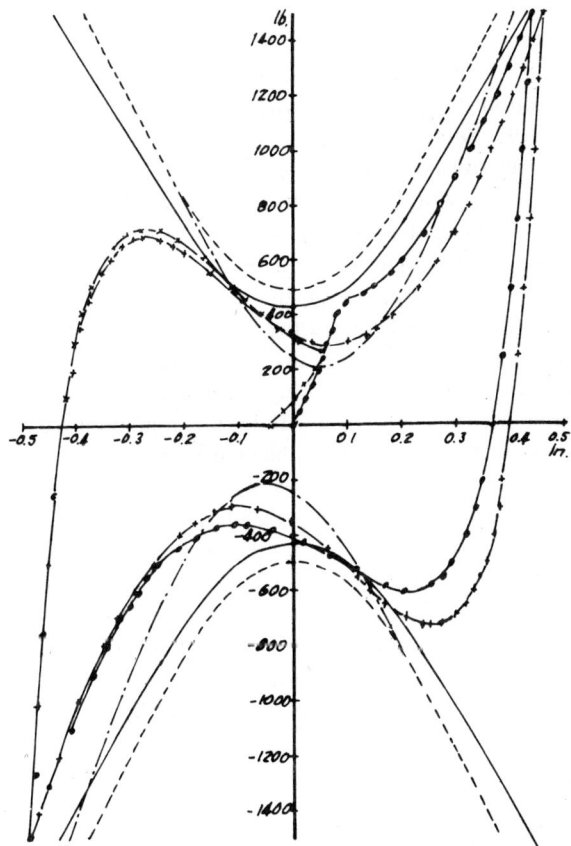

Fig. 6.55. Cyclic loading test of beam ($\frac{1}{4}$-in. depth).

2. At the first reversal in loading sign, the theoretical limit load recovers its significance of real collapse load, at which large deformations occur for very small load increments.

6.8. Minimum-weight design

6.8.1. *General theory*

We have so far considered the problem of limit analysis where the structure is given and the limit value of the load parameter is sought. The problem, however, that is most frequently encountered in practice is *designing* a structure to have a limit value of its load parameter not smaller than a given value which includes a safety factor against plastic collapse*. At the same time, some kind of optimization of the structure is usually desired. As already noted in Com. V., Section 9.1, the quality of a structure depends on so many variables (relative costs of labor and materials, aesthetic considerations, etc.) that an optimum is quite difficult to define. On the other hand, it may be of value to determine the *minimum-weight structure* for the given loads. Even if the theoretical design so obtained is not practical, it gives valuable information on the maximum saving in weight that can be achieved, with respect to other designs. It also gives the direction toward reaching this goal. It thus may suggest practical designs that come close to the optimum. For an homogeneous material**, minimum weight coincides with minimum volume. In the following we shall therefore study the minimum-volume problem.

Consider a perfectly plastic solid subjected to a one-parameter system of loads. The limit value P of the load parameter is given. Hence, the applied loads are completely known, and, in particular, they have given fixed points of application***. Furthermore, the solid is restricted to lie in a region R within a closed surface S_T, on which possible restraints are prescribed.

Suppose there exist a first structure, with volume V_1, located in R, with a limit load with value P. Hence, there exists, under this loading condition, a statically admissible stress field (Q) with which the normality law associates a kinematically admissible field of strain rates $(\dot{q})$ derived from a velocity field $(\mathbf{V})$.

The kinematic theorem of limit analysis enables us to write

$$\mathcal{P}(\mathbf{V}) = \int_{V_1} D(\dot{q}) dV_1 , \qquad (6.69)$$

where $\mathcal{P}(\mathbf{V})$ is the external power of dissipation and $D(\dot{q})$ is the internal power of dissipation per unit volume. Consider next a second structure, with volume V_2 also located within R, obtained by modifying the bounding surface of V_1 only in parts free from loads or kinematic restraints. Imagine moreover that the field $(\dot{q})$ can be extended in a kinematically admissible

---

* More than one loading case can also be dealt with: see [6.43, 6.44, 6.45].

** Nonhomogeneous materials, and design for maximum inertia, have been considered by Shield [6.46].

*** For movable loads see [6.44, 6.45, 6.47].

manner throughout R. Assume finally that the second structure does not collapse under the load P. We then have

$$\mathcal{P}(\mathbf{V}) \leqslant \int_{V_2} D(\dot{q}) dV_2 , \qquad (6.70)$$

for otherwise the use of the kinematic theorem (with collapse mechanism $\dot{q}$) would indicate collapse of the second structure for loads smaller than P.

Comparison of relations (6.69) and (6.70) furnishes

$$\int_{V_1} D(\dot{q}) dV_1 \leqslant \int_{V_2} D(\dot{q}) dV_2 . \qquad (6.71)$$

To obtain and inequality relating the volumes instead of the dissipation, it is *sufficient* that the dissipation per unit volume (which, as a rule, varies with position) be a constant α throughout R. If this condition is satisfied, eq. (6.71) gives

$$\alpha \int_{V_1} dV_1 \leqslant \alpha \int_{V_2} dV_2 , \qquad (6.72)$$

showing that volume V_1 is an *absolute minimum*. It may thus be stated [6.48, 6.49] that if, for a structure under given loads, it is possible to find (a) a statically admissible stress field and (b) a corresponding kinematically admissible mechanism with constant specific rate of dissipation throughout the domain R in which the structure must lie, the volume of the considered structure does not exceed that of any other structure in R that has the same limit load.

It may be useful to recall here a similar condition used for beams and frames (Com. V., Section 9.3). Indeed, we had there a total dissipation given by $\Sigma_{i=1}^{n} \alpha_i M_i$, where α_i is the sum of the absolute values of the hinge rotations in the "span" with plastic moment M_i and length l_i. As the weight function was

$$\sum_{i=1}^{n} l_i M_i ,$$

it was sufficient to have

$$\frac{\alpha_i}{l_i} = \text{const} \qquad (i=1,...,n)$$

to obtain direct proportionality of total dissipation and total weight.

We have so far neglected the body forces in our considerations. Before including them, we investigate whether it is possible to satisfy the sufficient condition for minimum weight just obtained. Obviously, if, from the very nature of the structure, it follows that the specific strain rate *must* vary, it will not be possible to fulfill the condition above. This situation occurs for nonsandwich structures subjected to bending. On the other hand, sandwich structures, membranes, solids in plane stress or plane strain will in general make it possible to satisfy the condition.

In order to study in further details the first of these two cases, it is necessary to reexamine the problem in terms of various types of structures.

6.8.2. *Structures*

Though our attention is presently directed toward plate problems, we shall consider the more general case of a shell with a given median surface A [6.46]. The applied loads are known. At the edge, forces and moments are given, except where the corresponding components of displacement or rotation are prescribed to vanish. Denoting by s the position of a generic point on the median surface, we specify a "design" by its thickness function $t(s)$.

Let D, $\mathcal{P}$, and $\mathcal{B}$ be the rate of dissipation per unit area of median surface, the power of applied loads, and the power of the body forces, respectively. Consider a design t_c which is collapsing under the given loads with a velocity field $(\mathbf{V})$. We have

$$\int_A \mathcal{P}(\mathbf{V})\,dA + \int_A \mathcal{B}(\mathbf{V})\,dA = \int_A D(\mathbf{V})\,dA . \qquad (6.73)$$

Eq. (6.73) may be rewritten

$$\int_A \mathcal{P}(\mathbf{V})\,dA = \int_A \Delta(\mathbf{V})\,dA , \qquad (6.74)$$

where

$$\Delta(\mathbf{V}) = D(\mathbf{V}) - \mathcal{B}(\mathbf{V}) \qquad (6.75)$$

is called the "modified dissipation rate".

We now consider a design t_s which is a neighboring design to t_c in the sense that $t_s = t_c + \delta t$, where δt is a small variation given to t. If we remark that D is a monotonic increasing function of the thickness, and that $\mathcal{B}$ is directly proportional to t, we can write the modified dissipation rate corresponding to the field $(\mathbf{V})$ in the design t_s as

$$\Delta(\mathbf{V}) + \delta t \frac{\partial}{\partial t} \Delta(\mathbf{V}), \tag{6.76}$$

neglecting second and higher-order terms in δt. We then have

$$\int_A \mathcal{P}(\mathbf{V})\, dA \leq \int_A \left[\Delta(\mathbf{V}) + \delta t \frac{\partial}{\partial t} \Delta(\mathbf{V}) \right] dA, \tag{6.77}$$

because $(\mathbf{V})$ is kinematically admissible for t_s under the given loads. From comparison of relations (6.74) and (6.77) we obtain

$$\int_A \delta t \frac{\partial}{\partial t} \Delta(\mathbf{V})\, dA \geq 0. \tag{6.78}$$

Hence, for the design t_c to be of *relative minimum-weight*, it is sufficient that

$$\frac{\partial}{\partial t} \Delta(\mathbf{V}) = \alpha, \tag{6.79}$$

where α is positive and constant (with respect to s). For plates subjected to bending, we have noted in Section 5.4.1 that the strain-rate vector has the same direction for all points through the thickness, and a magnitude proportional to the distance to the neutral layer. Hence, the dissipation per unit volume is proportional to that distance, and the dissipation per unit surface is proportional to t^2. Condition (6.79) is then interpreted as requiring the modified dissipation *per unit volume* to be constant over the surfaces of the plate. With no body forces present, the condition is

$$\frac{D}{t} = \text{const}. \tag{6.80}$$

Discussion by Mroz [6.50, 6.51] of the influence of second order terms neglected in (6.76) shows that, if the stress regime over the whole plate corresponds to a corner of the yield locus, a minimum of the weight is ensured. Otherwise, the nature of the extremum must be investigated.

In the case of sandwich shells, the core $H(s)$ is prescribed, and the unknown function is $t(s)$ where t is the total thickness of the two face sheets.

Here, D is proportional to t, and so is $\mathcal{B}$. Hence, for *any* design t_s supporting the loads, the modified rate of dissipation corresponding to the velocity field $\mathbf{V}$ is $\Delta(\mathbf{V})t_s/t_c$. Eq. (6.77) is replaced by

$$\int_A \mathcal{P}(\mathbf{V})\, dA \leq \int_A \frac{t_s}{t_c} \Delta(\mathbf{V})\, dA \ . \tag{6.81}$$

From comparison of relations (6.77) and (6.81), the following sufficient condition for *absolute* minimum weight is obtained:

$$\frac{\Delta(\mathbf{V})}{t_c} = \alpha\ , \qquad \alpha = \text{const} > 0\ . \tag{6.82}$$

Note that, when the minimum-weight design cannot be obtained (e.g., because of the mathematical complexity of the problem) it may be possible to bound it from above and below [6.49].

6.8.3. *Uniform-strength elastic design versus minimum-weight plastic design*
6.8.3.1. *Introduction*

In classical strength of materials, the desire to save as much material as possible has resulted in the concept of uniform-strength elastic solid, which must exhibit the same reserve of strength at all points. If the strength criterion is the yield limit, as usually accepted for ductile materials, the maximum admissible stress will be given by

$$\sigma_a = \frac{\sigma_Y}{s}\ , \tag{6.83}$$

where s is the chosen "safety factor". For uniform strength, we must have

$$\sigma_R(\sigma_x,...,\tau_{xy},...) = \sigma_a \tag{6.84}$$

in every point of the body under service loads P_s. Equivalently, one may prescribe

$$\sigma_R(\sigma_x,...,\tau_{xy},...) = \sigma_Y \tag{6.85}$$

at every point under loads sP_s, assuming elastic behavior. This latter definition will be used in the following. As long as $P < sP_s$, the body remains completely elastic. When $P = sP_s$, it is suddenly plastified throughout and flows plastically. The highest elastic load thus coincides with the limit load P_l and the "elastic" safety factor s is also the safety factor against plastic collapse. From this point of view, a uniform strength solid behaves as if its stress field were homogeneous.

Obviously, it will not always be possible to achieve truly uniform strength. Structures subjected to bending have normal stresses directly proportional to the distance to the neutral layer, and only a *restricted uniform strength* is possible, where all points of the upper and lower layers of the structure are simultaneously at the yield limit. On the other hand, statically determinate trusses are readily designed for uniform strength by mere choice of the cross sections of the bars to satisfy $|\sigma| = \sigma_Y$. A similar situation occurs with membranes (neglecting discontinuity effects that produce bending). The yield condition,

$$\sigma_R\left(\frac{Q_1}{t}, \frac{Q_2}{t}, ..., \frac{Q_n}{t}\right) = \sigma_Y,$$

where $Q_1, ..., Q_n$ are the generalized stresses (membrane forces), gives the thickness t at every point. Obviously, *uniform-strength statically determinate trusses and membranes are of minimum weight for the given loads*.

6.8.3.2. General theorems

It is of interest to investigate the relations between uniform-strength elastic designs and minimum-weight plastic design. The following theorems have been proved [6.52], assuming that yielding occurs when the elastic strain energy per unit volume attains a certain limit. If we denote by U the elastic strain energy per unit volume, it is thus assumed that the yield condition is

$$U(\sigma_x,...,\tau_{xy},...) = \frac{\sigma_Y^2}{K} \tag{6.86}$$

where K is a positive material constant. We immediately remark that, for incompressible materials, condition (6.86) reduces to the yield condition of von Mises. Indeed, we have

$$U = \frac{1}{2E}(\sigma_x^2+\sigma_y^2+\sigma_z^2) - \frac{\nu}{E}(\sigma_x\sigma_y+\sigma_y\sigma_z+\sigma_z\sigma_x) + \frac{1}{2G}(\tau_{xy}^2+\tau_{yz}^2+\tau_{zx}^2).$$

With $\nu = \frac{1}{2}$ (incompressibility), and $K = 2E$, eq. (6.86) reduces to eq. (1.34).

Also for any uniaxial state of stress, condition (6.86) is valid, whatever the material (otherwise ductile). Moreover, we immediately note that a *uniform-strength elastic design for condition (6.86) exhibits constant elastic strain-energy per unit volume*. We now proceed to prove the first theorem.

Theorem 1: With yield condition (6.86), *any uniform strength elastic design for given fixed loads is a minimum-weight plastic design for these loads*.

Elastic strains and stresses are related by

$$\epsilon_x = \frac{\partial U}{\partial \sigma_x}, \qquad \ldots (x,y,z), \tag{6.87}$$

$$\gamma_{xy} = \frac{\partial U}{\partial \tau_{xy}}, \qquad \ldots (x,y,z). \tag{6.88}$$

According to condition (6.86) and the associated normality law, plastic strain rates are given by

$$\dot{\epsilon}_x^p = \lambda K \frac{\partial U}{\partial \sigma_x}, \tag{6.89}$$

$$\dot{\gamma}_{xy}^p = \lambda K \frac{\partial U}{\partial \tau_{xy}}. \tag{6.90}$$

The elastic stress field of any uniform-strength design is statically admissible and at impending yield everywhere. The corresponding elastic strain field is compatible and satisfies eqs. (6.87) and (6.88). Because of the formal identity of eqs. (6.87) and (6.88) with eqs. (6.89) and (6.90), respectively, the elastic strain field is proportional to the plastic strain-rate field for the same stress field. Its rate of dissipation per unit volume

$$D = \sigma_x \dot{\epsilon}_x^p + \ldots + \tau_{xy}\dot{\gamma}_{xy}^p + \ldots, \tag{6.91}$$

is equal to $2U$ and thus constant. The conditions of Drucker and Shield [6.48] given in Section 6.8.1 are satisfied and a minimum-weight plastic design is thus obtained.

Theorem 2: *Any minimum-weight plastic design* for given fixed loads is a uniform strength elastic design for these loads.*

The stress field of a minimum-weight plastic design satisfies the equilibrium conditions and is at yield everywhere in order to allow a constant rate of dissipation. The constancy of the rate of dissipation also rules out discontinuities in the velocity field. The plastic strain-rate field is thus compatible in the sense of elasticity. It corresponds to the stress field by eqs. (6.89) and (6.90). Hence, it can be regarded as the elastic strain field corresponding to the considered stress field by relations (6.87) and (6.88) if λ may be taken equal to $1/K$. Indeed, using eqs. (6.89) and (6.90) in the expression (6.91) for the dissipation, and also the fact that U is quadratic homogeneous in the stress components, we obtain

$$D = 2\lambda KU. \tag{6.92}$$

Constancy of D actually occurs with λ = const.

The value of the constant is irrelevant (otherwise positive) and may be taken as $1/K$.

Theorem 3: *Any uniform-strength elastic design for given fixed loads is an elastic minimum-weight design for these loads.*

This statement means that, if W_e^{us} is the weight of the uniform-strength elastic design, no design with smaller weight can support the loads elastically. Indeed, let W_e be the weight of any elastic design for the given loads, and W_p^{min} the weight of the minimum-weight plastic design for the same loads. We have, according to Theorems 1 and 2 above

$$W_e^{us} = W_p^{min}. \tag{6.93}$$

But the design with weight W_e can support the loads at plastic collapse (or larger loads). Hence,

$$W_e \geqslant W_p^{min}. \tag{6.94}$$

From eqs. (6.93) and (6.94) we obtain

$$W_e \geqslant W_e^{us}, \tag{6.95}$$

---

* Obtained by the "constant dissipation method".

and the theorem is proved.

Similar general theorems can be established for structures. The reader is referred to the original paper [6.52] for further information. We here restrict ourselves to the case of plates.

6.8.3.3. *Plates in bending*

Consider a plate with variable thickness t, made of an incompressible elastic-plastic material that obeys the yield condition of von Mises. For elastic behavior we must have

$$M_1^2 + M_2^2 - M_1 M_2 \leq \left(\frac{\sigma_Y t^2}{6}\right)^2 \equiv M_e^2 , \qquad (6.96)$$

and elastic curvatures are [6.40]

$$\kappa_1 = \frac{3}{4F}\left(M_1 - \frac{M_2}{2}\right) ,$$

$$\kappa_2 = \frac{3}{4F}\left(M_2 - \frac{M_1}{2}\right) , \qquad (6.97)$$

where F is the flexural rigidity ($F = Et^3/12(1-\nu^2)$, with $\nu = \frac{1}{2}$). The yield condition is

$$M_1^2 + M_2^2 - M_1 M_2 = \left(\frac{\sigma_Y t^2}{4}\right)^2 \equiv M_p^2 = (1.5 M_e)^2 , \qquad (6.98)$$

and plastic curvature rates are

$$\dot{\kappa}_1^p = \lambda(2M_1 - M_2) ,$$

$$\dot{\kappa}_2^p = \lambda(2M_2 - M_1) . \qquad (6.99)$$

Consider an elastic design of restricted uniform-strength, for the value P of the load parameter. The condition (6.96) is strictly satisfied at every point. If we amplify the moment field and the loads by the factor $\phi = 1.5$, we obtain a statically admissible field satisfying the yield condition (6.98) at every point. The elastic curvature field (under P or under $1.5P$ indifferently) may be used as a corresponding kinematically admissible curvature rate field in view of the formal identity of relations (6.97) and (6.99). Its dissipation is

$$D = M_1\kappa_1 + M_2\kappa_2 = (M_1^2+M_2^2-M_1M_2)\frac{3}{4F}. \qquad (6.100)$$

Using relation (6.96) [or (6.98)] and the definition of F, in eq. (6.100), we obtain

$$D = \frac{3}{4Et^3} \cdot \frac{12.3}{4} \cdot \frac{\sigma_Y^2 t^4}{36} = \frac{3}{4}\sigma_Y^2 t. \qquad (6.101)$$

Hence

$$\frac{D}{t} = \frac{3}{4}\sigma_Y^2 = \text{const} > 0. \qquad (6.102)$$

Conditions of Section 6.8.2 are fulfilled and we conclude that *any von Mises plate of elastic restricted uniform strength for given loads is a relative minimum-weight plastic design* for these loads amplified by the shape factor* $\phi = 1.5$. It is readily proved that the converse property holds: *any von Mises plate that is a minimum-weight plastic design for given loads is an elastic uniform-strength design for the loads divided by the shape factor* $\phi = 1.5$.

It is sufficient to remark that constancy of D/t rules out discontinuities in the curvature rate field and also enforces $\lambda = $ constant in eqs. (6.99). The proof is then established along the same reasoning as for the preceding theorem.

The two theorems above are valid for sandwich plates if we take $\phi = 1$ instead of 1.5 and replace relative minimum weight by absolute minimum weight.

6.8.3.4. *Other yield conditions*

Consider for example the yield condition of Tresca.

A uniform-strength elastic design for that condition cannot be regarded as a minimum-volume plastic design for two reasons: (1) the elastic strain field does not correspond to the stress field by a normality law applied to the Tresca condition; (2) the elastic strain energy per unit volume is not in general constant.

We thus want to relate the volume of that uniform strength design to the minimum-volume plastic design for the same loads. To that purpose, we inscribe and circumscribe to the Tresca yield surface two yield surfaces of

---

* If not contradicted by consideration of second order terms.

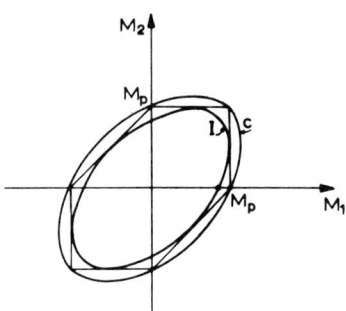

Fig. 6.56.

von Mises (fig. 6.56) that we denote surfaces I and C, respectively. We assume that Theorem 3 of Section 6.8.3.2 applies also to the Tresca condition (and, in general, to any yield condition):

$$V_e^{us} = V_e^{\min}.$$

Sacchi and Save have shown [6.52] that the following relation holds:

$$V_{e,\text{inscr}}^{\min} \geqslant V_e^{\min} \geqslant V_{e,\text{circum}}^{\min}.$$

In the present situation, the minimum-volume elastic designs for the inscribed and circumscribed Mises surfaces are also minimum-volume plastic designs and differ by a coefficient 1.15. We thus have

Volume uniform strength	=	Volume minimum volume	⩾	Volume uniform strength	⩾	Volume uniform strength	=	Volume minimum volume	=	$\dfrac{1}{1.15}$ Volume minimum volume
locus I		locus I		Tresca		locus C		locus C		locus I

But, from the static theorem applied as in Section 5.5.2, we know that the minimum-volume plastic design for the Tresca condition lies also between the same bounds. Hence, *for the Tresca condition, the maximum difference between the volume of a uniform-strength elastic design and that of the minimum-volume plastic design for the same loads is not more than 15%.*

6.8.4. Applications to plates

6.8.4.1. Circular isotropic plates loaded with the symmetry of revolution

Let $p(r)$ be the distributed load applied on a plate with radius R, either simply supported or built-in [6.54]. The fundamental equilibrium eq. (6.6) can be rewritten, after deriving both sides with respect to r, as

$$\frac{d^2}{dr^2}(rM_r) - \frac{dM_\theta}{dr} + rp = 0 . \tag{6.103}$$

Curvature rates are given by relations (6.7).
We first assume that the yield condition of Tresca is valid.

1. *Sandwich plate.* Let H be the core thickness, $t/2$ the (variable) thickness of each of the two face sheets. According to the results in Section 6.8.2, we want to satisfy the condition

$$\frac{D}{t} \equiv \frac{M_r \dot{k}_r + M_\theta \dot{k}_\theta}{t} = \text{const} > 0 . \tag{6.104}$$

A detailed discussion [6.54] shows that plastic regimes able to satisfy relations (6.7), (6.104), normality law and boundary conditions reduce to points A and C (or D and F). For example, regime AB, fig. 6.13 (b), corresponds to $\dot{k}_r = 0$, and condition (6.104) will enforce $M_\theta \dot{k}_\theta / t = \text{const}$. But $M_\theta = M_p = \sigma_Y(tH)/2$. We thus must have $\dot{k}_\theta = \alpha > 0$. Because $\dot{k}_\theta = (1/r)(dw/dr)$ (relation 6.7), we would obtain

$$\dot{w} = \alpha \frac{R^2 - r^2}{2} ,$$

and

$$\dot{k}_r = -\frac{d^2 w}{dr^2}$$

would not be zero, in contradiction with the original assumption of regime AB.

In regime A, we have $M_r = M_\theta = M_p = t(\sigma_Y H)/2$, and condition (6.104) becomes, using relation (6.7),

$$\frac{d^2\dot{w}}{dr^2} + \frac{1}{r}\frac{d\dot{w}}{dr} = -\alpha, \tag{6.105}$$

where α is a positive constant. Integration of eq. (6.105) gives

$$\dot{w} = \beta - \frac{1}{4}\alpha r^2 + \frac{1}{2}\alpha b^2 \ln\frac{b}{r}, \tag{6.106}$$

where β and b are integration constants. Because $\dot{\kappa}_r$ and $\dot{\kappa}_\theta$ must be positive, eq. (6.106) is valid only for $r \geqslant b$.

In regime C, where $M_r = -M_p$, $M_\theta = 0$, $-\dot{\kappa}_r \geqslant \dot{\kappa}_\theta \geqslant 0$, condition (6.104) becomes

$$\frac{d^2\dot{w}}{dr^2} = \alpha. \tag{6.107}$$

Hence, we obtain

$$\dot{w} = \tfrac{1}{2}\alpha r(r-2c) + \gamma \tag{6.108}$$

where c and γ are integration constants. For eq. (6.108) to be valid, we must satisfy $c \geqslant r \geqslant c/2$. We now successively consider two support conditions as follows:

(a) *Simple support*: regime A is acceptable throughout the plate. Eq. (6.106) becomes

$$\dot{w} = \tfrac{1}{4}\alpha(R^2 - r^2) \tag{6.109}$$

and equilibrium eq. (6.103) is here

$$r\frac{d^2M_p}{dr^2} + \frac{dM_p}{dr} = -rp(r), \tag{6.110}$$

from which we shall obtain $M_p(r) = \sigma_Y(H/2)t(r)$. Integration of eq. (6.110) with the boundary conditions

$$M_r = 0 \quad \text{for} \quad r = R, \quad V = 0 \quad \text{for} \quad r = 0,$$

will give, in the absence of central concentrated force,

$$M_p \equiv \sigma_Y \frac{H}{2} t = \int_r^R \frac{1}{\xi} d\xi \int_0^\xi \rho p(\rho) \, d\rho \,. \tag{6.111}$$

If there is a central concentrated force, a term $(P/2\pi) \ln(R/r)$ must be added to the right-hand side of eq. (6.111). We obtain t infinite at the center but the volume of the sheets remains finite.

(b) *Built-in support*: at the center, regime A applies because of the symmetry of revolution; at the boundary, $\dot{\kappa}_r < 0$ and regime C must be used. Hence, there will be a central region in regime A and a ring, adjacent to the support, in regime C. Let r_o be the radius of the central region.

For $0 \leqslant r \leqslant r_o$, relation (6.106) must be used. For $r_o \leqslant r \leqslant R$, relation (6.108) applies. For $r = R$, we must have $\dot{w} = 0$ and $d\dot{w}/dr = 0$ (built-in support, no hinge allowed, in order to obtain $D/t = \alpha$).

For $r = r_o$, $\dot{w}$ and $d\dot{w}/dr$ are continuous. We obtain $r_o = \frac{2}{3}R$ and

$$\dot{w} = \begin{cases} \dfrac{\alpha}{12}(2R^2 - 3r^2) & \text{for} \quad 0 \leqslant r \leqslant \tfrac{2}{3}R \,, \\[2ex] \dfrac{\alpha}{2}(r - R)^2 & \text{for} \quad \tfrac{2}{3}R \leqslant r \leqslant R \,. \end{cases} \tag{6.112}$$

For $0 \leqslant r \leqslant \tfrac{2}{3}R$, regime A, eq. (6.111) is valid. For $\tfrac{2}{3}R \leqslant r \leqslant R$, regime C, substitution of $-M_p$ and 0 for M_r and M_θ, respectively, in eq. (6.103) gives

$$\frac{d^2}{dr^2}(rM_p) = rp(r) \,. \tag{6.113}$$

Since M_r is continuous and hence vanishes at $r = \tfrac{2}{3}R$, and since V is continuous and vanishes at $r = 0$ (no central concentrated force) integration of eq. (6.113) gives

$$M_p \equiv \sigma_Y \frac{H}{2} t = \begin{cases} \displaystyle\int_r^{\tfrac{2}{3}R} \frac{1}{\xi} d\xi \int_0^\xi \rho p(\rho) \, d\rho & \text{for} \quad 0 \leqslant r \leqslant \tfrac{2}{3}R \\[3ex] \displaystyle\frac{1}{r} \int_{\tfrac{2}{3}R}^r d\xi \int_0^\xi \rho p(\rho) \, d\rho & \text{for} \quad \tfrac{2}{3}R \leqslant r \leqslant R \,. \end{cases} \tag{6.114}$$

For both types of support, the conditions of equilibrium, yield, and constant rate of dissipation are satisfied. The functions $t(r)$ defined by eqs. (6.111) and (6.112) thus furnish designs with absolute minimum volume of the face sheets for the considered loads. We now apply these results to two loading cases:

(a) *Circular loading*: $p(r) = p = $ const for $0 \leqslant r \leqslant a$, and $p(r) = 0$ for $a \leqslant r \leqslant R$.

For a *simple support* we directly obtain:

$$\sigma_Y \frac{H}{2} t = \begin{cases} \frac{p}{4}(a^2 - r^2) + \frac{pa^2}{2} \ln \frac{R}{a} & \text{for} \quad 0 \leqslant r \leqslant a, \\ \frac{p}{2} a^2 \ln \frac{R}{r} & \text{for} \quad a \leqslant r \leqslant R. \end{cases} \qquad (6.115)$$

The volume of the sheets is

$$V = \frac{1}{4} \left(2 - \frac{a^2}{R^2} \right) \frac{\pi p a^2 R^2}{\sigma_Y H}, \qquad (6.116)$$

to compare with the volume V_c of the sheets of the sandwich plate with constant sheet thickness, same radius and core, and same limit load. From results established in ref. [6.12], it can be deduced that

$$V_c = \left(1 - \frac{2}{3} \frac{a}{R} \right) \frac{\pi p a^2 R^2}{\sigma_Y H}. \qquad (6.117)$$

Saving in volume varies from 25% to 50% when a/R decreases from one to zero. With a concentrated central load P, one has

$$\sigma_Y \frac{H}{2} t = \frac{P}{2\pi} \ln \frac{R}{r}, \qquad (6.118)$$

and the volume of the sheets is

$$V = \frac{1}{2} \frac{PR^2}{\sigma_Y H}. \qquad (6.119)$$

For a *built-in plate* with $a < \frac{2}{3} R$, we obtain

$$\sigma_Y \frac{H}{2} t = \begin{cases} \frac{P}{4}(a^2 - r^2) + \frac{pa^2}{2} \ln \frac{2R}{3a} & \text{for} \quad 0 \leqslant r \leqslant a, \\ \frac{pa^2}{2} \ln \frac{2R}{3r} & \text{for} \quad a \leqslant r \leqslant \tfrac{2}{3}R, \\ \frac{pa^2}{2} \frac{r - \tfrac{2}{3}R}{r} & \text{for} \quad \tfrac{2}{3}R \leqslant r \leqslant R. \end{cases} \qquad (6.120)$$

If $a > \tfrac{2}{3}R$,

$$\sigma_Y \frac{H}{2} t = \begin{cases} \frac{p}{4}\left(\tfrac{4}{9}R^2 - r^2\right) & \text{for} \quad 0 \leqslant r \leqslant \tfrac{2}{3}R, \\ \frac{p}{6} \frac{r^3 - 8R^3/27}{r} & \text{for} \quad \tfrac{2}{3}R \leqslant r \leqslant a, \\ \frac{p}{6} \frac{3a^2 r - 8R^3/27 - 2a^3}{r} & \text{for} \quad a \leqslant r \leqslant R. \end{cases} \qquad (6.121)$$

With a concentrated load P at the center, we have

$$\sigma_Y \frac{H}{2} t = \begin{cases} \frac{P}{2\pi} \ln \frac{2R}{3r} & \text{for} \quad 0 \leqslant r \leqslant \tfrac{2}{3}R, \\ \frac{P}{2\pi} \frac{r - \tfrac{2}{3}R}{r} & \text{for} \quad \tfrac{2}{3}R \leqslant r \leqslant R, \end{cases} \qquad (6.122)$$

and a sheet volume of

$$V = \frac{1}{3} \frac{PR^2}{\sigma_Y H}. \qquad (6.123)$$

With a sandwich plate of uniform sheet-thickness the limit load is $P_l = 2\pi M_p$ (for both simple support and built-in edge) and the sheet volume is

$$V_c = \frac{PR^2}{\sigma_Y H}. \qquad (6.124)$$

Hence, minimum-volume design saves 67% of sheet material. It must however be kept in mind that the design (6.122) is highly theoretical because of its infinite thickness at the center.

The minimum-volume design for uniformly distributed load over the entire plate is obtained by substituting R for a in eqs. (6.121). It is represented in fig. 6.57 (a). The volume of the sheets is

$$V = 0.117 \frac{\pi p R^4}{\sigma_Y H}, \qquad (6.125)$$

whereas uniform thickness would have needed a volume

$$V_c = 0.178 \frac{\pi p R^4}{\sigma_Y H}. \qquad (6.126)$$

The saving here is 34%.

(b) *Annular loading*: $p(r) = 0$ for $0 \leqslant r \leqslant a$, and $p(r) = p = $ const for $a \leqslant r \leqslant R$.

Application of the general formulas (6.111) and (6.114) gives $t(r)$ and shows that, for both the simply supported and the built-in plates, it is constant in the central region. In particular, $t = 0$ in the central region of the built-in plate when $a > \frac{2}{3}R$.

2. *Solid plate*: We have seen in Section 6.8.2 that a relative minimum of the volume could be achieved by a design where $D/t = $ const > 0 (where D is the

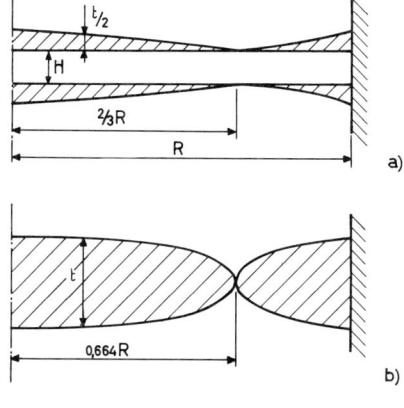

Fig. 6.57.

dissipation per unit area of the midplane), the plastic regime being that of a corner. We shall use regimes A and C, as above. In the present case

$$M_p = \sigma_Y \frac{t^2}{4}, \tag{6.127}$$

and D/t is no longer independent of t. Nevertheless, integration of equilibrium eq. (6.103), with the adopted plastic regimes, yields

$$M_p \equiv \sigma_Y \frac{t^2}{4} = \int_0^R \frac{1}{\xi} d\xi \int_0^\xi \rho p(\rho) \, d\rho \tag{6.128}$$

for the simply supported plate, and

$$M_p \equiv \sigma_Y \frac{t^2}{4} = \begin{cases} \int_r^{r_o} \frac{1}{\xi} d\xi \int_0^\xi \rho p(\rho) \, d\rho & \text{for} \quad 0 \leqslant r \leqslant r_o, \\ \frac{1}{r} \int_{r_o}^r d\xi \int_0^\xi \rho p(\rho) \, d\rho & \text{for} \quad r_o \leqslant r \leqslant R, \end{cases} \tag{6.129}$$

for the built-in plate. In relations (6.129), r_o is the radius at which the plastic regime changes from A to C. It is given by

$$\int_0^{r_o} \frac{\rho}{t(\rho)} d\rho = r_o \int_{r_o}^R \frac{d\rho}{t(\rho)}, \tag{6.130}$$

expressing continuity of $d\dot{w}/dr$ at $r = r_o$. It must also be verified that $r_o \geqslant R/2$. It remains to be shown that the condition of constant rate of dissipation is satisfied. For this proof, we refer the reader to the original paper [6.54]. We now consider some applications:
(a) *Circular loading*: $p(r) = p = \text{const} > 0$ for $0 \leqslant r \leqslant a$, and $p(r) = 0$ for $a \leqslant r \leqslant R$.

For a *simple support* we readily obtain

$$t = \begin{cases} \left(\dfrac{p}{\sigma_Y}\right)^{1/2} \left(a^2 - r^2 + 2a^2 \ln \dfrac{R}{a}\right)^{1/2} & \text{for} \quad 0 \leqslant r \leqslant a, \\ \left(\dfrac{p}{\sigma_Y}\right)^{1/2} a \left(2 \ln \dfrac{R}{r}\right)^{1/2} & \text{for} \quad a \leqslant r \leqslant R. \end{cases} \qquad (6.131)$$

When a/R decreases from 1 to 0, a saving in volume with respect to the constant thickness plate increases from 18% to 37%.

For a *built-in plate*, if $a < r_o$,

$$\sigma_Y \frac{t^2}{4} = \begin{cases} \dfrac{p}{4}(a^2 - r^2) + \dfrac{pa^2}{2} \ln \dfrac{r_o}{a} & \text{for} \quad 0 \leqslant r \leqslant a, \\ \dfrac{pa^2}{2} \ln \dfrac{r_o}{r} & \text{for} \quad a \leqslant r \leqslant r_o, \\ \dfrac{pa^2}{2} \dfrac{r - r_o}{r} & \text{for} \quad r_o \leqslant r \leqslant R. \end{cases} \qquad (6.132)$$

If $a > r_o$,

$$\sigma_Y \frac{t^2}{4} = \begin{cases} \dfrac{p}{4}(r_o^2 - r^2) & \text{for} \quad 0 \leqslant r \leqslant r_o, \\ \dfrac{p}{6} \dfrac{r^3 - r_o^3}{r} & \text{for} \quad r_o \leqslant r \leqslant a, \\ \dfrac{p}{6} \dfrac{3a^2 r - r_o^3 - 2a^3}{r} & \text{for} \quad a \leqslant r \leqslant R. \end{cases} \qquad (6.133)$$

In both cases ($a < r_o$ and $a > r_o$), the radius r_o is given by eq. (6.130) where expressions (6.132) and (6.133) for t have been used. The resulting transcendental equation has been solved numerically and results are given in fig. 6.58. The design corresponding to $a = R$ (uniformly distributed load throughout the plate) is represented in fig. 6.57 (b).

(b) *Annular loading, simply supported plate* [6.18]: $p(r) = 0$ for $0 \leqslant r \leqslant a$, and $p(r) = p = \text{const} > 0$ for $a \leqslant r \leqslant R$.

We remark that, for $0 \leqslant r \leqslant a$ we have $V = 0$, and hence, from eq. (6.5) with regime A, $dM_p/dr = 0$. The general formula (6.128) then gives

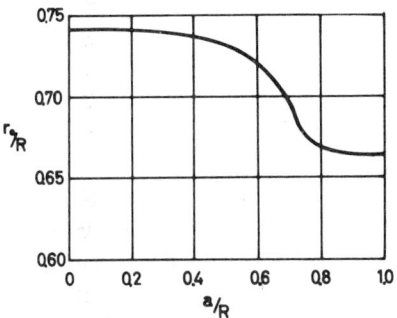

Fig. 6.58.

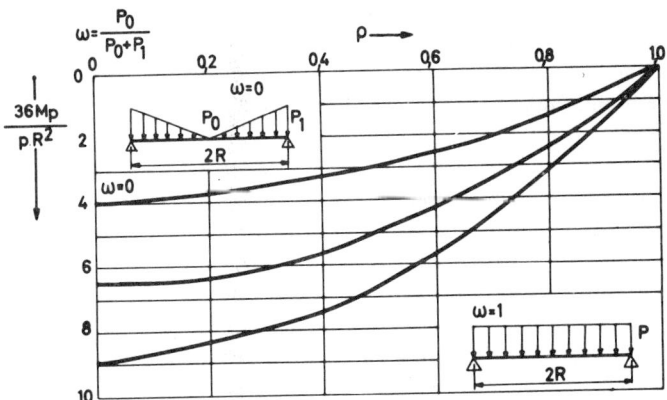

Fig. 6.59.

$$M_p \equiv \sigma_Y \frac{t^2}{4} = \begin{cases} \dfrac{p}{4}\left(R^2 - a^2 + 2a^2 \ln \dfrac{a}{R}\right) & \text{for} \quad 0 \leqslant r \leqslant a, \\[2ex] \dfrac{p}{4}\left(R^2 - r^2 + 2a^2 \ln \dfrac{r}{R}\right) & \text{for} \quad a \leqslant r \leqslant R. \end{cases} \quad (6.134)$$

(c) *Linearily distributed load, simply supported plate* [6.19]: Fig. 6.59 shows the loading condition and gives M_p as a function of $\rho = r/R$ for various loading cases. The pressure is p_0 at the center and $p_0 + p_1$ at the boundary. Let

$$\omega = \frac{p_0}{p_0 + p_1}.$$

The limit load is

$$p_l = \frac{36 M_{p0}}{R^2(4+5\omega)},$$

where M_{p0} is the plastic moment at the center ($r=0$).

3. *Remarks*: We have so far neglected the influence of the weight of the plate itself. It is possible to take this into account, at least in the case of a uniformly loaded sandwich plate, which is either simply supported [6.49] or built-in [6.54], but the mathematical work becomes complex and involves Bessel functions.

Minimum-volume plates obeying the yield condition of von Mises have been obtained by Freiberger and Tekinalp [6.55], and by Eason [6.56]. They have found that:

1. For the simply supported sandwich plate, results coincide with those obtained from the yield condition of Tresca, eqs. (6.111), (6.115) to (6.119), (6.131), (6.134) and fig. 6.59;

2. For the built-in sandwich plate subjected to circular loading, results differ slightly from those obtained with the Tresca condition: $r_o = 0.653R$ instead of $\frac{2}{3}R$, and expressions for $\sigma_Y(H/2)t(r)$ contain integrals that must be evaluated numerically. For a uniform load p, face-sheet volumes compare as follows:

von Mises $\quad 0.111 \dfrac{pR^4 \pi}{\sigma_Y H}$

Tresca $\quad 0.117 \dfrac{pR^4 \pi}{\sigma_Y H}$

For a concentrated load P at center, the face-sheet volumes are:

von Mises $\quad 0.324 \dfrac{PR^2}{\sigma_Y H}$

Tresca $\quad 0.333 \dfrac{PR^2}{\sigma_Y H}.$

In view of the rather theoretical character of the designs considered, use of Tresca's condition is very satisfactory.

6.8.4.2. Circular isotropic sandwich plates subjected to arbitrary loading

The yield condition of Tresca is used, and also the plastic regimes A and C and the corresponding mechanisms (6.109) and (6.112) satisfying the condition of constant rate of dissipation [6.57]. The directions of principal rates of curvature and of principal bending moments are radial and circumferential. Consequently, the equilibrium equations are

$$\frac{\partial^2}{\partial r^2}(rM_r) - \frac{\partial M_\theta}{\partial r} + \frac{1}{r}\frac{\partial^2 M_\theta}{\partial \theta^2} = -rp(r,\theta), \tag{6.135}$$

$$V_r = \frac{\partial M_r}{\partial r} + \frac{1}{r}(M_r - M_\theta), \tag{6.136}$$

$$V_\theta = \frac{1}{r}\frac{\partial M_\theta}{\partial \theta}. \tag{6.137}$$

1. *Simple support*: We have $M_r - M_\theta = M_p$ (regime A) throughout the plate, and $M_r = 0$ at $r = R$.
(a) Concentrated load P at a point with coordinates $r = \rho$, $\theta = \phi$.
   Integration of eq. (6.135) gives

$$M_p \equiv \sigma_Y \frac{H}{2} t = \frac{P}{2\pi} \ln \frac{r_2 \rho}{r_1 R}, \tag{6.138}$$

where

$$r_1^2 = \rho^2 + r^2 - 2\rho r \cos(\theta - \phi),$$

$$r_2^2 = \frac{R^4}{\rho^2} + r^2 - 2\frac{R^2}{\rho} r \cos(\theta - \phi). \tag{6.139}$$

(b) Distributed load $p(r,\theta)$.
   The resulting design is

$$\sigma_Y \frac{H}{2} t = \frac{1}{4\pi} \int_0^{2\pi} \int_0^R p(\rho,\phi) \ln \frac{R^2 + (r^2\rho^2/R^2) - 2\rho r \cos(\theta-\phi)}{\rho^2 + r^2 - 2\rho r \cos(\theta-\phi)} \rho \, d\rho \, d\phi \, .$$

(6.140)

2. *Built-in edge*: We take regime A for $0 \leq r \leq r_o = \frac{2}{3}R$ and regime C for $r_o \leq r \leq R$.
(a) Concentrated load with coordinates (ρ,ϕ), $\rho < \frac{2}{3}R$.
 Simple substitution of r_o for R in eq. (6.138) and (6.139) give the solution for $r \leq r_o$. For $r \geq r_o$, integration of eq. (6.135), taking account of regime C and continuity of $\partial M_p/\partial r$ for $r = r_o$, yields

$$\sigma_Y \frac{H}{2} t = \frac{P}{2\pi} \frac{r_o^2 - \rho^2}{r_o^2 + \rho^2 - 2\rho r_o \cos(\theta-\phi)} \left(1 - \frac{r_o}{r}\right) \quad \text{for } r_o \leq r \leq R . \quad (6.141)$$

(b) Distributed load $p(r,\theta)$, $r < r_o = \frac{2}{3}R$.
 For $r \leq r_o$, simple substitution of r_o for R in relation (6.140) gives the solution. For $r_o \leq r \leq R$, the design is

$$\sigma_Y \frac{H}{2} t = \frac{1}{2\pi} \left(1 - \frac{r_o}{r}\right) \int_0^{2\pi} \int_0^{r_o} p(\rho,\phi) \frac{r_o^2 - \rho^2}{r_o^2 + \rho^2 - 2\rho r_o \cos(\theta-\phi)} \rho \, d\rho \, d\phi \, .$$

(6.142)

(c) Distributed load $p(r,\theta)$, $r \geq r_o = \frac{2}{3}R$.
 With no load inside the circle of radius r_o, it is easily shown [6.57] that t vanishes everywhere except at the points of the radii between the loaded points and the built-in edge (fig. 6.60). The minimum-volume plate is thus a cantilever plate, with the sheet thickness given by

$$\sigma_Y \frac{H}{2} t = \int_{r_0}^r p(\rho,\theta)\rho \left(1 - \frac{\rho}{r}\right) d\rho \, . \quad (6.143)$$

(d) Distributed load $p(r,\theta)$.
 The design is obtained by simply adding the sheet thickness just obtained in (b) and (c) for the parts of the load respectively inside and outside the circle with radius r_o.

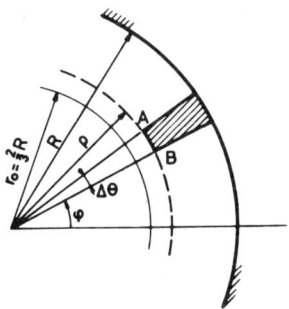

Fig. 6.60.

6.8.4.3. *Isotropic elliptic plate*

1. *Simply supported plate*: Prager has pointed out [6.58] that, if regime A is used in the equilibrium eq. (6.135), the resulting equation,

$$\frac{\partial^2 M_p}{\partial r^2} + \frac{1}{r}\frac{\partial M_p}{\partial r} + \frac{1}{r^2}\frac{\partial^2 M_p}{\partial \theta^2} = -p, \qquad (6.144)$$

is an equation of Poisson, the left-hand side being the laplacian of M_p, which we denote by ∇M_p. In cartesian rectangular coordinate axes x and y, the equilibrium equation is [see Section 6.5.2, eq. (6.37)]

$$\frac{\partial^2 M_x}{\partial x^2} + \frac{\partial^2 M_y}{\partial y^2} - 2\frac{\partial^2 M_{xy}}{\partial x \partial y} = -p.$$

Regime A means $M_x = M_y = M_p$, $M_{xy} = 0$ and we again obtain*

$$\nabla M_p = -p. \qquad (6.145)$$

Consider an elliptic plate (fig. 6.61), simply supported on an edge with equation

---

* A membrane analogy can thus be used to determine M_p. For p = const, the membrane problem is that of Prandtl's soap film analogy for the torsion problem.

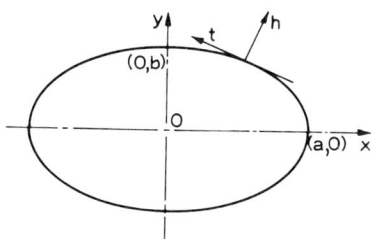

Fig. 6.61.

$$\frac{x^2}{a^2} + \frac{y^2}{b^2} = 1.$$

Denote by M_n and M_t the principal moments at the support, orthogonal and tangential to the edge, respectively. We have $M_n = 0$. But, for regime A, $M_n = M_t = M_p$. Hence, $M_p = 0$ at the edge. Integration of eq. (6.145) then gives, for *uniformly distributed load* p,

$$M_p = \frac{pa^2b^2}{2(a^2+b^2)} \left(1 - \frac{x^2}{a^2} - \frac{y^2}{b^2}\right). \tag{6.146}$$

It remains to associate to the moment field

$$M_x = M_y = M_p,$$

$$M_{xy} = 0,$$

a kinematically admissible mechanism with $D/t = $ const. We restrict our attention to the sandwich plate, for which

$$\frac{D}{t} = M_p \dot{k}_x + M_p \dot{k}_y = \sigma_Y \frac{H}{2} \left(\frac{\partial^2 \dot{w}}{\partial x^2} + \frac{\partial^2 \dot{w}}{\partial y^2}\right).$$

Constancy of the parenthesis is satisfied by

$$\dot{w} = \frac{\alpha a^2 b^2}{2(a^2+b^2)} \left(\frac{x^2}{a^2} + \frac{y^2}{b^2} - 1\right),$$

and the solution is completed.

2. *Built-in plate*: This problem has been treated by Shield [6.59] for a uniformly loaded sandwich plate. We refer the reader to the original publication.

6.8.5. *Final remarks and discussion*

A first approximation to minimum-volume plates can be obtained with stepped plates of segmentwise constant thickness. This problem has been first considered by Hopkins and Prager [6.69] and very recently in a more complete manner by König and Rychlewski [6.61]. An interesting practical result is that the optimum way of strengthening a simply supported uniformly loaded plate by increasing the thickness over a central circle is obtained with a ratio of 0.8 of the radius of the central circle to that of the plate and a ratio 0.65 of the thicknesses. Increase in carrying capacity with respect to the uniform plate with same volume of material is 15%. Obviously, for given limit load this solution furnishes minimum volume of the one-step plate. Increasing the number of steps to infinity furnishes the minimum-volume plate (6.131) for $a = R$. First intuitive approaches to minimum-volume design have been based on corner regime [6.19,6.58,6.62,6.63] to take advantage of mechanisms with an infinite degree of freedom. This "corner condition" is neither necessary (structures with regular yield surfaces also have minimum-volume designs) nor sufficient alone as shown by Hodge [6.64]. However, when used in conjunction with the condition of constant rate of dissipation, the corner regime is sufficient to furnish relative minimum volume of solid structures subjected to bending [6.50]. Indeed, it may occur that constant rate of dissipation is attainable only with corner regimes, as noted in Section 6.8.4.1.

Finally, we want to emphasize that, when minimum-volume designs of a sandwich structure for various loads are obtained from the same constant-dissipation mechanism, the minimum-volume design for a linear combination with nonnegative coefficients of the loads is simply obtained by the same linear combination of the thicknesses of the various designs. Indeed, equilibrium equations are satisfied because they are linear, the yield condition is also satisfied because plastic generalized stresses are proportional to the thickness, and the same constant-dissipation mechanism corresponds to the state of stress. This *restricted superposition method* applies to most examples treated above. It can be extended to structures subjected to movable loads [6.45].

6.9. Examples

6.9.1. Suppose a circular opening with 78.8-in. diameter must be made in a room subjected to internal pressure. The door for this opening will be made of a circular plate stiffened along the edge by a very rigid annulus on which the hinges and lock will be mounted. When it is closed, the door may be considered a built-in plate.

Let p_s be the service pressure, λ the load factor (or safety factor) and $p_l = \lambda p_s = 285$ psi the limit pressure.

To determine the needed thickness, fig. 6.16, with $a/R = 1$, gives

$$P_l = \pi R^2 p_l = 3.14 (39.4)^2 285 = 1.39 \times 10^6 \text{ lb}. \tag{a}$$

Hence $M_p \geqslant 39{,}200$ lb. But $M_p = \sigma_Y t^2/4$. With $\sigma_Y = 34{,}100$ psi, we deduce $t \geqslant 2.145$ in. We take $t = 2.20$ in. and thus obtain

$$M_p = 40{,}000 \text{ lbs}.$$

If elastic behavior had been prescribed up to $p_l = 285$ psi we would have had to write

$$\sigma_r \leqslant \sigma_Y \quad \text{at} \quad r = R. \tag{b}$$

Hence [6.40]

$$\frac{3pR^2}{4t^2} \leqslant \sigma_Y \quad \text{or} \quad t^2 \geqslant 0.75 p \frac{R^2}{\sigma_Y^2}. \tag{c}$$

On the other hand, eq. (a) gives

$$t^2 \geqslant 0.354 \, p \frac{R^2}{\sigma_Y^2}.$$

We see that prescribing no stress redistribution results in a 45.5% larger thickness.

We now assume that the door must be strengthened to support $p = 356$ psi without excess deformation. According to the results of Section 6.4, we shall strengthen exclusively in the circumferential direction. We now have

$$P_l = 1.25 \times 1.39 \times 10^6 \text{ lb} = 1.74 \times 10^6 \text{ lb}.$$

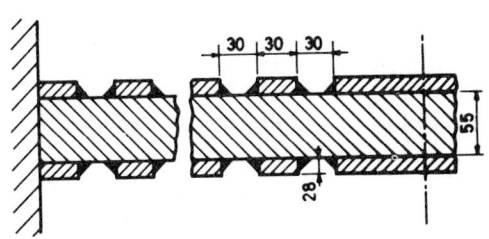

Fig. 6.62.

Hence, $P/\pi M_{rp} = 14$. On fig. 6.28 we read $k = 1.54$ and we successively have

$$M_{\theta p} = 1.54 \times 40{,}000 = 61{,}700 \text{ lb},$$

$$t_m^2 = 7.23 \text{ in.}^2.$$

To obtain the needed average $M_{\theta p}$ value, we need a supplement of thickness Δ such that, if present on half the radius only (fig. 6.62), we have

$$\frac{t^2 + (t+\Delta)^2}{2} \geqslant t_m^2.$$

We obtain $(t+\Delta)^2 \geqslant 2t_m^2 + t^2$, $t + \Delta \geqslant 4.4$ in.

Hence, $\Delta \geqslant 2.2$ in. We place annulus with 1.1×1.5 in.^{2} cross section, regularly distributed 1.5 in. apart (fig. 6.62), and a central plate for convenience.

6.10. Problems

6.10.1. A circular isotropic plate is loaded as shown in fig. 6.63. Find the formula relating the total loads P_1 and P_2 for collapse, on the basis of a statically admissible moment field and a corresponding mechanism. The yield condition of Tresca should be used. Verify that the theorem of Section 7.4.7 applies. *Answer*: $P_1 + P_2(3-2a/R) = 6\pi M_p$.

6.10.2. Find the analytic expressions of the moment field at collapse of a circular, isotropic, built-in plate with a central concentrated load P (Tresca condition). Show that a conical yield mechanism with a hinge circle at the

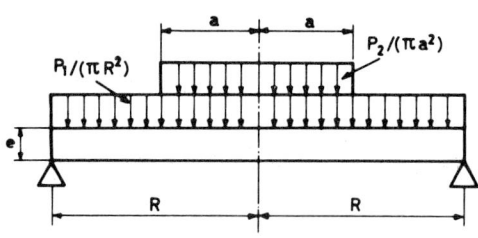

Fig. 6.63.

edge corresponds to this moment field. By extending the moment field in a statically admissible manner, show that $P = 2\pi M_p$ is the limit load for a built-in edge of arbitrary shape.

6.10.3. Find bounds for the limit value of the uniform pressure acting on a circular isotropic plate, simply supported and obeying the yield condition of von Mises. *Hint*: Lower bound – use $M_r = cp(R^2 - r^2)$, $M_\theta = cp(R^2 - \gamma r^2)$, and adjust the parameters c and γ.

Upper bound – use the result of Problem 3.8.1 to obtain the dissipation as a function of $\dot{w}$ and its derivatives. Use the field

$$\dot{w} = R - \frac{\rho}{2} - \frac{r^2}{2\rho} \quad \text{for} \quad 0 \leqslant r \leqslant \rho,$$

$$\dot{w} = R - r \quad \text{for} \quad \rho \leqslant r < R,$$

and adjust ρ.

6.10.4. Prove relation 6.28, fig. 6.19.

6.10.5. Find the lower bound $p_- = 20.6 M_p/l^2$ in relation (6.47). *Hint*: use the field

$$m_x = C(1-x^2),$$

$$m_y = C(1-y^2),$$

$$m_{xy} = (2C - 3p^*)xy,$$

with $p^* = pl^2/24M_p$, and adjust the parameter C.

6.10.6. Determine the upper bound $p_+ = 25.65 M_p/l^2$ for the problem of fig. 6.46.

6.10.7. Show that relation 6.27 gives the exact limit load. *Hint*: use a yield mechanism of the form of a truncated cone. To compute the rate of dissipation, see the fan mechanisms of Section 7.4.5.

References

[6.1] R.H.LANCE, E.T.ONAT, "A Comparison of Experiments and Theory in the Plastic Bending of Plates", *J. Mech. Phys. Solids*, **10**: 301, 1962.
[6.2] R.M.HAYTHORNTHWAITE, E.T.ONAT, "The Load-Carrying Capacity of Initially Flat Circular Steel Plates under Reversed Loading", *J. Aer. Sci.* **22**: 12 December 1955.
[6.3] R.M.HAYTHORNTHWAITE, "The Deflection of Plates in the Elastic-Plastic Range", *Proc. 2nd U.S. Nat. Cong. Appl. Mech.*, A.S.M.E., Ann Arbor, Mich., 1954.
[6.4] J.FOULKES, E.T.ONAT, "Tests on the Behavior of Circular Plates under Transverse Load", Brown Univ. Tech. Rep. 00R-3172/3, May 1955.
[6.5] E.T.ONAT, R.M.HAYTHORNTHWAITE, "The Load-Carrying Capacity of Circular Plates at Large Deflections", *J. Appl. Mech.*, **23**: 49, 1956.
[6.6] R.M.COOPER, G.A.SHIFRIN, "An Experiment on Circular Plates in the Plastic Range", *Proc. 2nd U.S. Nat. Cong. Appl. Mech.*, A.S.M.E., Ann Arbor, Mich., 1954.
[6.7] M.A.SAVE, "Vérification expérimentale de l'analyse plastique des plaques et des coques en acier doux". (Experimental verification of plastic limit analysis of mild steel plates and shells) *C.R.I.F. Report*, M.T.21, February 1966, Fabrimetal, 21, rue des Drapiers, Brussels.
[6.8] R.M.HAYTHORNTHWAITE, W.C.BOYCE, "The Load-carrying Capacity of Wide Beams at Finite Deflection", *Proc. 3rd U.S. Nat. Cong. Appl. Mech.*, pp. 541-550, June 1958.
[6.9] Y.OHASHI, S.MURAKAMI, "The Elastio-plastic Bending of a Clamped Thin Circular Plate", *Proc. 11th Int. Congr. Appl. Mech.*, Munich 1964, pp. 212-223, H.Görtler, ed., Springer, Berlin, 1966.
[6.10] Y.OHASHI, S.MURAKAMI, "Large Deflection in Elasto-plastic Bending of a Simply Supported Circular Plate under a Uniform Load", *Trans. A.S.M.E.*, ser. E, *J. Appl. Mech.*, **33**: 4, 866, 1966.
[6.11] Y.OHASHI, S.MURAKAMI, A.ENDO, "Elasto-plastic Bending of an Annular Plate at Large Deflection", *Ing. Archiv.*, **35**: 5, 340, 1967.
[6.12] H.G.HOPKINS, W.PRAGER, "The Load-carrying Capacity of Circular Plates", *J. Mech. Phys. Solids*, **2**: 1, 1953.
[6.13] C.A.ANDERSON, R.T.SHIELD, "On the Validity of the Plastic Theory of Structures for Collapse under Highly Localized Loading", *Trans. A.S.M.E., J. Appl. Mech.*, **23**: 629, September 1966.

[6.14] H.G.HOPKINS, A.J.WANG, "Load-carrying Capacities for Circular Plates of Perfectly-plastic Material with Arbitrary Yield Conditions", *J. Mech. Phys. Solids,* **3**: 117, 1954.
[6.15] V.V.SOKOLOVSKY, "Elasto-plastic Bending of Circular and Annular Plates", Brown Univ. Tech. Rep., 3, November 1955.
[6.16] A.SAWCZUK, T.JAEGER, *Grenztragfähigkeits-Theorie der Platten,* p. 522, Springer, 1963.
[6.17] D.C.DRUCKER, H.G.HOPKINS, "Combined Concentrated and Distributed Load on Ideally Plastic Circular Plates", *Proc. 2nd U.S. Nat. Cong. Appl. Mech.,* Ann Arbor, Mich., 1954.
[6.18] A.SAWCZUK, "Some Problems of Load-carrying Capacities of Orthotropic and Nonhomogeneous Plates", *Zakblad Mech. Osrod. Ciagl. Polsk. Akad. Nauk,* **VIII**: 4, Warsaw, 1956.
[6.19] W.OLSZAK, A.SAWCZUK, "Théorie de la capacité portante des constructions non-homogènes et orthotropes", *Suppl. to Annales de l'Inst. Tech. du Bat. et des Trav. Pub.,* 149, May 1960.
[6.20] A.SAWCZUK, "Linear Theory of Plasticity of Anisotropic Bodies and its Applications to Problems of Limit Analysis", *Zakblad, Mech. Osrod. Ciagl. Polsk. Akad Nauk,* **XI**: 5 Warsaw, 1959.
[6.21] J.MARKOWITZ, L.W.HU, "Plastic Analysis of Orthotropic Circular Plates", *Proc. A.S.C.E., J. Eng. Mech. Div.,* **90**: EM5, 251, October 1965.
[6.22] D.C.DRUCKER, "Plastic Design Methods, Advantages and Limitations", *Trans. Soc. Nav. Arch. Eng.,* **65**: 172, 1958.
[6.23] H.E.SHULL, L.W.HU, "Load-carrying Capacities of Simply Supported Rectangular Plates", *Trans. A.S.M.E., J. Appl. Mech.,* **30**: 617, December 1963.
[6.24] J.S.KAO, T.MURA, S.L.LEE, "Limit Analysis of Orthotropic Plates", *J. Mech. Phys. Solids,* **11**: 429, November 1963.
[6.25] A.SAWCZUK, S.KALISZKY, "On the Limit Analysis of Plates Supported by a Nonhomogeneous Plastic Subgrade under Rotational Symmetry conditions", *Acta Technica Ac. Sc. Hung, Tomus* 48, Fasc. 1-2, 1964.
[6.26] A.SAWCZUK, M.DUSZEK, "A Note on the Interaction of Shear and Bending in Plastic Plates", *Arch. Mech. Stosowanej,* **15**: 411, 1963.
[6.27] U.V.NEMIROVSKI, "Carrying Capacity of Ribbed Reinforced Circular Plates" (in Russian), *Izv. Nauk. U.S.S.R., Mekh. Mach.,* 2, 163, 1962.
[6.28] W.OLSZAK, W.URBANOWSKI, "Plastic Nonhomogeneity. A Survey of Theoretical and Experimental Research", *Nonhomogeneity in Elasticity and Plasticity, Proc. I.U.T.A.M. Symp.* Warsaw, 1958, W.Clszak ed., pp. 259-298, Pergamon Press, London, 1959.
[6.29] D.C.A.KOOPMAN, R.H.LANCE, "On Linear Programming and Plastic Analysis", *J. Mech. Phys. Solids,* **13**: 2, April 1965.
[6.30] C.MASSONNET, M.SAVE, *Calcul Plastique des constructions,* Vol. 1: One Parameter Structures, 2nd ed., Centre Belgo-Luxembourgeois d'Information de l'Acier, Brussels, 1967 (see Sections 10.4 and 10.5).
[6.31] W.PRAGER, "Programmation linéaire en théorie des constructions" (Linear programming in structural analysis) *Mémoires du C.E.R.E.S.,* **3**: 33, Liège 1962.
[6.32] C.GAVARINI, "I teoremi fondamentali del calcolo a rottura e la dulaita in programmazione linear" (Fundamental theorems of limit analysis and duality in linear programming), *Ingegneria Civile,* 18, 1966.

[6.33] P.VILLAGGIO, "Analisi limite di piastre sottili appoggiate plastico rigide" (Limit Analysis of Thin Simply Supported Rigid-Plastic Plates), *Giorn. Genio Civile*, **103**: 3, 133, March 1965.

[6.34] W.SCHUMANN, "On Isoperimetric Inequalities in Plasticity", *Quart. Appl. Math.*, **16**: 3, 303, 1958.

[6.35] W.SCHUMANN, "On Limit Analysis of Plates", *Quart. Appl. Math.* **16**: 1, 61, 1958.

[6.36] R.M.HAYTHORNTHWAITE, R.T.SHIELD, "A Note on the Deformable Region in a Rigid-Plastic Structure", *J. Mech. Phys. Solids*, **6**: 127, 1958.

[6.37] M.ZAID, "On the Carrying Capacity of Plates of Arbitrary Shape and Variable Fixity under a Concentrated Load", *J. Appl. Mech.*, **25**: 4, 598, 1958.

[6.38] B.TEKINALP, "Elastic-plastic Bending of a Built-in Circular Plate under a Uniformly Distributed Load", *J. Mech. Phys. Solids*, **5**: 135, 1957.

[6.39] R.M.HAYTHORNTHWAITE, "Mode Change During the Plastic Collapse of Beams and Plates", *Developments in Mechanics*, Plenum Press, New York, 1961.

[6.40] S.TIMOSHENKO, WOINOWSKY-KRIEGER, *Theory of Plates and Shells*, McGraw-Hill, New York, 1959.

[6.41] R.J.ROARK, *Formulas for Stress and Strain*, 3rd ed., McGraw-Hill, New York, 1954.

[6.42] R.M.HAYTHORNTHWAITE, "Plastic Behavior of Beams with Elastic End Constraints", *Proc. 9th Int. Cong. Appl. Mech.*, **VIII**: 59, Brussels, 1956.

[6.43] R.T.SHIELD, "Optimum Design Methods for Multiple Loading", *Zeit. Ang. Math. Phys., Z.A.M.P.*, **14**: 38, 1963.

[6.44] M.A.SAVE, W.PRAGER, "Minimum-weight Design of Beams Subjected to Fixed and Moving Loads", *J. Mech. Phys. Solids*, **II**: 255, 1963.

[6.45] M.A.SAVE, R.T.SHIELD, "Minimum-weight Design of Sandwich Shells Subjected to Fixed and Moving Loads", *Proc. 11th Int. Cong. Appl. Mech.*, Munich 1964, pp. 341-349, H.Gortler, ed., Springer, Berlin, 1966.

[6.46] R.T.SHIELD, "On the Optimum Design of Shells", *J. Appl. Mech.*, **27**: 316, June 1960, **27**: 316.

[6.47] O.GROSS, W.PRAGER, "Minimum-weight Design for Moving Loads", *Proc. 4th U.S. Nat. Cong. Appl. Mech.*, Berkeley, Calif., 1962, A.S.M.E., New York, 1963.

[6.48] D.C.DRUCKER, R.T.SHIELD, "Design for Minimum-weight", *Proc. 9th Int. Cong. Appl. Mech.*, pp. 212-222, Brussels, 1956.

[6.49] D.C.DRUCKER, R.T.SHIELD, "Bounds on Minimum-weight Design", *Quart. Appl. Math.*, **15**: 269, 1957.

[6.50] Z.MROZ, "On a Problem of Minimum-weight Design", *Quart. Appl. Math.*, **19**: 3, July 1961.

[6.51] Z.MROZ, "The Load-carrying Capacity and Minimum-weight Design of Annular Plate", *Rozpr. Inzyn* (Eng. Trans.), Warsaw, 1958.

[6.52] M.A.SAVE, "Some Aspects of Minimum-weight Design", *Engineering Plasticity*, pp. 611-626, Cambridge Univ. Press, Cambridge, 1968.

[6.53] G.SACCHI, M.A.SAVE, "A Note on Nonstandard Materials", *Meccanica*, **1**: 3, 1968.

[6.54] E.T.ONAT, W.SCHUMANN, R.T.SHIELD, "Design of Circular Plates for Minimum-weight", *Zeit. Ang. Math. Phys., Z.A.M.P.*, **8**: 485, 1957.

[6.55] W.FREIBERGER, B.TEKINALP, "Minimum-weight Design of Circular Plates", *J. Mech. Phys. Solids*, **4**: 294, 1956.

[6.56] G.EASON, "The Minimum-weight Design of Circular Sandwich Plates", *Zeit. Ang. Math. Phys. Z.A.M.P.*, **11**: 368, Zurich, 1960.

[6.57] W.PRAGER, R.T.SHIELD, "Minimum-weight Design of Circular Plates under Arbitrary Loading", *Zeit. Ang. Math. Phys., Z.A.M.P.*, **10**: 421, 1959.

[6.58] W.PRAGER, "Minimum-weight Design of Plates", *De Ingenieur*, **67**: 141, 1955.

[6.59] W.FREIBERGER, B.TEKINALP, "Minimum-weight Design of Circular Plates", *J. Mech. Phys. Solids*, **4**: 294, 1956.

[6.60] H.G.HOPKINS, W.PRAGER, "Limits of Economy of Material in Plates", *J. Appl. Mech.*, **22**: 3, 317, 1955.

[6.61] J.A.KONIG, R.RYCHLEWSKI, "Limit Analysis of Circular Plates with Jump Nonhomogeneity", *Int. J. Solids and Struct.*, **2**: 3, 493, July 1966.

[6.62] E.T.ONAT, W.PRAGER, "Limits of Economy of Materials in Cylindrical Shells", *De Ingenieur*, **67**: 46, 1956.

[6.63] W.FREIBERGER, "Minimum-weight Design of Cylindrical Shells", *J. Appl. Mech.*, **23**: 576, December 1956.

[6.64] P.G.HODGE Jr., Discussion of publication [6.63], *J. Appl. Mech.*, **24**: 486, 1957.

7

Reinforced Concrete Plates

7.1. Introduction

Mechanical behavior of reinforced concrete structures is much more complex than that of steel structures. In Chapter 11 of Com. V., we restricted ourselves to a very brief introduction to limit analysis and design of reinforced concrete beams and frames. This extremely short treatment of the question was justified by two reasons:

1. Many problems have not yet found satisfactory solutions, and the applicability of limit analysis and design to reinforced concrete beams and frames is still open to discussion [7.1];
2. A sufficiently exhaustive treatment of the subject, with due discussion of the conditions of applicability of the theory, would have required a volume by itself (see [7.2] for more information).

On the other hand, the situation is somewhat less complicated for plates and shells because the percentage of reinforcement is always relatively small and plastic behavior is more generally and more clearly established. Moreover, analysis and design of plates on the basis of the so-called "yield line method" [7.3] has long been used by practicing engineers and is now accepted by the

European Committee for Concrete [7.4]. As will be emphasized, this method is nothing but a special use of the kinematical approach of limit analysis [7.5, 7.6]. Hence, limit analysis and design of reinforced concrete plates appears to be of considerable practical importance.

Consider a reinforced concrete plate subjected to a system of loads that is defined to within a single load parameter P. Under increasing P, the following phenomena are observed: firstly, a very limited elastic range (OA in fig. 7.1); secondly, where the concrete is subjected to tensile stresses, it cracks on account of its small tensile strength (range AB in fig. 7.1); thirdly, plastic flow of the reinforcement: during this stage, the bending moment in the most stressed sections is kept at a constant value and a redistribution of bending moments will in general take place until sufficient reinforcement is plastified to allow further opening of cracks, resulting in a collapse mechanism of the plate. The value of the load parameter P reached at the end of the third stage is called the *limit load* (or sometimes, incorrectly, the fracture load).

The behavior described above implies that a reinforced concrete plate subjected to simple flexure may be regarded as a perfectly plastic structural ele-

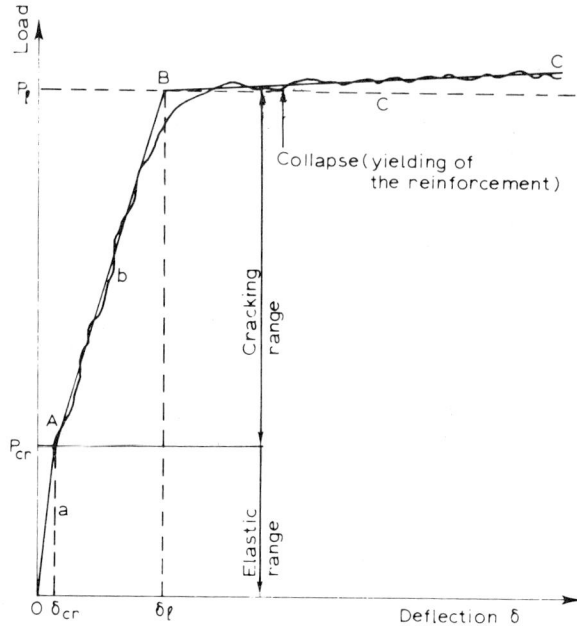

Fig. 7.1. Load versus deflection diagram.

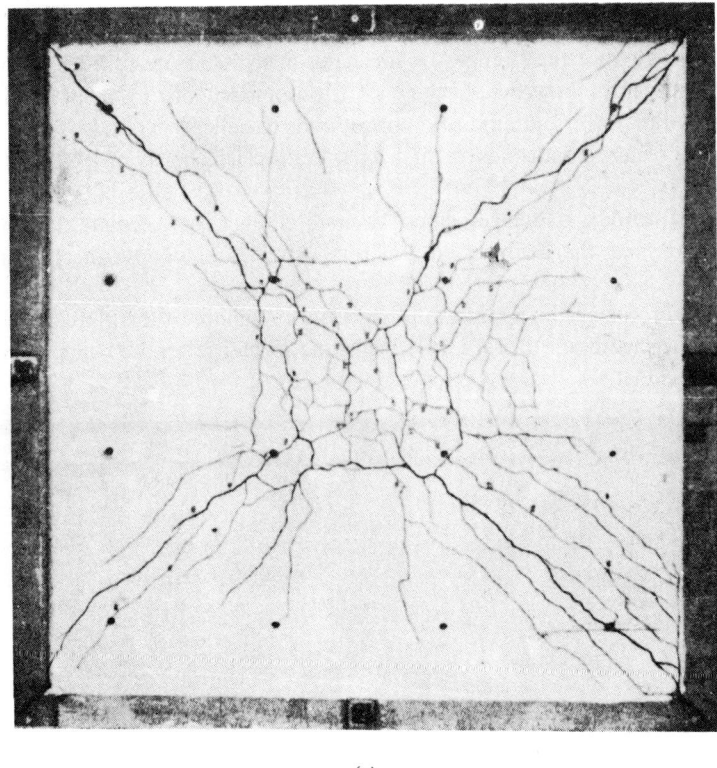

(a)

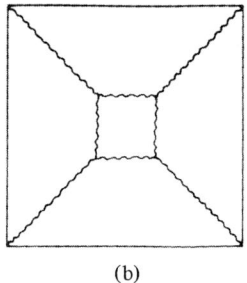

(b)

Fig. 7.2. (British Crown copyright reserved. Reproduced by permission of The Controller of Her Britannic Majesty's Stationary Office.)

ment. This is verified by the moment versus curvature diagram of fig. 7.1 [7.7], where the three ranges of mechanical behavior are indicated. In the present situation however, because of the homogeneous moment field, no redistribution occurs and point B corresponds to collapse.

The necessary conditions for such a perfectly plastic behavior are threefold:

1. The percentage of reinforcement is small enough for the plate not to fail by crushing of the compressed concrete before the collapse mechanism is formed;
2. Shear forces have no appreciable effect on the failure of the plate;
3. In-plane(membrane) forces have negligible influence on the collapse of the plate.

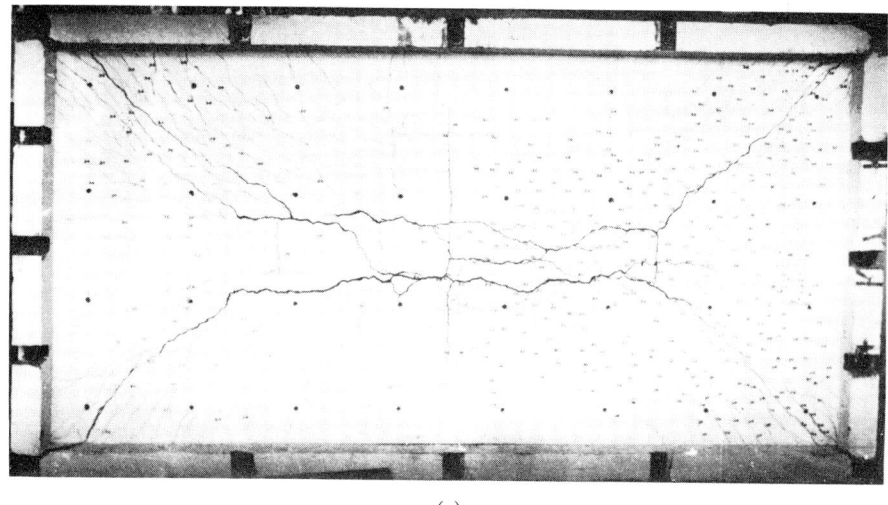

(a)

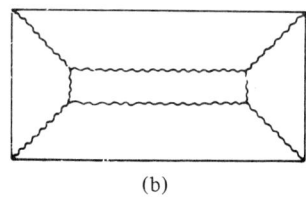

(b)

Fig. 7.3. (British Crown copyright reserved. Reproduced by permission of The Controller of Her Britannic Majesty's Stationary Office.)

The theory developed in Sections 7.3 to 7.6 will be based on the three conditions above. In Sections 7.7 and 7.8, we shall discuss of the effects of shear and membrane forces, when not negligible. Predictions of limit loads and collapse mechanisms have been widely verified experimentally [6.16,7.3]. Observed crack patterns can be idealized into "fracture lines" (according to Johansen [7.3]) that we shall preferably call "hinge lines" (see figs. 7.2 and 7.3). Our yield criterion will have to be compatible with this experimental fact.

7.2. Yield condition and flow rule

7.2.1. Introduction

Accepting the rigid-plastic moment versus curvature diagram of fig. 7.4, we assume that the reinforcement is placed in two orthogonal directions x and y in both the lower and the upper layers (fig. 7.5). If the reinforcement is not identical in both directions, the corresponding plastic moments will also differ and the plate will be orthotropic with regard to its ultimate strength.

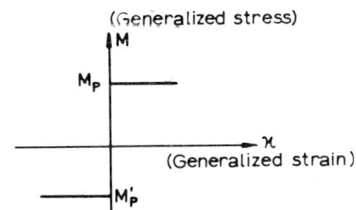

Fig. 7.4.

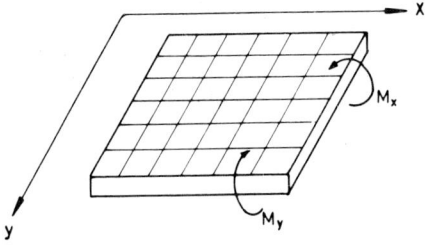

Fig. 7.5.

Ultimate moments for simple bending are denoted by M_{px}, M_{py} for positive bending (tension of the lower layer), M'_{px}, M'_{py} for negative bending (all positive values).

We accept the following physical criterion: Yielding occurs when the bending moment on the cross section with angle α (fig. 7.6) reaches a certain value that depends only on M_{px}, M_{py}, (M'_{px}, M'_{py}) and α.

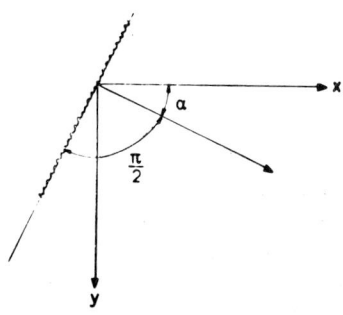

Fig. 7.6.

7.2.2. Yield condition

Let us describe the yield strength of the plate at a given point, for positive bending, by the *polar diagram* of fig. 7.7 (curve Y): the magnitude of segment

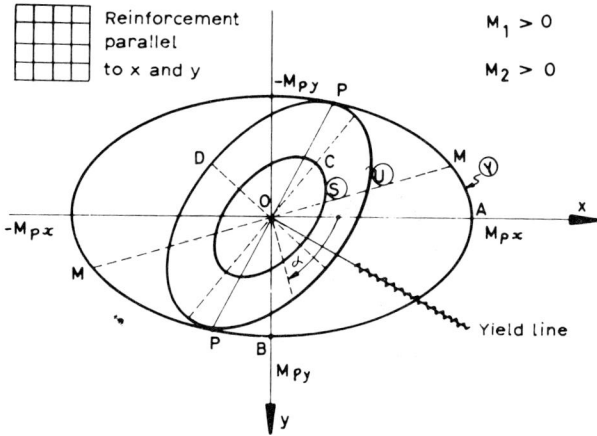

Fig. 7.7.

OA represents M_{px}, that of OB represents M_{py}, and generally, OM represents the ultimate bending moment on a cross section normal to the direction of OM (normal stresses on the section having the direction of OM). We denote this latter ultimate moment by $M_{p\alpha}$.

To describe the actual bending moments (as distinguished from the ultimate bending moments) at the same point of the plate, let us represent the bending moment M_α actually transmitted across a section through this point by a vector OS of magnitude $|M_\alpha|$ that is normal to the considered section, and let S be the locus of the endpoints of these vectors as the orientation of the considered section is changed. If all components of the moment tensor are increased proportionally, the curve S transforms into curve U which eventually becomes tangent at a certain point P with curve Y. We then have $M_\alpha = M_p$, and yielding occurs.

The curve Y_L for the lower layer must be used in the region of α where $|M| = M$ and the curve Y_U for the upper layer in the region where $|M| = -M$. Fig. 7.8 shows an example where yielding in negative bending takes place, contact at P corresponding to the condition $M_\alpha = -M_{p\alpha}$.

Following Johansen [7.3] and others (see, for instance, [7.8, 7.9, 7.10]), we adopt the following expressions for curves Y_L and Y_U:

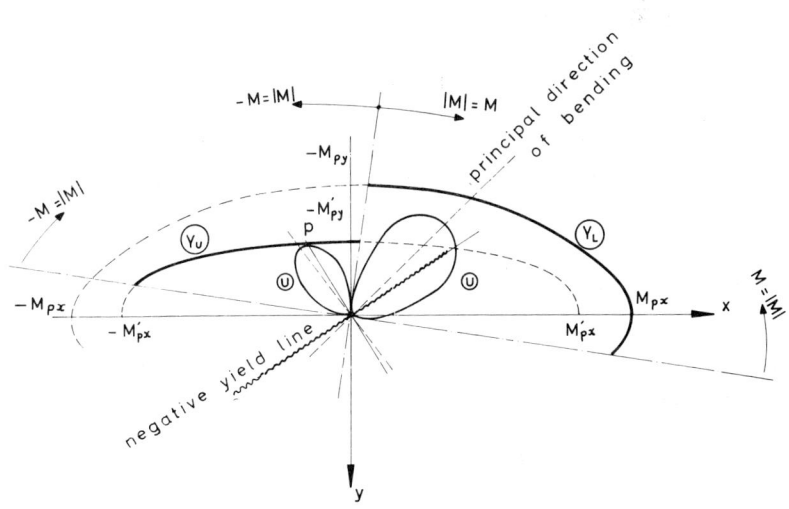

Fig. 7.8.

$$Y_L, \quad M_{p\alpha} = M_{px} \cos^2 \alpha + M_{py} \sin^2 \alpha, \tag{7.1}$$

$$Y_U, \quad M'_{p\alpha} = M'_{px} \cos^2 \alpha + M'_{py} \sin^2 \alpha. \tag{7.2}$$

We now want to obtain the yield condition in terms of the components M_x, M_y, M_{xy} of the moment tensor or, in other words, the yield surface in (M_x, M_y, M_{xy}) space. From the two-dimensional tensorial nature of the moment state it follows that (with convention of fig. 5.6 and with $M_{xy} = -M_{yx}$)*

$$M_\alpha = M_x \cos^2 \alpha + M_y \sin^2 \alpha - M_{xy} \sin 2\alpha. \tag{7.3}$$

Consider the case

$$M_\alpha = M_{p\alpha}. \tag{7.4}$$

Substitution of expressions (7.1) and (7.3) for $M_{p\alpha}$ and M_α into eq. (7.4), yields

$$M_x \cos^2 \alpha + M_y \sin^2 \alpha - M_{xy} \sin 2\alpha - M_{px} \cos^2 \alpha - M_{py} \sin^2 \alpha = 0. \tag{7.5}$$

With the use of elementary trigonometry**, eq. (7.5) may be rewritten as

$$\frac{M_{px} - M_x}{2 \tan \alpha} + \frac{M_{py} - M_y}{2} \tan \alpha + M_{xy} = 0. \tag{7.6}$$

This is the equation of a plane in (M_x, M_y, M_{xy})-space.

Varying α, we obtain a set of planes that will envelope a surface, the equation of which is derived by elimination of α from eq. (7.6) and the following equation:

$$-\frac{M_{px} - M_x}{\sin^2 \alpha} + \frac{M_{py} - M_y}{\cos^2 \alpha} = 0, \tag{7.7}$$

which is obtained by differentiating eq. (7.6) with respect to α.

We readily obtain

$$(M_{px} - M_x)(M_{py} - M_y) = M_{xy}^2. \tag{7.8}$$

* See Section 1.1.1 for a similar formula in plane stress [eq. (1.7)].
** Division by $2 \sin \alpha \cos \alpha$ would not be acceptable if $\sin \alpha = 0$ or $\cos \alpha = 0$. However, examination of these two cases easily reveals that the results obtained above remain valid.

Similarly, from $M_\alpha = -M'_{p\alpha}$ it follows that

$$(M'_{px}+M_x)(M'_{py}+M_y) = M^2_{xy} \ . \tag{7.9}$$

Eqs. (7.8) and (7.9) were obtained, in a slightly different manner, in [7.11] and can also be found in Nielsen [7.8], Wolfensberger [7.12], Kemp [7.9], and Morley [7.10]. These equations represent two cones (fig. 7.9). Their intersection lies on the surface with equation

$$(M_{px}-M_x)(M_{py}-M_y) = (M'_{px}+M_x)(M'_{py}+M_y) \ , \tag{7.10}$$

which is obtained by eliminating M_{xy} from (7.8) and (7.9). This surface actually reduces to a plane with equation

$$M_{px}M_{py} - M_y(M_{px}+M'_{px}) - M_x(M_{py}+M'_{py}) - M'_{px}M'_{py} = 0 \ . \tag{7.11}$$

This plane is parallel to the M_{xy}-axis and contains the straight lines with equations:

$$M_x = M_{px} \ ,$$
$$M_y = M'_{py} \ , \tag{7.12}$$

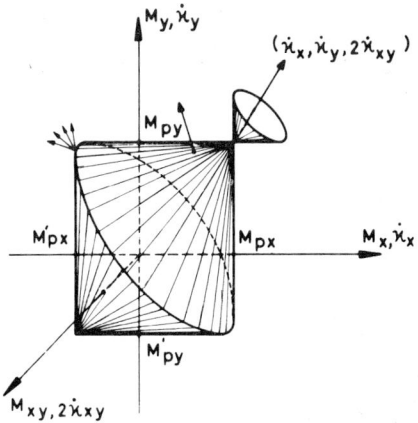

Fig. 7.9.

and

$$M_x = -M'_{px},$$

$$M_y = M_{py}. \tag{7.13}$$

Since the yield surface bounds the common part of the half spaces that are bounded by the tangent planes and contain the origin, it consists only of the parts of the cones [eqs. (7.8) and (7.9)] that are represented on fig. 7.9.

Every tangent plane touches the surface along a straight line that can be assigned a given value of α. Indeed, all points on such a contact line satisfy both eqs. (7.6) and (7.8) (or similar ones for negative yielding). Solving eq. (7.6) for $\tan \alpha$ and using eq. (7.8) we obtain

$$\tan \alpha = -\frac{M_{xy}}{M_{py} - M_y}. \tag{7.14}$$

Because

$$\tan 2\alpha = \frac{2 \tan \alpha}{1 - \tan^2 \alpha},$$

we have

$$\tan 2\alpha = \frac{-2 M_{xy}}{(M_{py}-M_y) - (M_{px}-M_x)}. \tag{7.15}$$

For the cone with eq. (7.9), we would obtain (with notation α' to indicate yielding in negative bending)

$$\tan \alpha' = \frac{M_{xy}}{M'_{py} + M_y}, \tag{7.16}$$

and

$$\tan 2\alpha' = \frac{2 M_{xy}}{(M'_{py}+M_y) - (M'_{px}+M_x)}. \tag{7.17}$$

Eq. (7.15) can be rewritten as

$$(M_{py}-M_{px}) \sin 2\alpha = (M_y-M_x) \sin 2\alpha - 2 M_{xy} \cos 2\alpha. \tag{7.18}$$

7.2] YIELD CONDITION AND FLOW RULE

But we know that, with $\beta = \alpha + \pi/2$

$$M_{\alpha\beta} = \tfrac{1}{2}(M_x - M_y) \sin 2\alpha + M_{xy} \cos 2\alpha . \tag{7.19}$$

Comparing eqs. (7.18) and (7.19), we deduce that

$$M_{\alpha\beta} = \tfrac{1}{2}(M_{px} - M_{py}) \sin 2\alpha \equiv T_{\alpha\beta} . \tag{7.20}$$

The twisting moment $M_{\alpha\beta}$ associated with the bending moment at yield $M_\alpha = M_{p\alpha}$ is thus determined if the stress point is to be on the yield surface or, equivalently, if curve U is to be *tangent* to curve Y_L (or Y_U). This was shown by Wolfensberger [7.12] and Kemp [7.9]. The twisting moment associated with $M_\alpha = -M'_{p\alpha}$ is

$$M_{\alpha\beta} = \tfrac{1}{2}(M'_{py} - M'_{px}) \sin 2\alpha \equiv T'_{\alpha\beta} . \tag{7.21}$$

7.2.3. Flow rule

After discussing the statical conditions for plastic flow, we must now investigate the precise manner in which this flow occurs. Consider a moment state represented by a point on the yield surface, with coordinates M_x, M_y, M_{xy}. According to Prager's terminology, M_x, M_y, M_{xy} are our generalized stresses (Section 5.2). The associated generalized strain rates must be chosen to furnish the specific rate of energy dissipated in plastic flow when the products of the corresponding generalized variables are added. Hence, these generalized strain rates are the rates of curvature $\dot{k}_x$, $\dot{k}_y$ and twice the rate of twist: $2\dot{k}_{xy}$.* The rate of energy dissipated per unit midsurface of the plate is:

$$D = M_x \dot{k}_x + M_y \dot{k}_y + 2M_{xy} \dot{k}_{xy} . \tag{7.22}$$

If we superimpose the systems of orthogonal cartesian coordinates M_x, M_y, M_{xy} and $\dot{k}_x$, $\dot{k}_y$, $2\dot{k}_{xy}$ and if we consider a point on the yield surface, the fundamental *flow rule* states that the corresponding generalized strain rates are components of a vector directed along the outward normal to the yield surface at the considered stress point. This is called the *normality rule*.

Because of perfect plasticity, the magnitude of the generalized strain rate vector remains arbitrary. Application of the normality law to the yield condition [eq. (7.8)] gives:

---

* Where $\dot{k}_{xy}$ is the twist component of the rate of curvature *tensor*.

$$\dot{\kappa}_x = \lambda(M_{py} - M_y),$$

$$\dot{\kappa}_y = \lambda(M_{px} - M_x),$$

$$2\dot{\kappa}_{xy} = 2\lambda M_{xy}, \qquad (7.23)$$

where λ is a nonnegative scalar. As the state of stress satisfies eq. (7.8), we have

$$\dot{\kappa}_x \cdot \dot{\kappa}_y = \dot{\kappa}_{xy}^2. \qquad (7.24)$$

Application of the normality rule to eq. (7.9) would result in the same relation [eq. (7.24)]. As can be seen on Mohr's circle, fig. 7.10, this relation holds when *one of the principal rates of curvature vanishes.* Hence, *local yielding consists of an elementary yield line* (of length dl). The angle γ of the normal to this yield line with the x-direction is given by

$$\tan 2\gamma = \frac{2\dot{\kappa}_{xy}}{\dot{\kappa}_y \cdot \dot{\kappa}_x}. \qquad (7.25)$$

Substituting the expressions (7.23) for $\dot{\kappa}_x$, $\dot{\kappa}_y$, $\dot{\kappa}_{xy}$ in eq. (7.25), we obtain

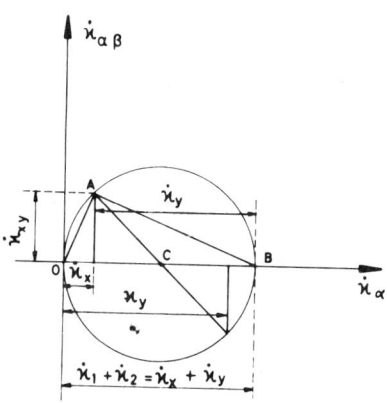

Fig. 7.10.

$$\tan 2\gamma = \frac{2M_{xy}}{(M_{px}-M_x) - (M_{py}-M_y)}. \tag{7.26}$$

Comparison of eq. (7.26) with eq. (7.15) gives $\gamma = \alpha \pm n\pi$. Similar results hold for negative yielding. Hence, *the direction of the yield line is that of the section subjected to the yielding moment*

From inspection of fig. 7.8, it is easily seen that:

1. For an isotropic plate ($M_{py}/M_{px} \equiv k = 1$, $M'_{py}/M'_{px} \equiv k' = 1$, but possibly $M'_{px} \neq M_{px}$) the directions of the yield lines coincide with the principal directions of the moment tensor;
2. For an orthotropic plate, yield lines no longer follow the principal directions of the moment tensor; when a positive and a negative yield line occur at a point, they are not orthogonal except when $(k'-1)/(1-k) = M_{px}/M'_{px}$ [7.9], and even in this case they differ from the principal directions of the moment tensor.

The last remark suggests that a discussion of the *singular points* and the *singular line* of the yield surface is in order.

Consider first a singular point characterized by $M_x = M_{px}$, $M_y = M_{py}$, $M_{xy} = 0$. The normality rule requires the strain-rate vector (with components $\dot{k}_x, \dot{k}_y, 2\dot{k}_{xy}$) to lie anywhere on or inside the cone of the outward normals of the yield surface at the singular point (fig. 7.9). Accordingly, we may have any combination of $\dot{k}_1 \geqslant 0$ and $\dot{k}_2 \geqslant 0$ with any principal directions. At the singular point $M_x = -M'_{px}$, $M_y = -M'_{py}$, $M_{xy} = 0$, we have any combination of $\dot{k}_1 \leqslant 0$ and $\dot{k}_2 \leqslant 0$. As shown in fig. 7.8, at both singular points the curve U coincides with the corresponding curve Y and all directions are potential yield lines (which may or may not become effective). Thus, the singular point corresponds to the intersection of any number of yield lines of the same sign. If this state of stress extends over a region, this region may flow arbitrarily as long as both principal curvatures are everywhere nonnegative.

At the intersection of the two cones we have a singular line. Any point of this line is the intersection of two straight contact lines of the two tangent planes with the two cones. Let $\dot{k}_\alpha > 0$ and $\dot{k}_{\alpha'} < 0$ denote the nonvanishing principal curvature rates corresponding to these two tangent planes. The normality rule requires that any nonnegative linear combination of these two vectors of curvature rate is admissible. The orientation of the two directions α and α' are well known when the moment tensor is given. Indeed, using eqs. (7.14) and (7.16), and the relation

$$\tan(a-b) = \frac{\tan a - \tan b}{1 + \tan a \cdot \tan b},$$

we have

$$\tan(\alpha-\alpha') = \frac{M_{xy}(M_{py}-M_y) + M_{xy}(M'_{py}+M_y)}{(M_{py}-M_y)(M'_{py}+M_y) - M^2_{xy}}. \tag{7.27}$$

Linear combination of the two curvature rate tensors results in a new tensor with principal directions that depend on the magnitudes of the coefficients of the combination.

It is worth recalling now that the specific rate of dissipation D (per unit area of midplane) can be viewed as the scalar product of vector $\boldsymbol{\sigma}$ with components M_x, M_y, M_{xy} (point P being on the yield surface) and vector $\dot{\boldsymbol{\varepsilon}}$ with components $\dot{\kappa}_x, \dot{\kappa}_y, 2\dot{\kappa}_{xy}$ (fig. 7.11).

$$D = \boldsymbol{\sigma} \cdot \dot{\boldsymbol{\varepsilon}}. \tag{7.28}$$

As is shown in Section 2.4, D is a *single-valued function of* $\dot{\boldsymbol{\varepsilon}}$, even if the yield surface is not regular (that is, if it has flats and vertices, fig. 7.12). In the general case, where there are two nonvanishing principal curvature rates $\dot{\kappa}_1$ and $\dot{\kappa}_2$ with orientation given by the angle α, we have (with $\beta = \alpha + \pi/2$);

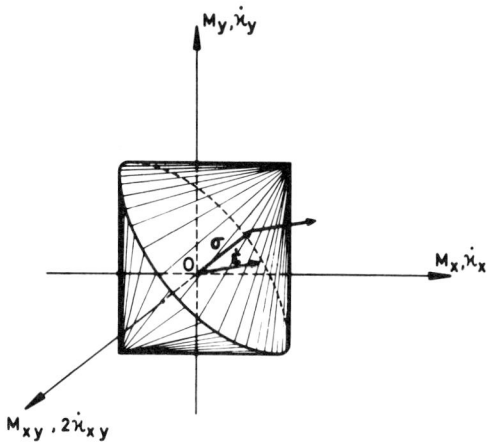

Fig. 7.11.

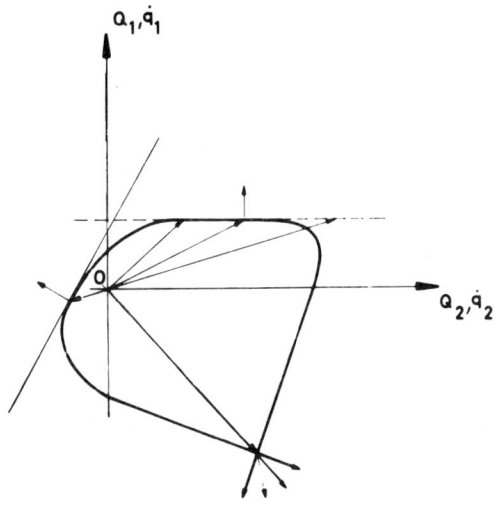

Fig. 7.12.

$$D = M_\alpha \dot{\kappa}_1 + M_\beta \dot{\kappa}_2 .\quad (7.29)$$

Denoting by $\dot{\kappa}_1$ the algebraically largest value of the principal curvature rates, we have

$$D = (M_{px} \cos^2 \alpha + M_{py} \sin^2 \alpha) \dot{\kappa}_1 + (M_{px} \cos^2 \beta + M_{py} \sin^2 \beta) \dot{\kappa}_2 \quad (7.30)$$

if $\dot{\kappa}_1 \geqslant \dot{\kappa}_2 \geqslant 0$. On the other hand, if $\dot{\kappa}_1 \geqslant 0 \geqslant \dot{\kappa}_2$, we have

$$D = (M_{px} \cos^2 \alpha + M_{py} \sin^2 \alpha) \dot{\kappa}_1 + (M'_{px} \cos^2 \beta + M'_{py} \sin^2 \beta) |\dot{\kappa}_2| . \quad (7.31)$$

Finally, if $0 \geqslant \dot{\kappa}_1 \geqslant \dot{\kappa}_2$, we have

$$D = (M'_{px} \cos^2 \alpha + M'_{py} \sin^2 \alpha) |\dot{\kappa}_1| + (M'_{px} \cos^2 \beta + M'_{py} \sin^2 \beta) |\dot{\kappa}_2| . \quad (7.32)$$

The various expressions for D clearly are single-valued functions of the components of the curvature rate tensor, which is given by $\dot{\kappa}_1$, $\dot{\kappa}_2$, and α.

7.3. Discussion of the yield condition

Theoretical justifications of the chosen yield criterion are presented by Nielsen [7.8], Wolfensberger [7.12], Kemp [7.9], and Morley [7.10], based on assumed behavior of steel and concrete.

Indirect experimental verification of the adequacy of the criterion is provided by numerous tests on transversally loaded plates (see, for example, [7.3,6.16,7.13]), but little direct experimental support exists for this [7.7], as quoted in [7.10,7.14, and 7.15].

We shall now summarize the results of experiments made by the coworkers of Massonnet at Liège University, Belgium [7.7].

Square plates 51.2 in. (1.3 m) wide were subjected to biaxial homogeneous bending by sixteen levers (four on each side), acted upon by hydraulic jacks and applying the edge moments by jaws with rubber protection (fig. 7.13). The eight jacks of set A are connected in parallel to the same pump, the eight jacks of set B to a second pump, and the sixteen jacks of set C to a third pump. Principal directions 1 and 2 are obviously parallel to the edges.

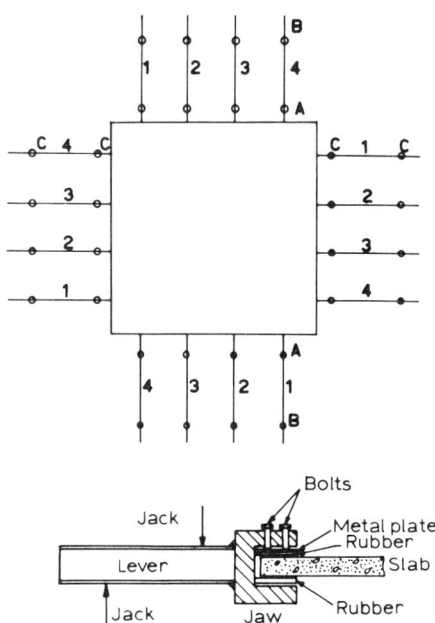

Fig. 7.13. Sketch of the test device.

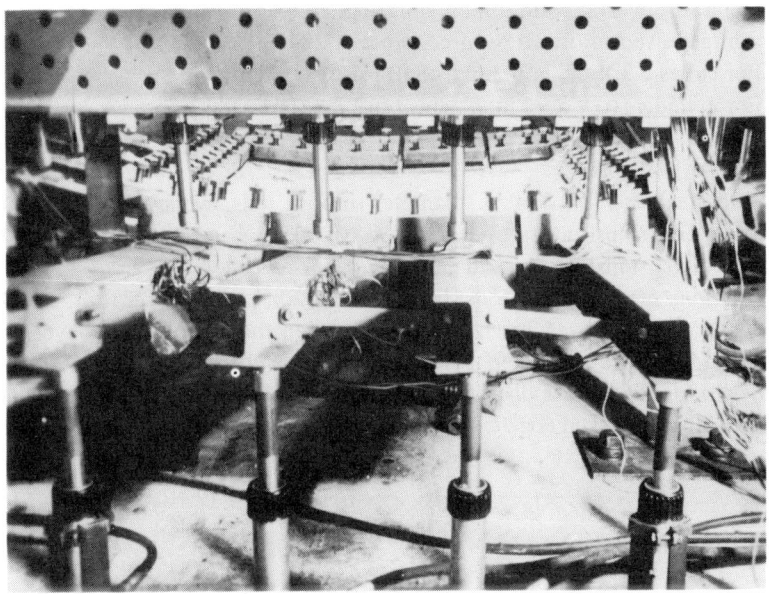

Fig. 7.14.

Jacks C apply the principal moment that will be labeled M_2, whereas jacks A and B apply the moment labeled M_1 and balance the dead weight of the slab and the levers. A photograph of the experimental device is shown in fig. 7.14, where levers and jacks are clearly seen. As is indicated by the photograph, free deformation of the edges is allowed. The moments M_1 and M_2 can be given arbitrary values of either the same signs or opposite signs. There is no shear force, except the negligible one caused by the dead weight. Bending moments are to all intents uniformly distributed along the edges, the largest discrepancies being less than 10%. From recorded pressure versus deflection diagrams analogous to the diagram in fig. 7.1, ultimate values of couples (M_1, M_2), corresponding to point B in fig. 7.1, have been determined. The characteristics of the tested plates are:

1. Composition of concrete:
   Crushed gravel 3/8 cm, 2760 lb
   Rhine sand 0/2 mm, 1660 lb
   High-strength Portland cement (PHR), 663 lb
   Water, 915 cm^3

220 REINFORCED CONCRETE PLATES [Ch. 7

2. Metal formworks, vibration by vibrating table.
3. Compressive rupture stress of concrete at 28 days:
   For cubes 6.3 × 6.3 × 6.3 in.: 4840 psi.
   For cylinders (ϕ = 5.94 in, h = 11.9 in.) 3970 psi.
4. Tensile rupture stress, on flexion prisms (3.94 × 3.94 × 19.7 in., span 15.8 in.) 497 psi.
5. Mild steel A 37, 0.394 in. diameter smooth rods, average tensile rupture stress 54400 psi, average yield stress 40200 psi (relative scattering 3.8%).
6. Slab dimensions: 51.2 in. side, 3.15 in. thickness.
7. Positions of reinforcing rods: layers of 0.802 in^2. of steel per foot, 2.36 in. apart: the effective* height is 2.58 in. when there is only one layer near each face of the slab; when two layers exist near the same face, the effective height is 2.58 in. for one and 2.18 in. for the other; edges of the slab are specially reinforced (fig. 7.15).

Fig. 7.15.

The reinforcing rods on top and bottom of any one slab are either unidirectional or follow two orthogonal directions. Their orientation with respect to the edges is defined by the smallest angle α they make with any edge.

---

* See Section 7.7, fig. 7.77 for definition of effective height.

DISCUSSION OF THE YIELD CONDITION

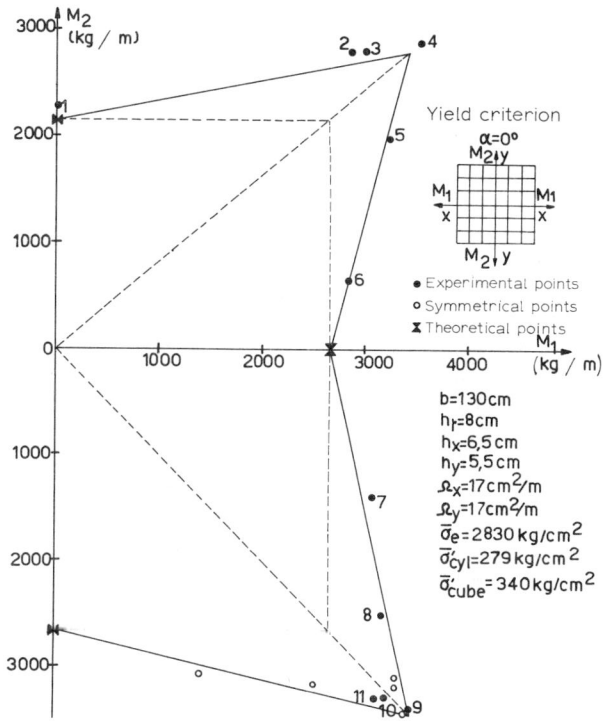

Fig. 7.16. Yield criterion.

The following cases have been considered:

1. $\alpha = 0°$; reinforcement parallel to the edges: 11 slabs. Experimental results are given in fig. 7.16. The theoretical points indicated by ✗ have been calculated in uniaxial bending, as for a beam. For $M_1 \geqslant 0$ and $M_2 \geqslant 0$, a difference between M_{p1} and M_{p2} is noted, arising from the difference in effective heights: it is moreover seen that, for $M_1 \cong M_2$ as well as for $M_1 \cong -M_2$, the ultimate strength is larger than in uniaxial bending by about 30%; finally, we observe that some parts of the yield curve turn their convexity toward the origin.

2. $\alpha = 45°$; results are shown in fig. 7.17. Slabs simply reinforced subjected to bending moments of opposite signs, have only one reinforcement layer near each face. Hence, they are extremely orthotropic, with coefficients $k = 0$ and $k' = \infty$. They are appreciably less strong than the other slabs, but appreciably

222 REINFORCED CONCRETE PLATES [Ch. 7

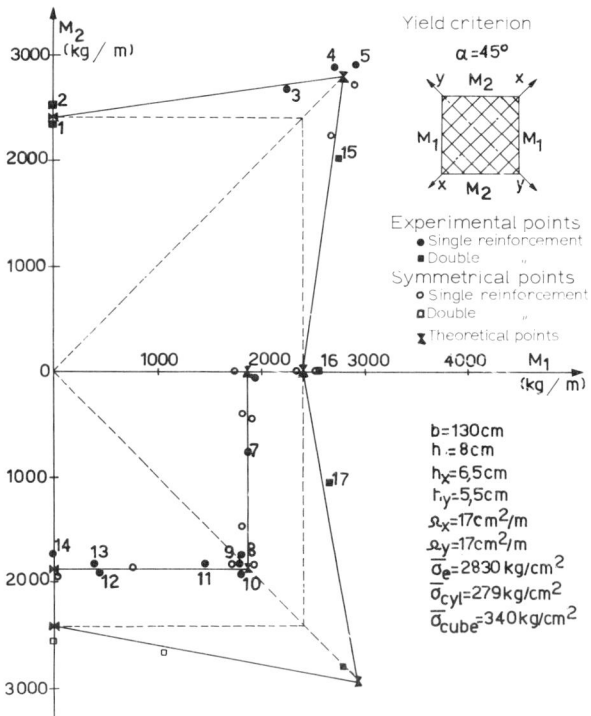

Fig. 7.17. Yield criterion.

stronger than predicted by the yield criterion of Johansen (relations 7.1 and 7.2).

All other slabs behave as isotropic plates because they have either two identical orthogonal layers near each face or two such layers near the sole face subjected to tension. Again, pointed corners are found in the diagonal directions of the M_1, M_2 diagrams (fig. 7.17).

It has been observed that cracks present a general orientation specified by an angle β but are locally formed of segments with two distinct orientations, both of which are different from β. Fig. 7.18 shows, for example, the tension face of the slab subjected to $M_1 = M_2$ with simple reinforcement together with the idealized cracking scheme, with average orientation of $18°30'$, made of segments parallel and at $45°$ to the side subjected to M_1.

Theoretical points and connecting segments have been obtained by interpreting the crack patterns and using some simple assumptions [7.7].

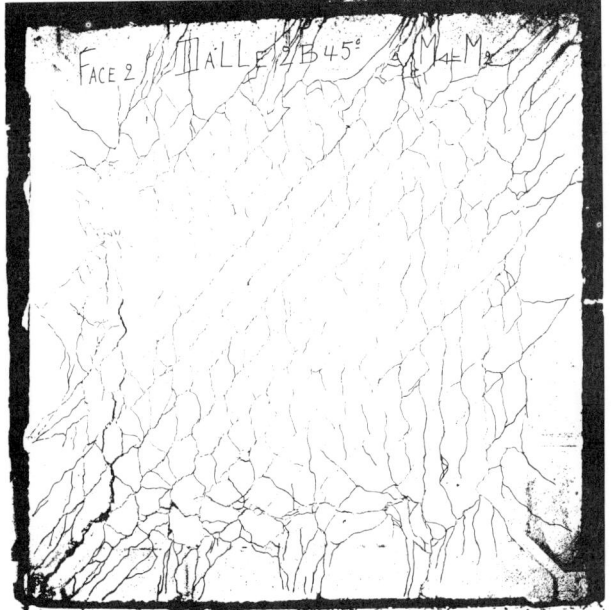
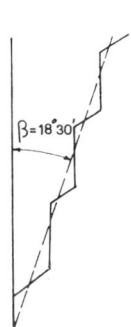

Fig. 7.18.

3. $\alpha = 22°30'$ and $\alpha = 30°$; results are shown in fig. 7.19. General features are the same as in the preceding figures. A detailed discussion can be found in the original paper [7.7].

Here, we only point out the following features:

1. *For isotropic plates*, the yield locus of Johansen is entirely inside the locus of the experimental points. Hence, if the ductility of the plate is sufficiently large, any solution based on Johansen's condition gives a lower bound for the limit load (see Section 5.5.2).

When using the condition of Johansen, one can allow for the influence of nonuniform effective height.

The margin of safety in the regions $M_1 \cong M_2$ and $M_1 \cong -M_2$ may be partially canceled by the fact that, as a rule, a kinematical approach is used. Accordingly, for practical use it seems justified to accept the considered yield condition.

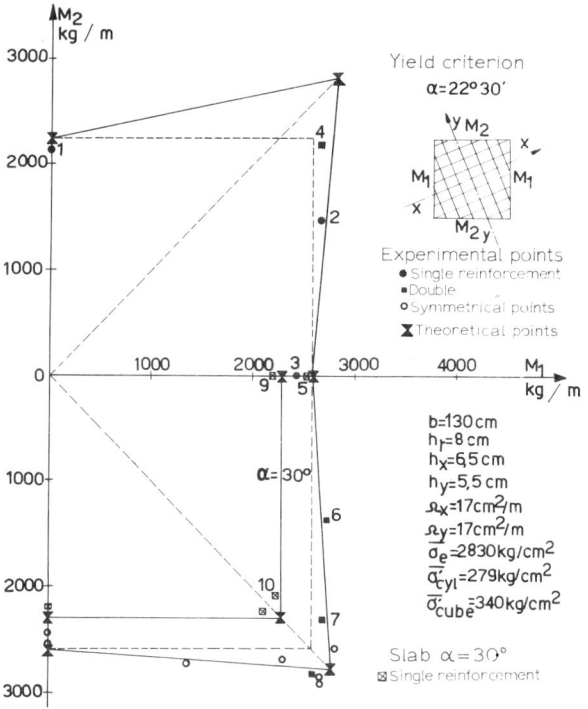

Fig. 7.19. Yield criterion.

2. *For orthotropic plates* (simple reinforcements):

$$M_{py} = 0, \qquad M_{px} \neq 0 \qquad (k=0),$$

$$M'_{py} \neq 0, \qquad M'_{px} = 0 \qquad (k'=\infty).$$

In the case where $M_1 \neq 0$, $M_2 = 0$, $\alpha = 45°$, we have $M_x = M_y = M_{xy} = M_1/2$. According to relation (7.8), we then find

$$M_1 = \frac{2M_{px}M_{py}}{M_{px} + M_{py}}.$$

With $M_{py} = 0$, we obtain $M_1 = 0$, and the normality law predicts $\dot{k}_x = \dot{k}_{xy} = 0$, $\dot{k}_y \neq 0$. Hence, yield lines parallel to the reinforcement rods should be obtained for arbitrarily small values of M_1, whereas experiments give $M_1 \cong$ 0.7 M_{px}. We conclude that Johansen's condition does not apply for very pronounced orthotropy.

Recent experiments by Lenschow and Sozen [7.14] do not show the increase of yield moments obtained at Liege University [7.7] when $|M_1| = |M_2|$. On the other hand, tests by Lenkei [7.15] not only have shown the existence of twisting moments in the yield line of an orthotropic slab when the principal directions of orthotropy differ from those of the moment tensor, but also indicate a smaller plastic moment than that predicted by Johansen's criterion.

In the absence of more precise information on the subject, it seems reasonable to continue to accept Johansen's yield condition, keeping in mind its approximate character.

7.4. The kinematic method (Johansen's fracture line theory)

7.4.1. *Kinematics of the mechanisms*

In accordance with our yield condition and with experimental evidence, we shall hereafter consider collapse mechanisms made of "hinge lines" (or "yield lines") about which adjacent parts of the plate will experience exclusively a relative rotation. Most often the yield lines will be straight because they will devide the slab into *rigid* parts hinged along the yield lines [7.3].

In the following, boundary conditions will be represented graphically as shown in fig. 7.20. At collapse, the plate is divided into various plane portions by the straight yield lines and thus bends into a polyhedron, or possibly into a ruled surface when an infinite number of infinitely close yield lines occur. A yield line will be called *positive* when subjected to a *positive* bending moment, and *negative* when subjected to a *negative* bending moment.

Consider first kinematically admissible mechanisms for the plates shown in fig. 7.21. The axes of rotation of the various portions of the plates coin-

Fig. 7.20.

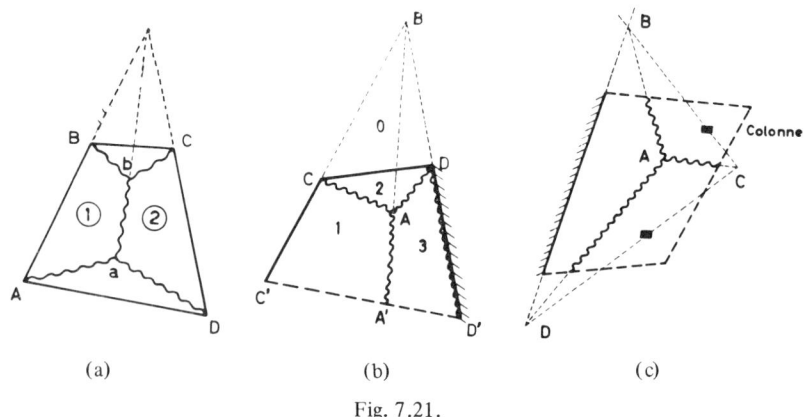

Fig. 7.21.

cide with either the simply supported edges, or the built-in edges, or are to pass through the points of support [fig. 7.21 (c)]. From inspection of fig. 7.21, we deduce that *a yield line separating two portions of the plate passes through the point of intersection of their axes of rotation.*

As we are only considering incipient collapse under constant load, all velocities of the collapse mechanism are only defined to within a common positive factor, the magnitude of which is irrelevant (except that it must be infinitesimal to justify disregarding changes of geometry). Thus, in the specification of a mechanism, one angle of rotation, or one deflection, can be assigned an arbitrary value*.

Conversely, *the mechanism is completely determined by giving the axes of rotation of the various portions of the plate and the ratios of the angles of rotation.* Indeed, as can seen from fig. 7.22, a (polygonal) contour line of arbitrary level h of the deflected plate consists of segments that are parallel to the axes of rotation at distances h/θ_i, these segments intersecting on the yield lines. The latter thus are the loci of the vertices of the polygonal contour lines. It follows directly from this property that the number p of geometric parameters needed completely to determine a mechanism is

$$p = n + a - 1, \qquad (7.33)$$

where n is the number of regions into which the plate is divided by the yield

---

* For brevity, we shall use the terms *rotation, deflection,* and *work* in the following, instead the more accurate terms *rate of rotation, rate of deflection,* and *rate of work.*

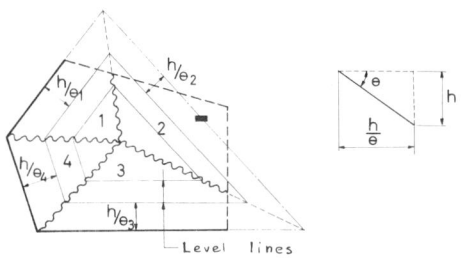

Fig. 7.22.

lines and a the degree of indeterminacy that remains in our knowledge of the axes of rotation. In fig. 7.21 (a), $n = 4$, $a = 0$ and hence $p = 3$. This actually constitutes a three-parameter "family" of mechanisms. In fig. 7.21 (b), $n = 3$, $a = 0$ and $p = 2$. In fig. 7.21 (c), $n = 3$ but $a = 2$ (two axes of rotation have only one fixed point each, namely, the supporting points) and thus $p = 3 + 2 - 1 = 4$.

The values of the angles of relative rotation in a given mechanism can be obtained by a graphical procedure, which is based on the following remark. On any hinge line or its extension, it is always possible to find two points through which pass either the axes of rotation of the adjacent rigid parts [points C, D, B in fig. 7.21 (b), for example] or two other hinge lines [point A, fig. 7.21 (b)]. With adequate sign conventions, the rotation vectors along the rotation axes and yield lines intersecting at the considered points form, at each point, a system equivalent to zero, by their very definition. Hence, intersections may be regarded as joints of a fictituous truss the bars of which are the axes of rotation and the hinge lines, the vectors of rotation playing the role of forces. Consequently, a Maxwell-Cremona diagram can be drawn for the rotation vectors. We shall call it "rotation diagram". To begin with, the magnitude of one rotation is assigned an arbitrary value. This rotation vector is represented by two forces in equilibrium, applied to the truss, equivalent to an internal force in the bar representing the considered axis.

In the example of fig. 7.21 (b), the bars will be AD, AC, BD, BC, and BA. The arbitrary rotation of axis CD, corresponding to a force in bar CD, will be represented by opposite external forces at C and D directed along CD (fig. 7.23). The regions of the plane of the truss are then labeled in correspondence with those determined by the plate boundary and its collapse mechanism [see fig. 7.21 (b) and fig. 7.23].

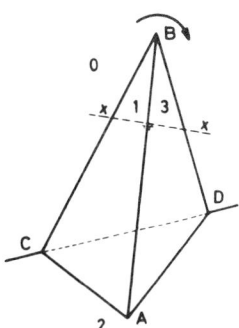

Fig. 7.23.

We then describe closed circular paths about each joint, in a sense of rotation that is fixed once for all.

Suppose for example that we begin at point B, fig. 7.23, moving in the clockwise sense, and successively cross segments BD, AB, and CB, that correspond to hinge lines DD' and AA', and to the axis of rotation CC', respectively. The intersections of the planes 0, 1, and 2 with a plane xx orthogonal to AB are shown in fig. 7.24. If the smallest angles θ_{xx} between the obtained intersection lines (fig. 7.24) are *positive* from one plane to the next *when turning about point B,* we have

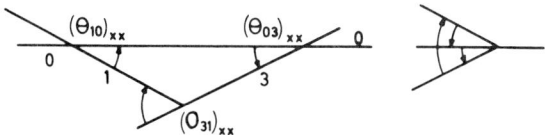

Fig. 7.24.

$$(\theta_{10})_{xx} + (\theta_{03})_{xx} + (\theta_{31})_{xx} = 0 . \tag{7.34}$$

Two similar relations are obtained from the consideration of intersecting planes normal to CB and DB. We thus arrive at the following result. With a given sign convention for rotation vectors (for example the right-hand screw rule) and if regions i, j, k are successively met when circulating about the joint where bars $i, j,$ and k meet, we have

$$\theta_{ij} + \theta_{jk} + \theta_{ki} = 0 . \tag{7.35}$$

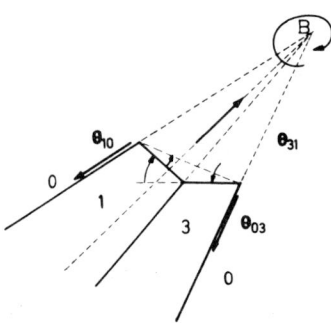

Fig. 7.25.

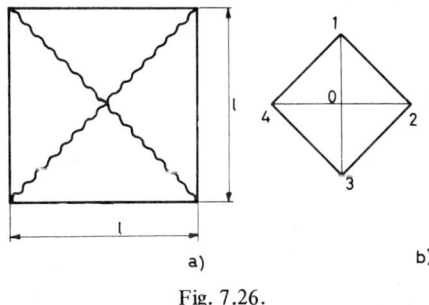

Fig. 7.26.

This property is illustrated in fig. 7.25. Consider, for example, a square, simply supported, and uniformly loaded isotropic plate. We adopt the collapse mechanism of fig. 7.26, which tends to deform the flat plate into a pyramid. If we give unit displacement to the center, the rotations about the edges are $2/l$. The rotation diagram is shown in fig. 7.26 (b). "Equilibrium" of joint A gives triangle 014, of joint B, 102, etc. In this manner, we obtain

$$D = M_p \cdot 2l\sqrt{2}(2l) \cdot \sqrt{2} = 8M_p . \qquad (7.36)$$

The work W of the applied load p is given by the product of p by the volume of the pyramid, that is

$$W = p \cdot \tfrac{1}{3}l^2 \cdot 1 . \qquad (7.37)$$

Equating D and W, we find

$$p_+ = \frac{24M_p}{l^2} .\tag{7.38}$$

With $P = pl^2$, eq. (7.38) can be written

$$P_+ = 24M_p .\tag{7.39}$$

If the plate was built-in and had the plastic moment M'_p in negative bending, relation (7.37) would remain unchanged, whereas the term

$$4l \cdot M'_p \frac{2}{l} = 8M'_p$$

would have to be added to the expression (7.36) of D. We therefore would obtain

$$P_+ = 24(M_p + M'_p) .\tag{7.40}$$

For a rectangular plate ($AB=DC=l$, $AD=BC=L>l$, fig. 7.27) it is readily verified that, if yield lines at corners make an angle of $45°$ with the edges, the rotations diagram is the same as that for the square plate (provided unit displacement is prescribed for segment EF, fig. 7.27). The terms $(L-l)4M_p/l$ and $pl/2(L-l)$ must be added to the expressions (7.36) and (7.37) of D and W, respectively. We then have

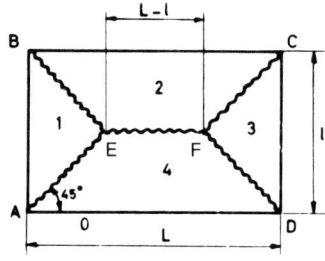

Fig. 7.27.

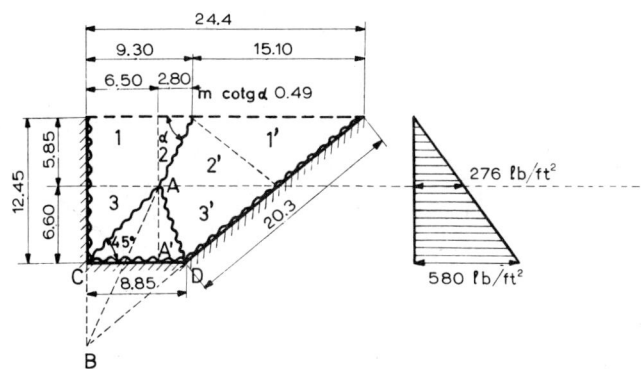

Fig. 7.28.

$$p_+ \left(\frac{l^2}{3} + \frac{lL}{2} - \frac{l^2}{2} \right) = 8M_p + 4 \frac{L \cdot l}{l} M_p ,$$

or

$$p_+ = \frac{24 M_p}{l^2} \frac{1 + L/l}{3L/l - 1} , \qquad (7.41)$$

which coincides with formula (6.35) obtained in Section 6.5.2.

As a third example, consider the isotropic plate of fig. 7.28, built-in on three sides, free on the fourth, and subjected to a linearly varying pressure as shown. Assume the mechanism shown in fig. 7.28 and defined by the position of point A and the angle α. Give to point A unit transversal displacement. The fictitious truss consists of the bars AC, CB, BD, DA, and AB, and is subjected to external forces at the joints C and D. The corresponding regions of the plate and the truss are labeled as shown in fig. 7.28 and 7.29. The label 0 refers to the region external to the plate. Forces at C and D are represented by segments 03 and 30, with common magnitude $1/AA' = \frac{1}{2}$. The following values of the rotations are obtained from the rotation diagram of fig. 7.29 (b): Rotations about

BC : $\theta_{BC} = |\overline{01}| = 0.500$,

BD : $\theta_{BD} = |\overline{02}| = 0.500$.

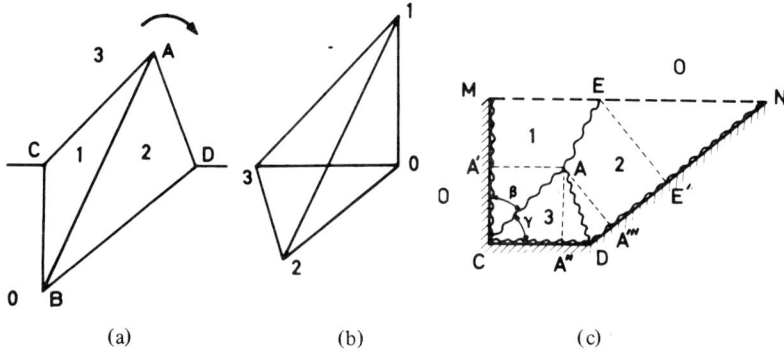

Fig. 7.29.

Rotations of adjacent parts about

BA : $\theta_{BA} = |\overline{12}| = 0.900$,

CA : $\theta_{CA} = |\overline{31}| = 0.706$,

AD : $\theta_{AD} = |\overline{32}| = 0.330$.

With $AC = 9.30$ ft, $AD = 6.90$ ft, $AF = 6.56$ ft, the work dissipated is

$D = M_p(0.706 \times 9.30 + 0.330 \times 6.90 + 0.9 \times 6.56)$

$+ M'_p(0.5 \times 12.45 + 0.5 \times 8.85 + 0.5 \times 20.3) = 14.7 M_p + 20.8 M'_p$.

To obtain the work of the load p, we must evaluate the resultant forces acting on parts 1, 2, and 3. For convenience, we compute partial resultant forces and their distances δ to the corresponding axes of rotation. One finds (fig. 7.29)

$p_{1a} = 5.7$ klb , $\qquad \delta_{p_{1a}} = 3.28$ ft ,

$p_{1b} = 0.852$ klb , $\qquad \delta_{p_{1b}} = 7.34$ ft ,

$p_{1c} = 6.28$ klb , $\qquad \delta_{p_{1c}} = 2.20$ ft ,

$p'_{1c} = 2.15$ klb , $\qquad \delta_{p'_{1c}} = 1.64$ ft .

(p_{1c} corresponds to the uniform part of the pressure, p'_{1c} to the linearily varying part.)

$$p_{2a} = 4.33 \text{ klb}, \quad \delta_{p_{2a}} = 2.33 \text{ ft},$$

$$p_{2b} = 6.06 \text{ klb}, \quad \delta_{p_{2b}} = 4.73 \text{ ft},$$

$$p_{2c} = 10.05 \text{ klb}, \quad \delta_{p_{2c}} = 2.17 \text{ ft},$$

$$p'_{2c} = 3.41 \text{ klb}, \quad \delta_{p'_{2c}} = 1.64 \text{ ft},$$

$$p_3 = 8.47 \text{ klb}, \quad \delta_{p_3} = 2.20 \text{ ft},$$

$$p'_3 = 5.76 \text{ klb}, \quad \delta_{p'_3} = 1.64 \text{ ft}.$$

The work of the pressure thus is

$$W = \Sigma p_i \delta_i \theta_i = 0.5(5.7 \times 3.28 + 0.852 \times 7.34 + 6.28 \times 2.20 + 2.15 \times 1.64)$$

$$+ 0.5(4.33 \times 2.33 + 6.04 \times 4.73 + 10.05 \times 2.17 + 3.41 \times 1.64)$$

$$+ 0.5(8.47 \times 2.20 + 5.76 \times 1.64) = 64.1 \text{ klb·ft}.$$

Equation $W = D$ gives

$$14.7 M_p + 20.8 M'_p = 64.1 .$$

With $M'_p = 0$ (simple supports) we would obtain

$$M_p = 4.35 \text{ klb}.$$

7.4.2. *Work equation*

Consider (fig. 7.30) a yield line with length l, adjacent to the parts ① and ② of the plate that rotate about the axes i and j, respectively. Denote by $\boldsymbol{\theta}_i$ and $\boldsymbol{\theta}_j$ vectors with magnitude equal to the angles of rotations of the two parts about their respective axes i and j, and directed along these axes, both pointing away from the intersection I of the axes.

Denote by $\boldsymbol{\theta}$ a vector with a magnitude equal to the *relative* rotation of parts ① and ② about the yield line (rotation "in" the yield line), and directed along this yield line. We have [fig. 7.30 (b)]

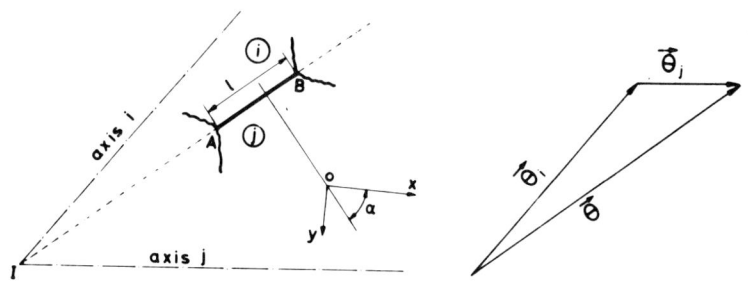

Fig. 7.30.

$$\theta = \theta_i + \theta_j . \quad (7.42)$$

Denote by α the angle of the normal to the yield line with the x-axis, the latter coinciding with the direction of one of the reinforcing bars. According to relations (7.1) or (7.2), the plastic moment is constant along the whole yield line. In a positive yield line we thus have

$$M_\alpha = M_{p\alpha} = M_{px} \cos^2 \alpha + M_{py} \sin^2 \alpha , \quad (7.43)$$

and the work dissipated in this yield line is

$$D_\alpha = |M_\alpha| \, |\theta| \, |l| \quad (7.44)$$

where l is a vector directed along the yield line, oriented as θ, and with a magnitude equal to the length of the line. We have

$$|\theta||l| = \frac{l_x \theta_x}{\sin^2 \alpha} = \frac{l_y \theta_y}{\cos^2 \alpha} , \quad (7.45)$$

where l_x, l_y, θ_x, θ_y are the projections of l and θ on the x- and y-axes, respectively.

Substitution of expressions (7.45) and (7.43) for $|\theta||l|$ and $|M_\alpha|$ into eq. (7.44), and use of $M_{py} = kM_{px}$, give

$$D_\alpha = M_{px}(l_y \theta_y + kl_x \theta_x) . \quad (7.46)$$

A negative yield line would have furnished

$$D'_\alpha = M'_{px}(l_y\theta_y + k'l_x\theta_x) \ . \tag{7.47}$$

Summing the works dissipated in all the yield lines we obtain the total work dissipated as

$$D = M_{px}\underset{+}{\Sigma}(l_y\theta_y + kl_x\theta_x) + M'_{px}\underset{-}{\Sigma}(l_y\theta_y + k'l_x\theta_x) \ , \tag{7.48}$$

where the first sum is extended to all positive yield lines and the second to all negative yield lines.

When the plate is isotropically reinforced, we have $k = k' = 1$. Hence, eq. (7.48) reduces to

$$D = M_p\underset{+}{\Sigma}(l_y\theta_y + l_x\theta_x) + M'_p\underset{-}{\Sigma}(l_y\theta_y + l_x\theta_x) \ . \tag{7.49}$$

As we know that

$$l_x\theta_x + l_y\theta_y = \mathbf{l} \cdot \boldsymbol{\theta} \ ,$$

eq. (7.42) gives

$$l_x\theta_x + l_y\theta_y = \mathbf{l} \cdot \boldsymbol{\theta}_i + \mathbf{l} \cdot \boldsymbol{\theta}_j \ , \tag{7.50}$$

and eq. (7.49) can be rewritten:

$$D = M_p\underset{+}{\Sigma}(\mathbf{l}\cdot\boldsymbol{\theta}_i + \mathbf{l}\cdot\boldsymbol{\theta}_j) + M'_p\underset{-}{\Sigma}(\mathbf{l}\cdot\boldsymbol{\theta}_i + \mathbf{l}\cdot\boldsymbol{\theta}_j) \ . \tag{7.51}$$

If we denote by l_i and l_j the projections of $\mathbf{l}$ on $\boldsymbol{\theta}_i$ and $\boldsymbol{\theta}_j$, eq. (7.51) becomes

$$D = M_p\underset{+}{\Sigma}(l_i|\theta_i| + l_j|\theta_j|) + M'_p\underset{-}{\Sigma}(l_i|\theta_i| + l_j|\theta_j|) \ . \tag{7.52}$$

Eq. (7.52) may be expressed as follows: *in an isotropic plate, the work dissipated in the positive yield lines separating various portions of the plate is the product by M_p of the sum of the products of the rotations of these portions about their respective axes by the projections on these axes of the considered yield lines; the work dissipated in the negative yield lines is obtained similarly, replacing M_p by M'_p and summing along negative yield lines.* Note that some values l_i or l_j might be negative. In the example of fig. 7.31 of a plate built-in along the edges MCDN and free along MN, we have

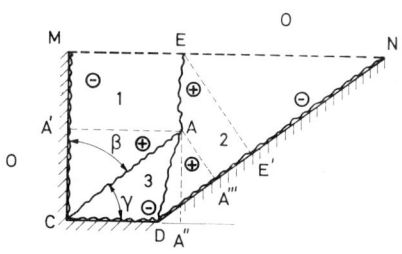

Fig. 7.31.

$$D = M_p[\theta_1(\overline{CA'} + \overline{A'M}) + \theta_2(\overline{DA'''} + \overline{A'''E}) + \theta_3(\overline{CA''} + \overline{A''D})]$$

$$+ M'_p(\theta_1 \cdot \overline{CM} + \theta_3 \cdot \overline{CD} + \theta_2 \cdot \overline{DN}).$$

$\overline{CA''}$ and $\overline{A''D}$ have opposite signs and their sum is CD. The work dissipated is

$$D = M_p(\theta_1 \cdot \overline{CM} + \theta_2 \cdot \overline{DE} + \theta_3 \cdot \overline{CD}) + M'_p(\theta_1 \cdot \overline{CM} + \theta_2 \cdot \overline{DN} + \theta_3 \cdot \overline{CD}). \quad (7.53)$$

7.4.3. *Orthotropic plates: affinity method*

The use of eq. (7.48) implies the calculation of the projections θ_x and θ_y on the axes (parallel to the reinforcing bars) of the *relative* rotations of the portions adjacent to the yield lines whereas in eq. (7.51) applicable to isotropic plates, *absolute* rotations only are needed. These latter being much easier to evaluate, we want to reduce the study of an orthotropic plate to that of a *fictitious equivalent isotropic plate*, that is an isotropic plate with same limit load P_l (or with same plastic moment M_p under the given loads). We must therefore define the fictitious isotropic plate such that the work dissipated in its mechanism and the work of the loads be the corresponding works in the actual plate multiplied by a common factor. Consider a plate obtained multiplying all dimensions of the original plate parallel to O_y by an "affinity" coefficient γ, without altering the dimensions along O_x. Call the resulting plate the "affine plate", with respect to the given one [7.3]. A point P with coordinates x_1 and y_1 on a segment of yield line becomes P_γ with coordinates x_1 and γy_1 (fig. 7.32).

Suppose that one of the two rigid parts to which point P belongs rotates about an axis RR, fig. 7.32, by an angle θ. The displacement w of P is given by the magnitude of the vectorial product of vectors $\mathbf{\theta}$ and $\mathbf{d} \equiv \mathbf{PN}$. We thus have

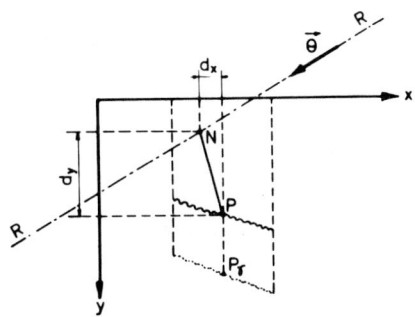

Fig. 7.32.

$$w_p = \theta_y d_x - \theta_x d_y .$$

Similarly, in the affine plate

$$w_{p_\gamma} = (\theta_y)_\gamma d_x - (\theta_x)_\gamma d_y \cdot \gamma$$

because $(d_x)_\gamma = d_x$ and $(d_y)_\gamma = \gamma d_y$.

If we let

$$w_{P_\gamma} = w_P , \quad \text{or} \quad \gamma y_1 (\theta_x)_\gamma = \gamma_1 \theta_x ,$$

in order, as will be seen later, to maintain the same distributed load q (lb/in^2) on the affine plate as on the given plate, we obtain

$$(\theta_x)_\gamma = \frac{\theta_x}{\gamma} , \tag{7.54}$$

and

$$(\theta_y)_\gamma = \theta_y . \tag{7.55}$$

The affine plate is isotropically reinforced to exhibit moments M_p and M'_p equal to the moments M_{px} and M'_{px} of the given plate, respectively. Application of eq. (7.48) to the affine plate, setting $k = k' = 1$, gives

$$D_\gamma = \sum_+ \left(M_{px} \gamma l_y \theta_y + M_{px} l_x \frac{\theta_x}{\gamma} \right) + \sum_- \left(M'_{px} \gamma l_y \theta_y + M'_{px} l_x \frac{\theta_x}{\gamma} \right) . \tag{7.56}$$

We now restrict ourselves to the study of orthotropic plates with the same orthotropy coefficients in both the upper and the lower layers of bars. Hence $k = k'$. Let

$$\frac{1}{\gamma^2} = k . \tag{7.57}$$

Substitution of expression (7.57) of γ in eq. (7.56) and comparison with eq. (7.48) yield:

$$D_\gamma = \gamma D . \tag{7.58}$$

The work w of the loads is the sum of the contributions of the distributed load $q(x,y)$ on a certain area A and of the concentrated loads P_i at various points i. Hence

$$W = \int_A qw \, dA + \sum_i P_i w_i . \tag{7.59}$$

For the affine plate, $w_\gamma = w$, $A_\gamma = \gamma A$. If we let $q_\gamma = q$, we have

$$\int_{A_\gamma} q_\gamma w_\gamma \, dA_\gamma = \gamma \int_A qw \, dA .$$

Besides, if we let $(P_i)_\gamma = \gamma P_i$, we finally obtain

$$W_\gamma = \gamma W . \tag{7.60}$$

Applying the kinematic theorem to the affine plate we write $W_\gamma = D_\gamma$. Substitution of expressions (7.58) and (7.60) for W_γ and D_γ in this relation gives $\gamma W = \gamma D$, and consequently, the affine plate will exhibit the same limit load (or the same plastic moment M_{px}) as the given orthotropic plate. To sum up, the affine plate of an orthotropic plate with given M_{px}, M'_{px}, k and with $k' = k$, is obtained by multiplying the dimensions in the Oy-direction and the concentrated loads by the affinity coefficient $\gamma = 1/\sqrt{k}$. Dimensions in the ox-direction and distributed loads are unaltered. If a load is distributed along a line, its resultant must be multiplied by γ.

Consider for example the plate in fig. 7.33, simply supported along the edges $EABF$ and free along EF, uniformly loaded with $q = 1$ (unit of load per unit of area), and assumed to have the collapse mechanism represented on

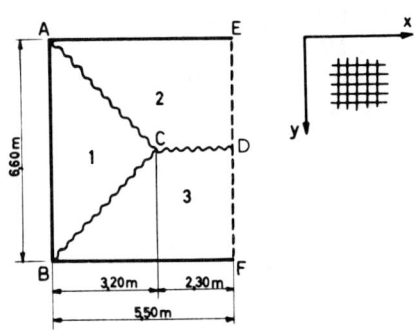

Fig. 7.33.

the drawing. The reinforcement will ensure $k = M_{py}/M_{px} = 2$, and we want to determine M_{px}.

Let the displacement of point C be unity. We have immediately $|\theta_1| = 1/3.2$, $|\theta_2| = |\theta_3| = 1/3.3$. Obviously $|\theta_{CD}| = 2/3.3$ and $(\theta_{CD})_y = |\theta_{CD}|$, whereas $(\theta_{CD})_x = 0$. $\boldsymbol{\theta}_{AC}$ is obtained graphically in fig. 7.34. Obviously, $(\theta_{AC})_x = 1/3.3 = (\theta_{BC})_x$, $(\theta_{AC})_y = 1/3.2 = (\theta_{BC})_y$. Application of eq. (7.49) (with no negative yield line) gives

$$D = M_{px}\left[2\times\frac{3.3}{3.2} + 2\left(2.3\times\frac{2}{3.3} + 2\times\frac{3.2}{3.3}\right)\right] = 8.73 M_{px}.$$

The work of the loads is

$$W = q\left(\frac{3.2\times 6.6}{2}\times\frac{1}{3} + 2\times 3.2\,\frac{3.3}{2}\times\frac{1}{3} + 2\times 2.3\times 3.3\times\frac{1}{2}\right) = 14.64.$$

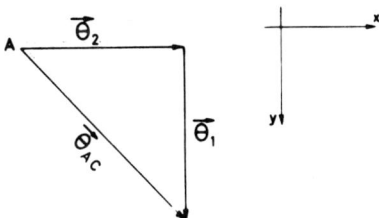

Fig. 7.34.

Hence, from $W = D$ we obtain $M_{px} = 1.68$.

Let us now apply the affinity method to the same problem. The affinity coefficient is $\gamma = 1/\sqrt{2} = 0.707$. Consequently, we obtain the plate and the mechanism in fig. 7.35, where $|\theta_1| = 1/3.2$, $|\theta_2| = |\theta_3| = 1/2.33$. Applying eq. (7.52) we obtain

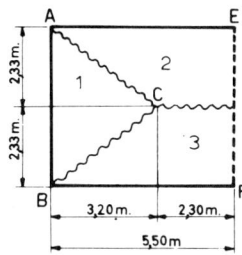

Fig. 7.35.

$$D_\gamma = M_{px}\left(\frac{4.66}{3.2} + 2\times\frac{5.5}{2.33}\right) = 6.162 M_{px}.$$

The work of the loads is

$$W = q\left(4.66\times 3.2\times\frac{1}{2}\times\frac{1}{3} + 2\times 3.2\times\frac{2.33}{2}\times\frac{1}{3} + 2\times 2.33\times 2.3\times\frac{1}{2}\right) = 10.435,$$

whence $M_{px} = 1.692$. The difference between the two results is less than the error due to the use of the slide rule. Note also that the value obtained for M_{px} is a lower bound because we have used exclusively the kinematic theorem.

In their recent book cited in ref. [7.16], Wood and Jones have extended the affinity method to plates with nonorthogonal reinforcing bars.

7.4.4. Obtaining the actual collapse mechanism

Application of the kinematical theorem essentially consists of assuming a type of collapse mechanism, that is a p-parameter family of mechanisms.

For this family, the corresponding load P_+ is readily obtained from the work equation, as a function of the p parameters $\alpha, \beta, \gamma, \delta$ ... which, according to the kinematic theorem, must be chosen to minimize the limit load. This minimum is not necessarily analytic. However, if it is analytic, and if the function $P_+ (\alpha,\beta,\gamma,...)$ is continuous, the minimum condition is

$$\frac{\partial P}{\partial \alpha} = 0, \quad \frac{\partial P}{\partial \beta} = 0, \quad \frac{\partial P}{\partial \gamma} = 0, \quad \ldots. \tag{7.61}$$

The system of p eqs. (7.61) furnishes the "best" values of the parameters, and the corresponding mechanism is the actual collapse mechanism provided the assumed family does contain the actual collapse mechanism.

Consider, for example, an infinitely long rectangular plate of width a, that is simply supported along its infinitely long parallel edges and loaded by a concentrated load P at a point of its axis. The plate is isotropic, with plastic moments M_p and M'_p. If we first suppose that the collapse mechanism is of the type represented in fig. 7.36 (a), and if we give a unit displacement to the point of application of the load P, the equation $W = D$ takes the form

$$1 \cdot P = M_p \left(8 \frac{\alpha}{a} + \frac{2a}{\alpha} \right) + M'_p 2 \frac{a}{\alpha}. \tag{7.62}$$

The system (7.61) reduces here to the single equation $\partial P/\partial \alpha = 0$, which furnishes

$$\alpha = \frac{a}{2} \sqrt{1 + \frac{M'_p}{M_p}} \tag{7.63}$$

Substitution of eq. (7.63) for α into (7.62) yields

$$P = 8 M_p \sqrt{1 + \frac{M'_p}{M_p}}. \tag{7.64}$$

Expression (7.64) furnishes the smallest value of P that can be obtained from the assumed type of collapse mechanism, but we are by no means certain that this family is the correct one.

This certainty may be achieved only if we are able to apply the combined theorem, indicating for the mechanism which gives the minimum value [eq. (7.64)] of P a corresponding statically admissible moment field. On the other hand, to ascertain that the mechanism obtained above is *not* the correct collapse mechanism, it is *sufficient* to find one mechanism of another family yielding a smaller load than eq. (7.64). Consider, for example, the mechanism shown in fig. 7.36 (b). This mechanism consists of two "fans" of infinitely close positive yield lines, bounded by two negative circular yield lines. This kind of fan may be considered as the limit of the mechanism shown in fig. 7.37, in which the number of positive yield lines is finite, and the negative

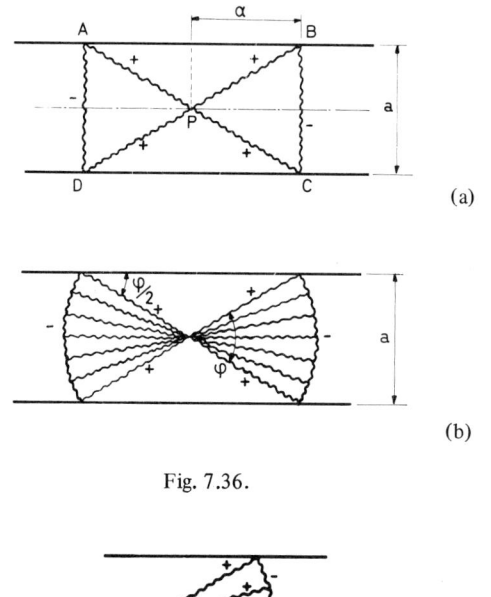

Fig. 7.36.

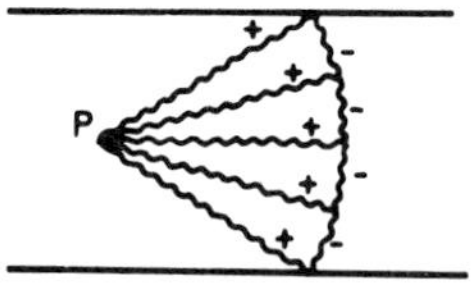

Fig. 7.37.

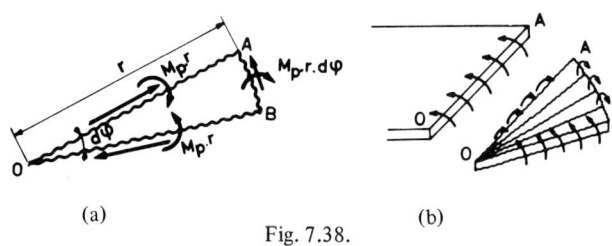

(a)   (b)

Fig. 7.38.

yield lines form a polygon. The work dissipated in a fan, for a unit displacement of its center point, is obtained as follows. The positive moments M_p acting on the sides OA and OB of an elementary triangle [fig. 7.38 (a)] have the resultant $M_p \cdot r \cdot d\phi$ with the direction AB of the tangent at A to the circular negative yield line. Its angle of rotation is $1/r$. Hence the elementary work dissipated by the moments applied to OA and OB is $M_p r\, d\phi/r = M_p \cdot d\phi$.

In the negative yield line AB, it is $M'_p r \, d\phi/r = M'_p \cdot d\phi$. Summing and integrating, we obtain

$$D = \int_\phi (M_p + M'_p) \, d\phi = (M_p + M'_p)\phi . \tag{7.65}$$

In considering the mechanism of fig. 7.36 (b), we must also take account of the work done by the moments applied by the fan to its radial edges, as shown in fig. 7.38 (b). Applying the work equation, we obtain

$$1. \; P = 2(M_p + M'_p)\phi + 4M_p \cot\frac{\phi}{2} . \tag{7.66}$$

The only parameter is ϕ and the minimum condition $\partial P/\partial \phi = 0$ yields

$$\cot\frac{\phi}{2} = \sqrt{M'_p/M_p} . \tag{7.67}$$

In the particular case where $M_p = M'_p$, eq. (7.67) yields $\phi = \pi/2$; substituting this value of ϕ into eq. (7.66), we obtain

$$P = 4M_p + 2\pi M_p = 10.28 M_p ,$$

whereas eq. (7.64) gives

$$P = 8\sqrt{2} M_p = 11.3 M_p .$$

Comparison of the two values of P reveals that the mechanism of fig. 7.35 cannot be the actual collapse mechanism. A number of collapse mechanisms for various loading conditions are found in refs. [6.16, 7.3, 7.16, 7.17, and 7.18]. We shall restrict ourselves to briefly describing some particularly interesting simple cases.

7.4.5. *Some examples of fan mechanisms caused by concentrated loads*
7.4.5.1. *Circular isotropic plate, simply supported and loaded in its center*

The circular fan mechanism transforms the plate into a cone. It gives the limit load

$$P_+ = 2\pi M_p . \tag{7.68}$$

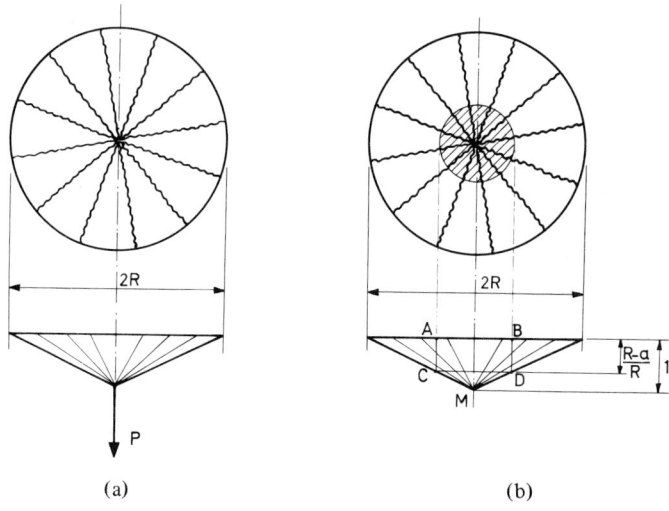

Fig. 7.39.

If the load is distributed over a small central circle of radius a (fig. 7.39), the same mechanism gives the same dissipated work $2\pi M_p$. But the work of the load here is the product of the intensity p by the combined volumes of the cylinder $ABDC$ and the cone CMD. Thus,

$$W = p\left[\frac{R-a}{R}\pi a^2 + \pi a^2 \tfrac{1}{3}\left(1-\frac{R-a}{R}\right)\right],$$

or

$$W = p\pi a^2 \left(\frac{R-a+\tfrac{1}{3}a}{R}\right).$$

Introducing the notation $p\pi a^2 = P$, the equation $W = D$ yields

$$P_+ = \frac{2\pi M_p}{1-\tfrac{2}{3}a/R}. \tag{7.69}$$

7.4.5.2. Circular isotropic plate, loaded in its center, and built-in along the edge

We now have a circular negative yield line at the edge. We only have to add the term $2\pi M'_p$ in the expression for D, and thus obtain

$$W = 2\pi(M_p + M_p'),\qquad(7.70)$$

and

$$P_+ = \frac{2\pi(M_p + M_p')}{1 - \tfrac{2}{3}a/R}.\qquad(7.71)$$

7.4.5.3. Isotropic plate of arbitrary shape, built-in along the edge, and subjected to a load distributed over a small circle of radius a

A collapse mechanism corresponding to a cone whose axis, normal to the plate, passes through the center of the circle where the load is applied, and limited by a negative circular yield line of radius R, furnishes the load [eq. (7.71)]. It is easily seen that the smallest value of P_+ is obtained for R as large as possible, i.e., for a yield circle that is tangent to the edge of the plate (fig. 7.40).

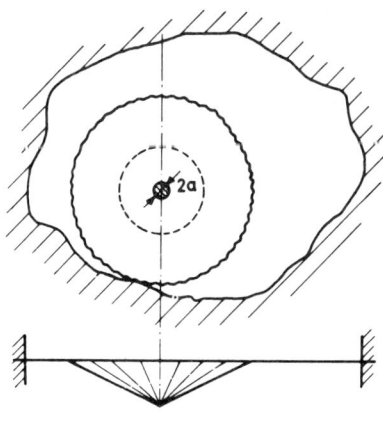

Fig. 7.40.

7.4.5.4. Circular isotropic plate loaded at the center and simply supported along a part of its edge

For $\beta < 116°$, the collapse mechanism is conical in the supported part of the edges, with two planes in the free part, which are separated by a yield line along the axis of symmetry (fig. 7.41, [7.18]). One obtains

$$P_+ = 2M_p(\pi - \beta + \sin\beta).\qquad(7.72)$$

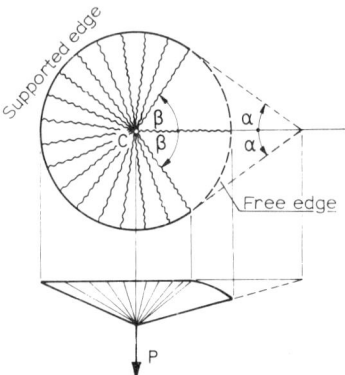

Fig. 7.41.

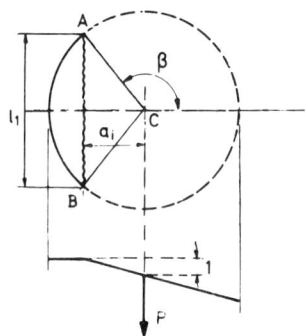

Fig. 7.42.

For $\beta > 116°$, this load is larger than that corresponding to a single straight line joining the extreme points of the supported part of the edge (fig. 7.42). Eq. (7.72) must then be rejected and replaced by

$$P_+ = 2M_p \tan \beta . \tag{7.73}$$

7.4.5.5. *Concentrated load applied near the straight edge of an isotropic plate*

If the edge is simply supported, the smallest load is obtained with

$$\tan \phi = \sqrt{M'_p/M_p} , \tag{7.74}$$

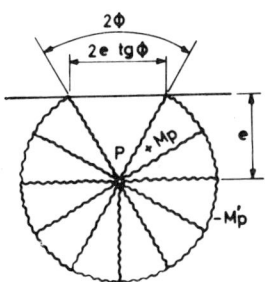

Fig. 7.43.

which, for $M'_p = M_p$, yields

$$P_+ = (3\pi+2)M_p ,\qquad(7.75)$$

with $\phi = 45°$ (fig. 7.43). When M'_p approaches zero, we obtain

$$P_+ = 2\pi M_p ,\qquad(7.76)$$

with $\phi = 0$. If the edge is built-in, one returns to the case discussed in Section 7.4.2.3.

7.4.6. Corner and edge effects
7.4.6.1. Corner effects: introduction

Consider a yield line passing through the intersection of two straight simply supported edges (fig. 7.44). If is well known that, when the plate is loaded, the corner point C tends to lift from the support. *If it is free* to do this (unilateral support), a corner region 3 will actually rotate about a certain axis aa (fig. 7.44) and the point C will move upwards. Thus, the yield line

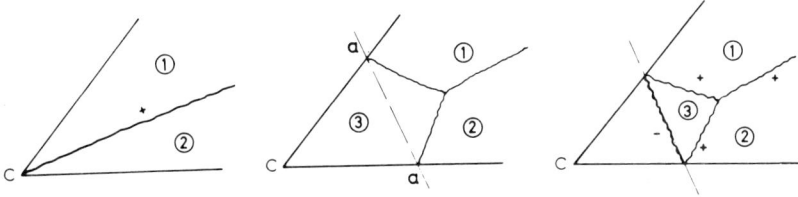

Fig. 7.44.

divides into two new branches passing through the intersections of *aa* and the supporting edges. *If the corner point C is prevented from leaving the support* ("anchored" corners or "bilateral" support), the axis *aa* is replaced by a negative yield line, and the positive initial yield line must branch into two positive yield lines as shown in fig. 7.44. In both cases, the modification of the mechanism results in a decrease of the load P_+ below that given by the original mechanism. The magnitude of the reduction depends on the shape of the plate and on the ratio M_p/M'_p. Detailed discussion of this problem can be found in the book by Wood [7.16]. We shall restrict ourselves to illustrating these considerations by two examples.

7.4.6.2. Square isotropic plate, simply supported and uniformly loaded

Assume bilateral supports. Neglecting corner effects, the collapse mechanism is of the type shown in fig. 7.45. Formula (7.38) furnishes the load $p_+ = 24M_p/l^2$. The corner effect transforms the mechanism of fig. 7.45 into that of fig. 7.46 (a) which contains two parameters a and b. Provided that the center O is given a unit displacement, one has, for part 1, $\theta_y = 0$ and $\theta_x = 2/l$.

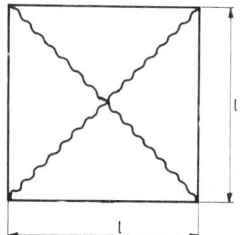

Fig. 7.45.

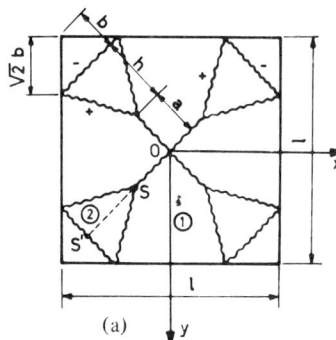

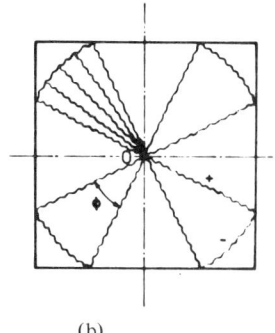

Fig. 7.46.

The triangular part 2 rotates around the negative yield line, whereas the corner itself remains at rest. The vertex S of 2 has the displacement

$$1 - \frac{a/\sqrt{2}}{l/2} = 1 - \frac{\sqrt{2}a}{l},$$

and the height SS' of this triangle has the value $l/\sqrt{2} - (a+b)$. The angle of rotation about the negative yield line thus is

$$\frac{1 - \sqrt{2}a/l}{l/\sqrt{2} - (a+b)}.$$

The work dissipated is

$$D = 4M_p \frac{2}{l}(l - 2b\sqrt{2}) + 4(M_p + M'_p)2b \frac{1 - \sqrt{2}a/l}{l/\sqrt{2} - (a+b)}.$$

Since the corner triangles do not move, the work of the loads is

$$W = \frac{1}{3}pl^2 - \frac{4}{3}ph^2\left(1 - \sqrt{2}\frac{a}{l}\right).$$

Equating W and D, one obtains p as a function of the parameters a and b, which must then be calculated to minimize p. Thus,

$$\frac{\partial p}{\partial a} = 0, \qquad \frac{\partial p}{\partial b} = 0.$$

These nonlinear equations must be solved numerically.

Introducing the values of a and b evaluated in this manner in the expression of p, we obtain, if $M'_p = 0$,

$$P \equiv pl^2 = 22M_p, \tag{7.77}$$

which compares with $P = 24M_p$. If we replace the mechanism of fig. 7.46 (a) by that of fig. 7.46 (b), where the corner effect is obtained by a fan mechanism, we obtain

$$P = 21.7M_p \tag{7.78}$$

for $M'_p = 0$.

7.4.6.3. *Triangular plate*

The case of the equilateral triangular plate (fig. 7.47) is particularly interesting. For a concentrated load at the center, the simple mechanism of fig. 7.47a furnishes

$$P_+ = 6\sqrt{3}M_p = 10.4M_p . \tag{7.79}$$

For $\phi = 30°$, the mechanism with three fans, shown in fig. 7.47b, yields

$$P_+ = 3\left[\frac{\pi}{6}(M_p + M'_p) + M_p \frac{1}{d}2d\right].$$

With $M'_p = M_p$ this furnishes

$$P_+ = (\pi+6)M_p = 9.14M_p . \tag{7.80}$$

The load reduction thus is $[(10.4-9.14)/9.14] \times 100 = 13.8\%$, even if the negative reinforcement is identical to the positive one.

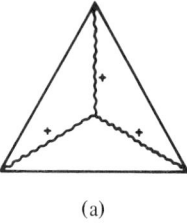

Fig. 7.47.

7.4.6.4. *Intersection of yield lines and the edges*

When the kinematic theorem is used, only considerations of kinematical admissibility are relevant [7.6, 7.19].

Because it derives from considerations of statical admissibility, the condition (given, for example, by Wood in ref. [7.16])

$$\Psi \geqslant \cot^{-1}\sqrt{\frac{M'_p}{M_p}} \tag{7.81}$$

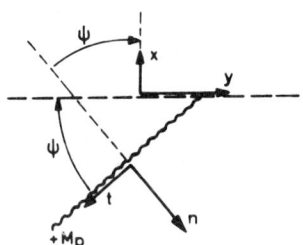

Fig. 7.48.

(see fig. 7.48 for notations) should be disregarded except when one aims at a complete solution (based on the combined theorem). In this case, and for an isotropic plate, when $\Psi \neq \pm \pi/2$, twisting moments must exist along the edge. Because n and t are principal directions, it follows from $M_x = 0$ that

$$M_{xy} = M_n \cot \Psi . \qquad (7.82)$$

It is sometimes useful to replace M_{xy} by two shear forces V as shown in fig. 7.49. Since $M_n = M_p$, we have

$$V = M_p \cot \Psi . \qquad (7.83)$$

Positive values of M_{xy} and V are shown in fig. 7.49.

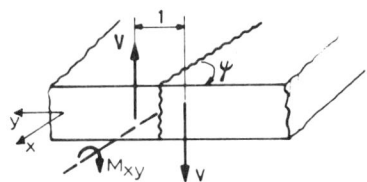

Fig. 7.49.

7.4.7. *Superposition method*

The principle of superposition, extensively used in elasticity, does not hold in plasticity. However, if we want to construct solutions of more complex problems from those of simpler problems, a (restricted) superposition method may be used.

Consider a plate with given geometry and boundary conditions and assume that the exact plastic moments $M_{p1}, M_{p2}, ..., M_{pn}$ are known for n distinct loading conditions symbolically represented by $P_1, P_2, ..., P_n$. The plate is then subjected to the simultaneous action of the various loads each of which is multiplied by a positive scalar factor a_i. The resulting loading condition will be symbolically denoted by

$$P_\Sigma = a_1 P_1 + a_2 P_2 + ... + a_n P_n . \tag{7.84}$$

We shall prove the following theorem: *the plate with plastic moment*

$$M_{P\Sigma} = a_1 M_{p1} + a_2 M_{p2} + ... + a_n M_{pn} , \tag{7.85}$$

has a collapse load $P_{\Sigma l}$ larger than or equal to P_Σ. Suppose first that the mechanisms that have yielded M_{p1}, M_{p2}, ..., M_{pn}, all differ from the collapse mechanism for P_Σ. For this mechanism, denote by $\mathcal{P}'$ the powers of the loads and by D' the rates of dissipation. We have, by definition,

$$\mathcal{P}'_{P\Sigma} = a_1 \mathcal{P}'_{P1} + a_2 \mathcal{P}'_{P2} + ... + a_n \mathcal{P}'_{Pn} . \tag{7.86}$$

For the individual loads $P_1, P_2, ..., P_n$, this mechanism is kinematically admissible. Hence, the plastic moments M_i defined by

$$\mathcal{P}'_{P_i} = D'(M_i) , \qquad i = 1, 2, ..., n , \tag{7.87}$$

are such that

$$M_i \leqslant M_{pi} , \qquad i = 1, 2, ..., n . \tag{7.88}$$

From relation (7.88) we immediately obtain

$$a_1 M_1 + a_2 M_2 + ... + a_n M_n \leqslant a_1 M_{p1} + a_2 M_{p2} + ... + a_n M_{pn} . \tag{7.89}$$

We also know that, for a given mechanism, the total rate of dissipation is proportional to the value of the plastic moment. We thus have

$$a_1 D'(M_1) + a_2 D'(M_2) + ... + a_n D'(M_n) = D'(a_1 M_1 + a_2 M_2 + ... + a_n M_n) . \tag{7.90}$$

The exact plastic moment under P_Σ is M_Σ given by

$$\mathcal{P}'_{P_\Sigma} = D'(M_\Sigma) . \tag{7.91}$$

From eqs. (7.86), (7.87), and (7.90), it follows that

7.4] THE KINEMATIC METHOD

$$M_\Sigma = a_1 M_1 + a_2 M_2 + ... + a_n M_n .\tag{7.92}$$

The relations (7.89), (7.85), and (7.92) thus furnish

$$M_\Sigma \leqslant M_{P_\Sigma} .\tag{7.93}$$

It follows from relation (7.93) that the exact plastic moment for P_Σ is not larger than M_{P_Σ}. Alternatively, the limit load $P_{\Sigma l}$ of the plate with the plastic moment M_{P_Σ} is not smaller than P_Σ, and the theorem is proved.

Suppose now that for the loads $P_1, P_2, ..., P_n$ the plates with the plastic moments $M_{P1}, M_{P2}, ..., M_{Pn}$ have the same collapse mechanism, and denote the powers of the loads for this mechanism by $\mathcal{P}$ and the rate of the dissipation by D. This mechanism is only kinematically admissible for P_Σ. Hence, the work equation

$$\mathcal{P}_{P\Sigma} = D(X) \tag{7.94}$$

furnishes a plastic moment satisfying

$$X \leqslant M_\Sigma .\tag{7.95}$$

On the other hand,

$$\mathcal{P}_{P\Sigma} = a_1 \mathcal{P}_{P1} + a_2 \mathcal{P}_{P2} + ... + a_n \mathcal{P}_{Pn} ,$$

where

$$\mathcal{P}_{Pi} = D(M_{Pi}) , \qquad i = 1, 2, ..., n ,$$

because the mechanism is exact for all P_i. From the two preceding equations we have

$$\mathcal{P}_{P\Sigma} = a_1 D(M_{Pi}) + a_2 D(M_{P2}) + ... + a_n D(M_{Pn}) ,$$

or,

$$\mathcal{P}_{P\Sigma} = D(a_1 M_{P1} + a_2 M_{P2} + ... + a_n M_{Pn}) .\tag{7.96}$$

Comparison of eqs. (7.94) and (7.96) yields

$$X = M_{P\Sigma} \, . \tag{7.97}$$

From eqs. (7.97) and (7.95), we have

$$M_{P\Sigma} \leqslant M_\Sigma \, . \tag{7.98}$$

Comparison of relations (7.98) and (7.93) gives

$$M_\Sigma = M_{P\Sigma} \, . \tag{7.99}$$

We thus have proved the following theorem: *If the collapse mechanisms for the individual loads are identical, the collapse mechanism for the load P_Σ is the same and the plastic moment M_Σ is equal to $M_{P\Sigma}$.*

If we finally consider a given plate with plastic moment M_p, we must have $M_{P\Sigma} = M_p$ or

$$a_1 + a_2 + \ldots + a_n = 1 \, . \tag{7.100}$$

Hence, the load P_Σ defined by eq. (7.84) which satisfies condition (7.100) is smaller than or at most equal to the limit value $P_{\Sigma l}$. It is the exact limit load when all collapse mechanisms for $P_1, P_2, \ldots, P_n$ are identical. The collapse mechanism for P_Σ then coincides with this common mechanism.

An example is provided by the circular isotropic simply supported plate of Section 7.4.5.1. When subjected to a concentrated load P at the center, its plastic moment is

$$M_{p1} = \frac{P}{2\pi} \, . \tag{7.101}$$

When the plate is subjected to a load p uniformly distributed on a central circle with radius a, the plastic moment is

$$M_{p2} = \frac{p\pi a^2 (1 - \tfrac{2}{3} a/R)}{2\pi} \, . \tag{7.102}$$

In both cases, the mechanisms correspond to the transformation of the plate into a cone. If the plate is subjected to $k_1 P + k_2 p$, its exact plastic moment therefore is

$$M_{p(1+2)} = k_1 \frac{P}{2\pi} + \frac{k_2 p\pi a^2}{2\pi} \left(1 - \frac{2}{3}\frac{a}{R}\right) . \tag{7.103}$$

7.4.8. *Interaction between plate and edge beams*

In almost all practical cases the plates are monolithic with edge beams. The corresponding problem of collapse is exceedingly difficult. To simplify it, we first consider the simpler problem of a plate simply supported by edge beams which are themselves supported on columns. The edge conditions influence the collapse mode of the plate in a manner that we shall study for the frequently encountered cases of square and rectangular plates [7.16].

Consider first a square plate resting on edge beams with axes in the middle surface of the plate, the torsional rigidity of the beams being neglected. These beams are in turn supported by columns at the four corners of the plate. If the beams are very strong, their deformations will remain elastic (and therefore negligible in the rigid-plastic scheme up to the collapse of the plate). This collapse will take place according to the mechanism of fig. 7.45, called "diagonal mechanism", under the total load

$$P \equiv pl^2 = 24M_p . \qquad (7.104)$$

However, if the beams are less strong, they may participate in the collapse of the plate, which presents, then, two median yield lines terminating in plastic hinges in the beams.

This mechanism is represented in fig. 7.50 (a). If the center is given a unit displacement, the hinges displacements are $\frac{1}{2}$. The relative rotation at the yield hinges is $2\frac{1}{2}/(l/2) = 2/l$, which is also the relative rotation of two portions adjacent to a yield line. We thus have

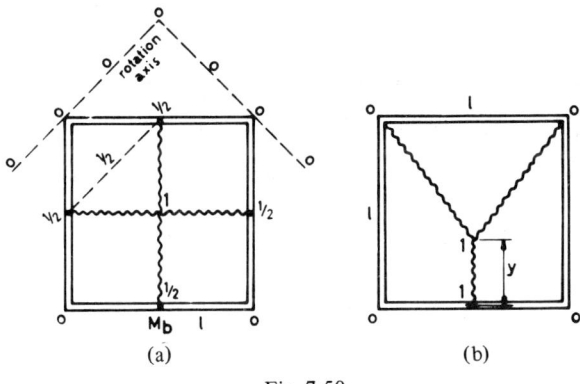

Fig. 7.50.

$$D = (2M_p l + 4M_{pb}) \frac{2}{l} = 4M_p + 8 \frac{M_{pb}}{l},$$

where M_{pb} represents the plastic moment of the edge beams. The work of the loads is $W = pl^2/2$, whence

$$(pl^2)_+ = 8M_p \left(1 + \frac{2M_{pb}}{M_p l}\right). \qquad (7.105)$$

If we introduce

$$\gamma_p = \frac{M_{pb}}{M_p l/2},$$

which will be called the coefficient of plate-beam interaction, eq. (7.105) may be written in the form

$$(pl^2)_+ = 8M_p(1+\gamma_p). \qquad (7.106)$$

Comparing eq. (7.104) with eq. (7.106), we see that, for $\gamma_p = 2$, the "diagonal" collapse mode (plate only) and the "median" mode (plate and beams) result in the same load, under which both of them may occur. All combinations of these two modes are possible, in particular those shown in fig. 7.51, which all yield the same load $(pl^2)_+ = 24M_p$. If $\gamma_p > 2$, then the collapse mode is the diagonal one, and the applicable equation is eq. (7.104). If, on the other hand, $\gamma_p < 2$, the "median" mode is relevant, and the governing equation is eq. (7.106). For $\gamma_p = 0$, one returns to the case of the plate

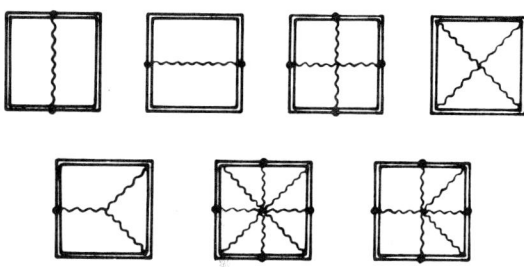

Fig. 7.51.

supported at the four corners. Eq. (7.106) then gives $(pl^2)_+ = 8M_p$. If one of the beams is weaker than the others, the mechanism of fig. 7.50 (b) furnishes

$$\frac{pl^2}{6M_p} = \frac{4 + \frac{1}{1 - y/l} + 2\gamma_p}{2 + y/l}.$$

The value of y/l corresponding to the smallest load is

$$\frac{y}{l} = \frac{10 + 4\gamma_p}{8 + 4\gamma_p}\left[1 - \sqrt{1 - \left(\frac{3 + 2\gamma_p}{4 + 2\gamma_p}\right)\left(\frac{8 + 4\gamma_p}{10 + 4\gamma_p}\right)^2}\right],$$

γ_p referring to the weak beam ($\gamma_p = 0$ corresponds therefore to a free edge; it gives $(pl^2)_+ = 14.15M_p$).

In the case of a rectangular plate, we shall denote by M_{pb} and M_{pB} the plastic moments of the short and long beams, respectively. If all beams are sufficiently strong, the plate will collapse according to the mechanism represented in fig. 7.52. Under a uniformly distributed load, the work of the load is

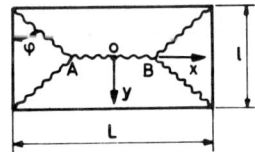

Fig. 7.52.

$W = \frac{1}{3}pl^2 \tan\phi + \frac{1}{2}p(L - l \tan\phi)l$,

if AB takes a unit displacement, and the work dissipated is

$$D = M_p\left[\frac{2l}{\tan\phi/2} + 2L\frac{2}{l}\right] = 4M_p\left(\frac{1}{\tan\phi} + \frac{L}{l}\right).$$

The equation $W = D$ furnishes the expression of p as a function of ϕ. The minimum condition $\partial p/\partial\phi = 0$ finally yields $\tan\phi = \sqrt{3 + (l/L)^2} - l/L$, and hence

$$(plL)_+ = 24M_p\frac{L}{l}\frac{1}{[\sqrt{3 + (l/L)^2} - l/L]^2}. \qquad (7.107)$$

It is worth noting that, for a built-in plate with negative yield lines along the edges, the influence of the additional term in the dissipation is simply to substitute $(M_p + M'_p)$ for M_p in eq. (7.107).

If the long beams are weak in comparison with the short ones, we have the mechanism of fig. 7.53 (a), which gives

$$(plL)_+ = \frac{8}{L}(M_p l + 2M_{pB}) . \qquad (7.107)'$$

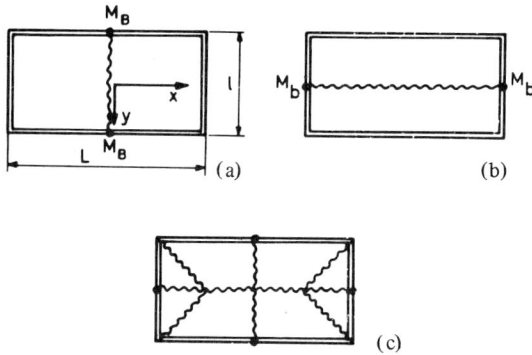

Fig. 7.53.

If the short beams are weak, we have mechanism (b) of fig. 7.53, which yields

$$(plL)_+ = \frac{8}{l}(M_p L + 2M_{pb}) . \qquad (7.107)''$$

To produce one mechanism rather than the two others, the corresponding load $(plL)_+$ must be the smallest of the three, which imposes two conditions to be satisfied by M_{pb} and M_{pB}. Introducing the notations

$$\gamma_B = \frac{M_{pB}}{M_p l/2}, \qquad \gamma_b = \frac{M_{pb}}{M_p L/2}, \qquad (7.108)$$

these conditions are easily obtained, by using eqs. (7.107) to (7.108), in the following form:

1. Combined collapse: plate and long beams [fig. 7.53 (a)] if

$$1 + \gamma_B \leqslant \left(\frac{L}{l}\right)^2 (1+\gamma_b),$$

and   Governing eq. (7.107)'

$$\gamma_B \leqslant \left(\frac{L}{l}\right)^2 - 1 + \frac{2L/l}{\sqrt{(l/L)^2 + 3} - l/L}.$$

2. Combined collapse: plate and short beams [fig. 7.53 (b)] if

$$1 + \gamma_B \geqslant \left(\frac{L}{l}\right)^2 (1+\gamma_b),$$

and   Governing eq. (7.107)''

$$\gamma_b \leqslant \frac{2l/L}{\sqrt{(l/L)^2 + 3} - l/L}.$$

3. Collapse of the plate only (fig. 7.52) if

$$\gamma_B \geqslant \left(\frac{L}{l}\right)^2 - 1 + \frac{2L/l}{\sqrt{(l/L)^2 + 3} - l/L},$$

and   Governing eq. (7.107)

$$\gamma_b \geqslant \frac{2l/L}{\sqrt{(l/L)^2 + 3} - l/L}.$$

If certain of the preceding inequalities are replaced by equalities, mixed collapse modes are obtained such as the one in fig. 7.53 (c). The conditions determining the type of collapse have been represented graphically in fig. 7.54.

7.4.9. *General fan mechanisms*

In Sections 7.4.4 and 7.4.5, the only fan mechanisms used were circular fans. In order to obtain smaller upper bounds, more general fan patterns can be used. Janas has recently discussed this question [7.19]. It must first be pointed out that there is no kinematic objection to using a bounding negative yield line that is not orthogonal to the rays of a polar fan, as done by Mansfield [7.20], Sawczuk [7.21], and Johansen [7.3]; the lowest upper bounds

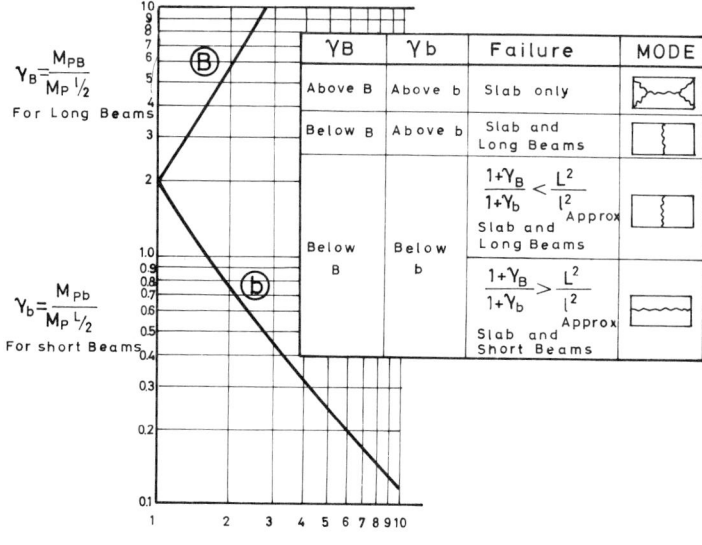

Fig. 7.54.

are obtained from a bounding logarithmic spiral. Numerous examples of applications can be found in the book by Sawczuk and Jaeger [6.16]. Obviously, for isotropic plates there exist no statically admissible moment fields corresponding to such fans, because principal directions of curvature rates and moments coincide and positive and negative lines must therefore intersect orthogonally. The better upper bounds obtained from fans with noncircular boundaires simply prove that the complete solutions do not belong to the considered family of mechanisms with polar fans. Indeed, the actual collapse mechanisms might even not belong to the whole yield line class; they might need completely plastified regions corresponding to the vertices and edges of the yield locus. For this reason, orthogonality of bounding yield line to rays, and in general any condition of statical origin, should be excluded from a purely kinematic approach. Only when a complete solution is desired, should these conditions be allowed to affect the choice of mechanisms.

Improved upper bounds may be obtained from the consideration of more general, nonpolar fans, the straight yield lines of which are tangent to some (curved) envelope (see fig. 7.55, for example). Janas [7.19] has given dissipation formulas for fans of this type, and for their combinations. The best mechanism of the family is obtained by the differentiation process applied to

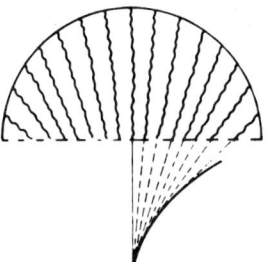

Fig. 7.55.

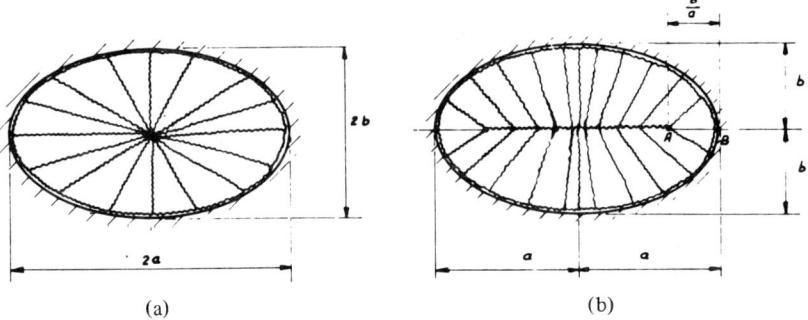

Fig. 7.56.

the choice of the envelope line, or by using the nodal force method (see Section 7.4.10), as done by Nielsen [7.22], who obtained $p_+ = 4.40 M_p/R^2$ for the simply supported uniformly loaded, semicircular slab in fig. 7.55. In the case of the elliptic built-in isotropic plate in fig. 7.56, the best collapse pattern is the simple nonorthogonal polar fan of fig. 7.56 (a) when the plate is subjected to a uniform load. It gives

$$p_+ = 3(M_p + M'_p) \frac{a^2 + b^2}{a^2 b^2}. \tag{7.109}$$

For a uniform line load $\bar{p}$(lb/in.) along the longer axis, the mechanism of fig. 7.56 (b) gives a lower load. This load $\bar{p}_+$ is given in fig. 7.57.

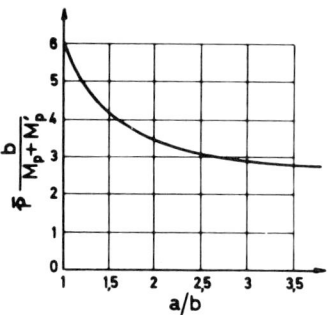

Fig. 7.57.

7.4.10. *Nodal forces*

According to the kinematic theorem, the best mechanism of a given family is the mechanism that corresponds to the smallest load P_+. This smallest value might well not be an analytic minimum as shown by the example of the built-in beam, fig. 7.58. In this case, $\partial P_+/\partial \alpha = 0$ yields $\alpha = \infty$. Smallest P_+ is obtained for $\alpha = l/2$, largest physically admissible value of the parameter α.

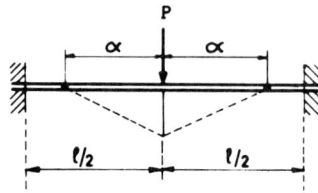

Fig. 7.58a.

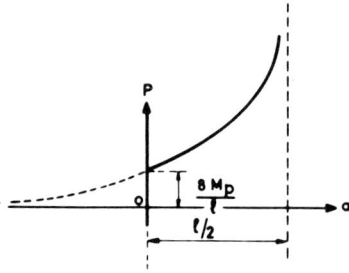

Fig. 7.58b.

In the following, we *assume* that the differentiation with respect to the parameters will effectively furnish the best mechanism. The resulting set of equations, eqs. (7.61), will sometimes be very hard to solve because these equations are nonlinear. Hence, some substitute method might be searched. One method is the so-called "equilibrium method". If we remark that computing the dissipation corresponding to a given mechanism imposes *only* the value of the bending moment in the yield lines, we can imagine several moment fields that are in equilibrium with the loads and exhibit the imposed moments in the yield lines. These fields will in general violate the yield condition, even in the yield lines where only M_n is determinate.

The work equation can be viewed as an equation of virtual work for a field of this kind. The shear forces, twisting and bending moments acting in the yield lines must be such that every rigid part of the collapsing plate is in equilibrium. Torque and shear forces can be reduced to equivalent transversal forces by the well-known formula of Thomson-Tait, eq. (7.119). Now, if these transversal forces are defined by formulas expressing mathematically that M_{p+} (or P_+) is an analytic extremum, the equilibrium of the rigid parts *acted upon by these transverse forces* will force the parameters to assume values that correspond to vanishing derivatives. Note that this method is nothing but a convenient mathematical substitute for equating the derivatives to zero. Equilibrium of rigid parts is assumed to be possible under *imposed* boundary bending moments and *arbitrary* transversal forces, the latter being then determined from stationarity of M_p (or P_+). The approach is essentially kinematical. No statical admissibility is considered here, despite the name of the method. This has been clearly understood only recently (see [7.23–7.28], and [7.6]). All necessary details of the method can be found in [7.25] and [7.29]. We shall restrict ourselves to an illustrative example. Consider a rectangular built-in plate subjected to a uniformly distributed load p (fig. 7.59). As indicated in fig. 7.59, lower reinforcement is orthotropic, and negative restraining moments along the edge may have different values $-C_1 M_p, -C_2 M_p, -C_3 M_p, -C_4 M_p$.

We consider the three-parameter family of mechanisms (x_1, x_2, y_1) of fig. 7.59. The general formulas [7.25, 7.29] show that all nodal forces K_1 to K_6 vanish in this particular case.

Rotational equilibrium of the part 1 about the edge gives

$$M_p L(1+C_1) = \frac{p x_1 y_2^2}{6} + \frac{p x_2 y_2^2}{6} + \frac{p}{2}(L-x_1-x_2) y_2^2 \tag{a}$$

and rotational equilibrium of 2 about its edge gives

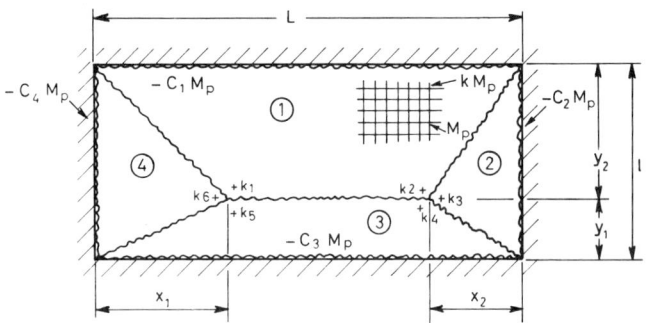

Fig. 7.59. (Reprinted by permission of Thames and Hudson Ltd. from R.H.Wood, *Plastic and Elastic Design of Slabs and Plates*, London, 1961.)

$$M_p l(1+C_2) = \frac{plx_2^2}{6}. \tag{b}$$

Similarly,

$$M_p L(1+C_3) = \frac{px_1 y_1^2}{6} + p\frac{x_2 y_1^2}{6} + \frac{p}{2}(L-x_1-x_2)y_1^2, \tag{c}$$

$$M_p l(1+C_4) = p\frac{lx_1^2}{6}. \tag{d}$$

We also have (fig. 7.59)

$$y_1 + y_2 = l. \tag{e}$$

We thus have five equations with the five unknowns: x_1, x_2, y_1, y_2, and M_p. Let $A = \sqrt{1+C_2} + \sqrt{1+C_4}$, $B = \sqrt{k+C_1} + \sqrt{k+C_3}$, and solve the system (a) to (e). We obtain [7.16]

$$M_p = \frac{p}{24}\left(\frac{2l}{B}\right)^2 \left[\sqrt{3+\left(\frac{2l/B}{2L/A}\right)^2} - \frac{2l/B}{2L/A}\right]^2. \tag{f}$$

It is easily seen that, in the case of the isotropic rectangular plate where $k = 1$ and $C_1 = C_2 = C_3 = C_4 = 1$, relation (f) reduces to eq. (7.107) where $2M_p$ is substituted for M_p.

THE STATIC METHOD

Space limitations prevent us from further developing the nodal force theory. The reader is referred to [7.25, 7.28, and 7.29].

7.5. The static method

7.5.1. Introduction

The static method consists of finding a statically admissible moment field covering the entire plate. If this field corresponds to a mechanism in accordance with the flow rule, the combined theorem shows that the load is the exact limit load; otherwise, it is a lower bound P_- for the limit load.

It should be noted that the equilibrium of the rigid parts of a plate mechanism (with no violation of the yield condition in the yield lines) is a necessary but not a sufficient condition for the existence of a statically admissible field. Equivalent transverse forces in the yield lines should be determined from statical admissibility. Verification of the equilibrium of rigid parts, however, is, in general, not a useful step towards obtaining a statically admissible moment field. On the other hand, use of the so-determined transverse forces (nodal forces) in the kinematic method could result in an unnecessarily high upper bound, without at all insuring the existence of a corresponding statically admissible moment field. Consequently nodal forces should not be used in a statical approach.

7.5.2. Circular plate, simple support

We first note that, when both principal moments have same sign, Tresca's yield condition (fig. 7.60) is identical to Johansen's yield condition [eqs.

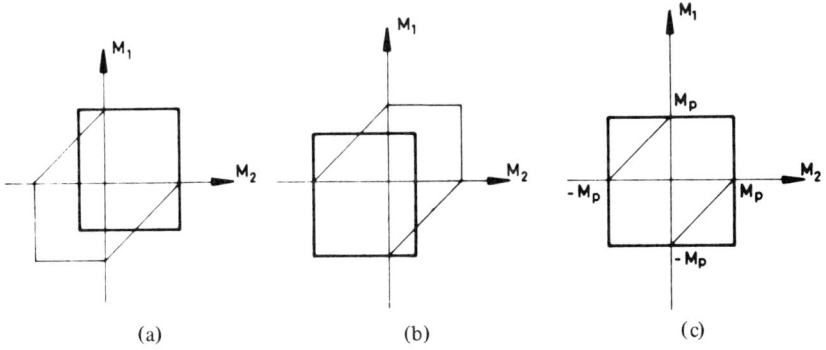

Fig. 7.60.

266   REINFORCED CONCRETE PLATES   [Ch. 7

(7.8), (7.9)] for $M_{xy} = 0$. This remark applies to simply supported circular plates symmetrically loaded, for which the following exact solutions are known:

1. "Circular loading" of total magnitude P on a central circle with radius a concentric to the circular edge. See Section 6.3.2, relations (6.11) to (6.14). The limit concentrated central load is given by eq. (6.16).
2. Annual loading: see Section 6.3.3, fig. 6.15;
3. Line load: see relation (6.23);
4. Annular plate subjected to uniform load: see curve a in fig. 6.17;
5. Orthotropic plates: see Section 6.4.2, where all curves in figs. 6.30, 6.31, 6.33, 6.35, and 6.37 are applicable.

Note that all preceding solutions are complete, consisting of both a statically admissible moment field and a corresponding mechanism.

7.5.3. *Built-in isotropic circular plate, circular loading*

Due to plane polar symmetry, the yield condition in fig. 7.61 can be used. On the basis of the existing solution for the simply supported plate, we consider the moment field.

$$\left.\begin{array}{l} M_\theta = M_p \\ M_r = M_p - (M_p + M'_p) \dfrac{r^2}{a^2} \cdot \dfrac{1}{3 - 2a/R} \end{array}\right\} 0 \leqslant r \leqslant a,$$

$$\left.\begin{array}{l} M_\theta = M_p \\ M_r = M_p - (M_p + M'_p) \dfrac{3 - 2a/r}{3 - 2a/R} \end{array}\right\} a \leqslant r \leqslant R, \qquad (7.110)$$

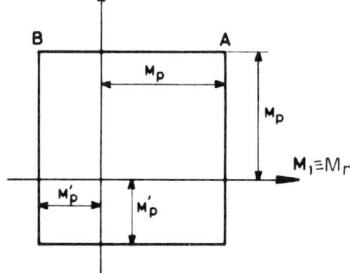

Fig. 7.61.

where a is the radius of the central loaded circular area. The plastic regime is AB, M_θ being kept constant and equal to M_p, whereas M_r varies in a parabolic manner from M_p at the center to $-M'_p$ at the edge. Substitution of relations (7.110) into the fundamental equilibrium eq. (6.103) gives

$$p_- = \frac{2(M_p + M'_p)}{a^2(1 - \frac{2}{3}a/R)}. \tag{7.111}$$

With the notation $P = p\pi a^2$, formulas (7.111) and (7.71) give identical values. Hence, we have obtained the exact limit load

$$P_l = p_l \pi a^2 = \frac{2\pi(M_p + M'_p)}{1 - \frac{2}{3}a/R}. \tag{7.112}$$

Note that the fields (7.110) and the field (6.12), (6.13) can both be given by

$$\left. \begin{array}{l} M_\theta = M_p \\ M_r = M_p - \dfrac{pr^2}{6} \end{array} \right\} 0 \leqslant r \leqslant a,$$

$$\left. \begin{array}{l} M_\theta = M_p \\ M_r = M_p - \dfrac{pa^2}{2} + \dfrac{pa^3}{3r} \end{array} \right\} a \leqslant r \leqslant R, \tag{7.113}$$

where the expressions (7.111) or (6.11) for p must be used for the built-in plate and the simply supported plate, respectively. From eq. (7.112) we obtain the exact limit value of the central concentrated load

$$P_l = 2\pi(M_p + M'_p). \tag{7.114}$$

7.5.4. Simply supported isotropic square plate uniformly loaded

Assume equal plastic moment for both positive and negative bending (square yield condition, fig. 7.62). In the polar coordinate system of fig. 7.63, with origin at the center of the plate, consider the field [7.5]

$$\begin{aligned} M_\theta &= M_p, \\ M_r &= M_p \left(1 - \frac{r^2}{R^2}\right), \\ M_{r\theta} &= 0, \end{aligned} \tag{7.115}$$

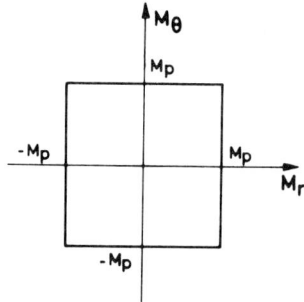

Fig. 7.62.

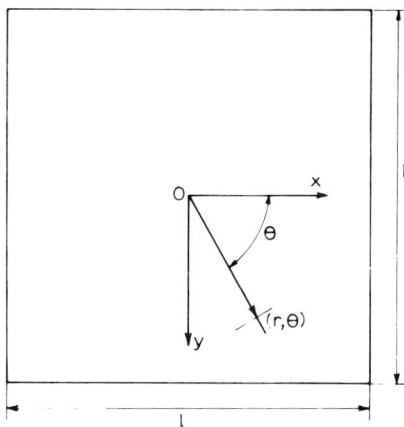

Fig. 7.63.

obtained by replacing a by R in eq. (6.12). The radial and circumferential directions are thus assumed to be principal directions. We then change our coordinate system from r, θ to x, y with the formulas

$$r^2 = x^2 + y^2, \quad \cos\theta = \frac{x}{r}, \quad \sin\theta = \frac{-y}{r}, \quad R = \frac{l}{2},$$

and use the relations (1.10) and (1.11) where M_r, M_θ, M_x, M_{xy}, θ are substituted for σ_1, σ_2, σ, $-\tau$, α, respectively. We transform the field (7.115) into

$$M_x = M_p \left(1 - \frac{4x^2}{l^2}\right),$$

$$M_y = M_p \left(1 - \frac{4y^2}{l^2}\right),$$

$$M_{xy} = 4M_p \frac{xy}{l^2}. \tag{7.116}$$

As is readily verified, the yield condition (7.8) is satisfied at every point of the square plate. The boundary conditions also are obviously satisfied.

Substitution of expressions (7.116) into the equilibrium eq. (6.37) (where $p = c^t$) yields the lower bound

$$p_- = \frac{24M_p}{l^2}. \tag{7.117}$$

Comparing with the upper bound, eq. (7.38), we find the exact limit load as

$$p_l = \frac{24M_p}{l^2}. \tag{7.118}$$

The edge reaction (per unit of length), statically equivalent to the shear force and the torque, is given by [7.16]

$$Q_x = \frac{\partial M_x}{\partial x} - 2\frac{\partial M_{xy}}{\partial y}. \tag{7.119}$$

Eq. (7.119) applied at $x = \pm l/2$ furnishes $Q_x = pl/3$, which turns out to be constant. Each corner of the plate is subjected to a concentrated vertical downward force:

$$R = \frac{1}{4}\left|pl^2 - \frac{4}{3}pl^2\right| = \frac{pl^2}{12}.$$

Assume the plate to be simply supported on edge beams with the constant plastic moment M_{pb}, which are in turn simply supported on corner columns. The maximum bending moment in one of these beams is then $M_{max} = pl^3/24$. The beam will remain rigid as long as

$$\gamma_p \equiv \frac{M_{pb}}{M_p(l/2)} > \frac{M_{max}}{(pl^2/24)(l/2)} = 2 , \qquad (7.120)$$

a result already found in Section 7.4.8. For $\gamma_p > 2$, though the plate is completely plastic it collapses according to diagonal mechanism of fig. 7.45.

For $\gamma_p < 2$, we generalize eqs. (7.116) by simply modifying the last equation into

$$M_{xy} = 4M_p \frac{xy}{l^2} (\gamma_p - 1) . \qquad (7.121)$$

It is readily verified that the obtained moment field is statically admissible for all γ such that $0 < \gamma < 2$, and that the plate is at yield along its median lines ($x=y=0$). Eq. (7.119) gives the edge reaction

$$Q_x = \frac{4M_p}{l} \cdot \gamma_p ,$$

and the supporting beams are subjected to a maximum bending moment

$$M_{max} = Q_x \frac{l^2}{8} = M_{pb} .$$

The beams are thus at collapse together with the plate, as shown in fig. 7.50.

We once again have a complete solution, and both equilibrium eq. (6.37) and work equation furnish the exact limit load, namely

$$p_l = \frac{8M_p}{l^2} (1+\gamma_p) . \qquad (7.122)$$

Note that, for $\gamma_p = 0$ (vanishing beams, plate supported at the corners), the plate is plastic at every point as well as for $\gamma_p = 2$ (plate on rigid supports).

7.5.5. *Simply supported isotropic rectangular plate uniformly loaded*

As in Section 7.4.8, denote by l and L the lengths of the short and long edges, respectively. We generalize the field, eqs. (7.116), into

$$M_x = M_p \left(1 - \frac{4x^2}{L^2}\right),$$

$$M_y = M_p \left(1 - \frac{4y^2}{l^2}\right),$$

$$M_{xy} = \frac{4M_p xy}{lL}. \qquad (7.123)$$

Substitution of expressions (7.123) for M_x, M_y, M_{xy} into equilibrium eq. (6.37) furnishes

$$p_- = \frac{8M_p}{l^2}\left(1 + \frac{l}{L} + \frac{l^2}{L^2}\right). \qquad (7.124)$$

Though the field (7.123) satisfies the yield condition (7.8) at every point of the plate, there is no corresponding mechanism for rigid support conditions because the trajectories of principal moments are curved lines (except the median lines). Nevertheless, the upper bound, eq. (7.107), does not differ by more than about 1.5% from the lower bound, eq. (7.124), just obtained.

Assume now that the plate is simply supported on edge beams with plastic moment M_{pb} and M_{pB} for the short and long beam, respectively. Substitution of expressions (7.123) for M_x and M_y in eq. (6.37), integration, and use of symmetry, give

$$M_{xy} = p\frac{xy}{2} - 4M_p xy \left(\frac{1}{L^2} + \frac{1}{l^2}\right). \qquad (7.125)$$

Reaction on the long beam is [eq. (7.119)]

$$|Q_{l/2}| = p\frac{l}{2} - 4M_p \frac{l}{L^2},$$

and the load corresponding to the collapse of the long beams is determined from condition

$$M_{\max} = |Q_{l/2}| \frac{L^2}{8} = M_{pB}.$$

We obtain

$$p_-(Ll) = \frac{8}{L}(M_p l + 2M_{pB}) . \tag{7.126}$$

Similarly, the load corresponding to collapse of the short beams is

$$p_-(lL) = \frac{8}{l}(M_p L + 2M_{pb}) . \tag{7.127}$$

With definitions (7.108) of symbols γ_b and γ_B, eqs. (7.126) and (7.127), respectively become

$$p_- = \frac{8M_p}{L^2}(1+\gamma_B) , \tag{7.128}$$

$$p_- = \frac{8M_p}{l^2}(1+\gamma_b) . \tag{7.129}$$

If

$$\frac{1+\gamma_B}{1+\gamma_b} < \frac{L^2}{l^2} , \tag{7.130}$$

the load (7.128) is reached before the load (7.129). Long beams are at collapse whereas short beams are still rigid.

Introducing expressions (7.123) of M_x and M_y and expression (7.125) of M_{xy} in the yield condition (7.8), and using eq. (7.128), it is easily seen that the moment field in the plate is statically admissible as long as

$$\gamma_B < \frac{L}{l} + \frac{L^2}{l^2} \tag{7.131}$$

(to compare with a similar condition in the kinematical approach of Section 7.4.8).

When both conditions (7.130) and (7.131) are satisfied, expression (7.128) is the exact limit load because it is identical with the upper bound, eq. (7.107)'.

Similarly, eq. (7.129) is the exact limit load (with collapsing plate and short beams) when

$$\frac{1+\gamma_B}{1+\gamma_b} > \left(\frac{L}{l}\right)^2 \tag{7.132}$$

and

$$\gamma_b < \frac{l}{L} + \frac{l^2}{L^2}. \tag{7.133}$$

When equality signs must be used in relations (7.130) to (7.133), combined mechanisms occur.

7.5.6. *Continuous isotropic rectangular plates subjected to uniform load*

We consider an isotropic rectangular plate continuous over many transversal simple supports (fig. 7.64) [7.30]. A typical part of this continuous plate, spanning between two adjacent supports apart from the length L, is to be regarded as a plate built-in across the supports, free along the two other edges and subjected to uniform load p.

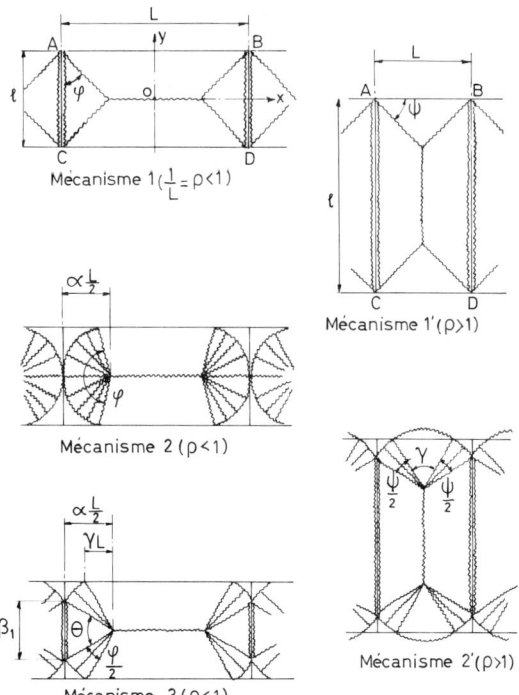

Fig. 7.64.

In order to check the quality of our static solution, we first establish upper bounds as developed in Section 7.4. The assumed mechanisms are shown in fig. 7.64. Upon optimization they give the upper bounds in table 7.1. The plastic moment is M_p for both positive and negative bending.

Table 7.1. Minimum values of pl^2/M_p for various values of ρ

$\rho = \dfrac{l}{L}$	Minimum values of pl^2/M_p (upper-bound solution)	Relevant mode
$\to \infty$	$16p^2$	
8	1.125.75	
7	873.63	
6	653.52	$2'$
5	465.42	
4	309.36	
3	185.37	
2	93.56	
1.5	59.90	
1	34.09	
2/3	21.95	3
1/2	17.02	
1/3	13.27	
1/4	11.67	
1/5	10.82	
1/6	10.28	
1/7	9.92	
1/8	9.65	2
1/12	9.06	
1/16	8.78	
1/20	8.62	
1/24	8.51	
1/36	8.34	
1/48	8.25	

The static solution is constructed by taking polynomials of the second, sixth, and sixth degree for M_y, M_x, and M_{xy}, respectively. After boundary and equilibrium conditions are satisfied, the remaining three parameters are chosen so as not to violate the yield condition (7.8) and (7.9) while maximizing p. This is achieved, by trial and error, for the square plate, where

$\rho \equiv l/L = 1$. The resulting values of the parameters turn out to give good lower bounds for other ratios ρ. The final solution is

$$M_x = M_p \left[1 + \frac{1}{l^2} \left(4 - \frac{pl^2}{2M_p} + \gamma \rho \right) \frac{x^2}{L^2} + \frac{7\gamma x^4}{\rho L^4} - 24 \frac{\gamma x^6}{\rho L^6} \right],$$

$$M_y = M_p \left(1 - \frac{4y^2}{l^2} \right),$$

$$M_{xy} = \gamma M_p \frac{xy}{Ll} \left(1 - \frac{4x^2}{L^2} \right) \left(1 + \frac{18x^2}{L^2} \right), \qquad (7.134)$$

Table 7.2. Upper- and lower-bound solutions for various values of ρ

$\rho = \dfrac{l}{L}$	Values of pl^2/M_p		$100 \left(\dfrac{\text{UB}-\text{LB}}{\text{UB}} \right)$
	Upper bound	Lower bound	
$\to \infty$	$16\rho^2$	$16\rho^2$	0
8	1.125.0	1.093	2.8
7	873.60	845.00	3.3
6	653.50	630.00	3.6
5	465.40	446.00	4.2
4	309.40	289.00	6.6
3	185.40	175.00	5.6
2	93.60	87.80	6.2
1.5	59.90	56.00	6.5
1	34.09	32.10	5.8
2/3	21.95	20.40	7.1
1/2	17.02	15.88	6.6
1/3	13.27	12.23	7.8
1/4	11.67	10.74	8.0
1/5	10.82	9.96	7.9
1/6	10.28	9.49	7.7
1/7	9.92	9.18	7.5
1/8	9.65	8.96	7.2
1/12	9.06	8.50	6.2
1/16	8.78	8.31	5.4
1/20	8.62	8.21	4.8
1/24	8.51	8.15	4.2
1/36	8.34	8.07	3.2
1/48	8.25	8.04	2.5
$1/\infty$	8.00	8.00	0

where

$$\gamma = \frac{pl^2/M_p - 16\rho^2 - 8}{2.5\rho}. \tag{7.135}$$

The corresponding values of pl^2/M_p are given in table 7.2 for various values of ρ, together with the upper bound for the same ρ and the difference between the bounds.

The regions where negative reinforcement is needed can be determined from fig. 7.65 in which the locus where the smallest principal moment changes sign is drawn for various values of ρ.

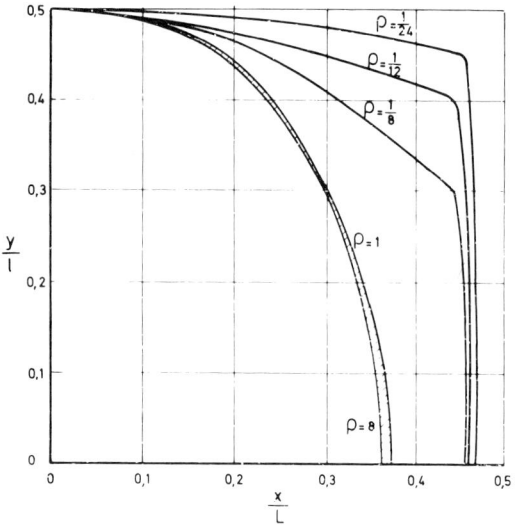

Fig. 7.65.

7.5.7. *Isotropic rectangular built-in plate*

Even for a square plate with uniformly distributed load, this apparently very simple problem turns out to be extremely difficult. No complete solution is known so far, and presently available good lower bounds were obtained by numerical procedures [7.12,7.31]. Recently, Şobotka [7.32] has attempted to give a statically admissible moment field with corresponding loads quite close to the upper-bound values given by eq. (7.107) modified by substituting $(M_p+M'_p)$ for M_p. Unfortunately, his moment fields cannot be accepted because they contain inadmissible discontinuities in shear force.

7.5.8. Orthotropic plates: affinity method

The affinity method of Section 7.4.3 will be extended to the statical approach to limit analysis [7.8]. We thus assume that the same orthotropy coefficient k applies to ultimate strength in positive and negative bending. If this situation did not occur, the complete statical procedure should be applied, as was discussed for metal plates in Chapter 6.

If a field M_x, M_y, M_{xy} does not violate the yield condition of an isotropic plate with plastic moments M_p and M'_p, it is readily verified that the field M_x, kM_y, and $\sqrt{k}M_{xy}$ does not violate the yield condition (7.8) and (7.9) of the plate with plastic moments M_p, M'_p, and orthotropy coefficient k. This is a direct consequence of the very nature of the yield condition (7.8) and (7.9). Now, if so changing the moment field with the orthotropy avoids violating the yield condition, this change does not in general maintain equilibrium if loads are unaltered. To maintain equilibrium, the functions M_x, kM_y and $\sqrt{k}M_{xy}$ should be used in a plate obtained from the original one by changing its dimensions by the affine transformation $x^* = x$, $y^* = \sqrt{k}y$, where stars refer to the new plate. We then successively have

$$\frac{\partial^2 M_x}{\partial x^{*2}} = \frac{\partial^2 M_x}{\partial x^2}, \quad \frac{\partial kM_y}{\partial y^*} = k\frac{\partial M_y}{\partial y} \cdot \frac{\partial y}{\partial y^*} = \sqrt{k}\frac{\partial M_y}{\partial y},$$

$$\frac{\partial kM_y}{\partial y^{*2}} = \sqrt{k}\frac{\partial^2 M_y}{\partial y^2} \cdot \frac{\partial y}{\partial y^*}, \quad \frac{\partial^2 \sqrt{k}M_{xy}}{\partial x^* \partial y^*} = \sqrt{k}\frac{\partial}{\partial x}\left(\frac{\partial M_{xy}}{\partial y} \cdot \frac{\partial y}{\partial y^*}\right) = \frac{\partial^2 M_{xy}}{\partial x \partial y}.$$

Because the equilibrium equation

$$\frac{\partial^2 M_x}{\partial y^2} + \frac{\partial^2 M_y}{\partial y^2} - 2\frac{\partial^2 M_{xy}}{\partial x \partial y} = -p$$

is satisfied in the initial plate, the equilibrium equation

$$\frac{\partial^2 M_x}{\partial x^{*2}} + \frac{\partial^2 kM_y}{\partial y^{*2}} - 2\frac{\partial^2 \sqrt{k}M_{xy}}{\partial x^* \partial y^*} = -p$$

will be satisfied in the affine plate with the same load per unit area. Consequently, to apply a lower-bound solution for an isotropic plate to an orthotropic but otherwise identical plate the following steps are sufficient:

1. Multiply the functions M_y and M_{xy} by k and $\sqrt{k}$, respectively, M_x being unaltered;

2. Multiply by $1/\sqrt{k}$ all dimensions in the direction of the y-axis, and use the lower-bound value of the load parameter for the resulting isotropic plate.

Consider for example a rectangular, simply supported plate subjected to uniformly distributed load. We immediately transform the field (7.123) of the isotropic plate into

$$M_x = M_p \left(1 - \frac{4x^2}{L^2}\right),$$

$$M_y = kM_p \left(1 - \frac{4y^2}{l^2}\right),$$

$$M_{xy} = \sqrt{k} \, \frac{4M_p xy}{Ll}, \qquad (7.136)$$

and the limit load (7.124) into

$$p_- = \frac{8M_p k}{l^2} \left(1 + \frac{l}{\sqrt{k}L} + \frac{l^2}{kL^2}\right). \qquad (7.137)$$

The upper bound, eq. (7.107), can be transformed in a similar way to give

$$p_+ = \frac{24M_p}{l^2} \cdot k \cdot \frac{1}{\left[\sqrt{3 + \frac{l^2}{kL^2}} - \frac{l^2}{\sqrt{k}L}\right]^2}. \qquad (7.138)$$

The two bounds just obtained do not differ by more than 1.5% [7.33, 7.34].

Obviously, the affinity method is not restricted to rectangular plates.

7.5.9. *Other lower bounds and complete solutions*

Upper-bound solutions, based solely on a yield line pattern, have two main disadvantages:

1. They give too large a value of the limit load and, hence, are unsafe; this fact is counterbalanced to some extent by the neglected work hardening and the effects of changes of geometry;
2. Except along the yield lines, they do not give any information on the required reinforcement; when nonuniform reinforcement is desirable, this drawback is very serious and empirical formulas must be used.

On the other hand, lower-bound solutions (and, obviously, complete solutions) are safe and give all necessary indications concerning economic reinforcement. They are therefore extremely useful but, unfortunately, statically admissible moment fields are much more difficult to obtain than collapse mechanisms. Still more difficult to find are complete solutions, giving the exact limit load.

Presently known complete solutions are few in number. Most of them are found in the books by Sawczuk and Jaeger [6.16] and by Nielsen [7.8].

Nielsen [7.8] also points out an analogy between some moment fields and slip line fields in plane plastic strain. This analogy furnishes some additional complete solutions, but only for very special problems.

As Prager has shown [6.31], the limit analysis of a framed structure can be formulated as a linear programming problem. Wolfensberger [7.12] has recently applied this remark to find statically admissible solutions in reinforced concrete plates. This method implies suitable linearization of the yield condition and piecewise linearization of the moment field, and requires the use of an electronic computer. For more details see ref. [7.12]. Ceradini [7.35], Gavarini [7.36], and Sacchi [7.37] have worked along similar lines.

Massonnet has given recently [7.38] some new complete solutions based on combination of radial fields corresponding to plastic regime AB (fig. 7.61).

Following Nielsen [7.8], we give a summary of most (if not all) complete solutions known so far in fig. 7.66 (a) to (t). Moment fields not given in the present book should be found in the references given with the figures. Further lower-bound solutions for rectangular plates with various edge conditions can be found in [7.8].

7.5.10. *Remarks on uniqueness of solution*

As was seen in Sections 7.4.8, one can obviously find different collapse mechanisms corresponding to the same load. When this load is the exact limit load, the moment fields corresponding to the various mechanisms may possibly differ only in the common rigid regions (see Section 4.4).

For example, if, for the simply supported, uniformly loaded square plate, we want to find a statically admissible stress field different from eqs. (7.116), we must cause the principal positive moment acting on the diagonals to reach the value M_p. Such a field was specified by Vallance [7.16]. It gives non-uniform edge reactions. If the plate is supported by edge beams, these reactions cause a maximum moment of $pl^3/21$, instead of $pl^3/24$ with the field (7.116).

If the plate is to be *uniformly* reinforced for $M_p = pl^2/24$, the edge beams can be designed for the smaller maximum moment (here $pl^3/24$). According to the static theorem, the structure (plate+beams) will support the load.

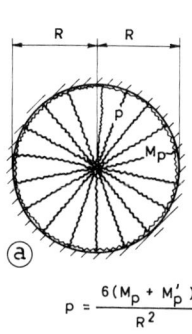
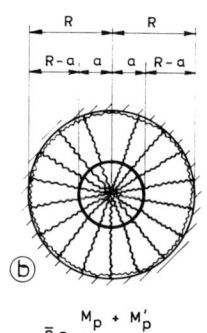

(a) $p = \dfrac{6(M_p + M'_p)}{R^2}$

(b) $\bar{p} = \dfrac{M_p + M'_p}{a\left(1 - \dfrac{a}{R}\right)}$

$2\pi a \bar{p} = P \rightarrow 2\pi(M_p + M'_p)$
pour $a \rightarrow 0$

ref. [7·3]

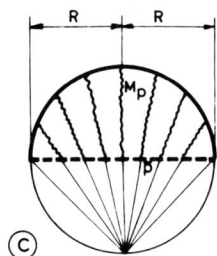
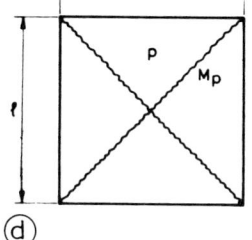

(c) $\bar{p} = \dfrac{2 M_p}{R}\ (M_p = M'_p)$

ref. [7·3]

(d) $p = \dfrac{24 M_p}{\ell^2}\ (M_p = M'_p)$

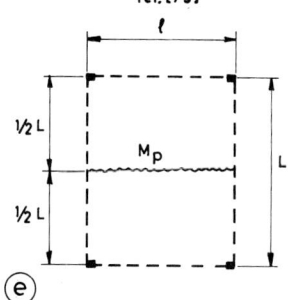
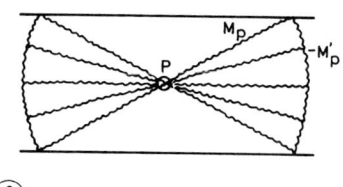

(e) $p = \dfrac{8 M_p}{L^2}\ \left(M'_p = \dfrac{\ell}{L} M_p,\ L \geqq \ell\right)$

ref. [7·16]

(f) $P = 4\left(\sqrt{M_p M'_p}\right) + (M_p + M'_p)\mathrm{Arctg}\sqrt{\dfrac{M_p}{M'_p}}$

ref. [7·16]
(solution non complète)

7.5] THE STATIC METHOD

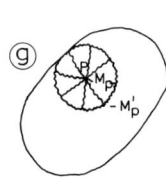

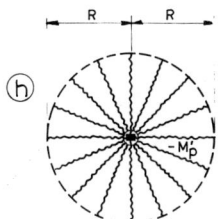

$P = 2\pi(M_p + M'_p)$
ref. [6·16]

$P = \dfrac{3 M'_p}{R^2} (M_p = \dfrac{1}{2} M'_p)$
ref. [7·8]

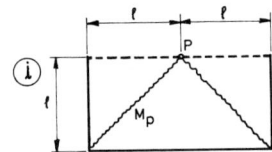

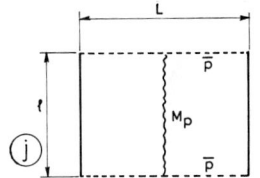

$P = 4 M_p (M_p = M'_p)$
ref. [7·8]

$\bar{p} = \dfrac{4 M_p \ell}{L^2} (M'_p = \dfrac{\ell}{L} M_p , L \geqq \ell)$
ref. [7·8]

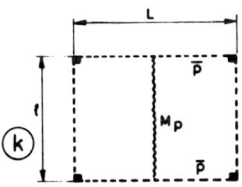

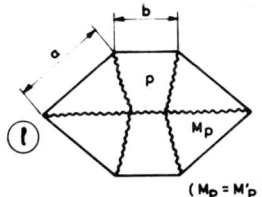

$\bar{p} = \dfrac{4 M_p \ell}{L^2} (M'_p = \dfrac{\ell}{L} M_p , L \geqq \ell)$
ref. [7·8]

$p = \dfrac{6 M_p}{a^2} \dfrac{1}{\left[\sqrt{2(\frac{a}{b})^2 + \sqrt{2}\frac{a}{b} + \frac{3}{2}} - \sqrt{2}\frac{a}{b}\right]^2}$
ref. [7·8]

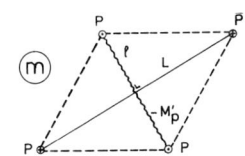

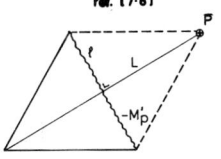

$P = \dfrac{2 M_p \ell}{L} \cdot (M'_p = M_p , L \geqq \ell)$
ref. [7·8]

$P = \dfrac{2 M_p \ell}{L} \cdot (M'_p = M_p , L \geqq \ell)$
ref. [7·8]

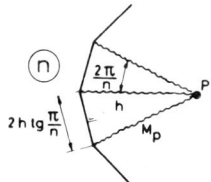

$P = 2nM_p \, tg\frac{\pi}{n} \, (M_p = M'_p \cot\frac{\pi}{n})$

ref. [7-8]
[7-38]

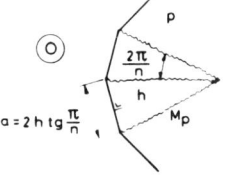

$p = \frac{6M_p}{h^2} = \frac{24 M_p \, tg^2\frac{\pi}{n}}{a^2}$

$(M_p = M'_p \cot^2\frac{\pi}{n})$

ref. [7-8]
[7-38]

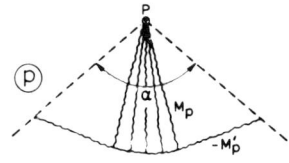

$P = 2\sqrt{M_p \, M'_p} + (M_p + M'_p)(\alpha - 2\,\text{Arctg}\sqrt{\frac{M_p}{M'_p}})$

$(tg^2\frac{\alpha}{2} \leq \frac{M_p}{M'_p})$

ref. [7-8]

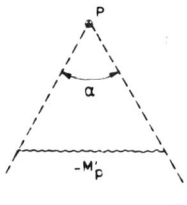

$P = 2 M'_p \, tg\frac{\alpha}{2}$

$(tg^2\frac{\alpha}{2} \leq \frac{M_p}{M'_p})$

ref. [7-8]

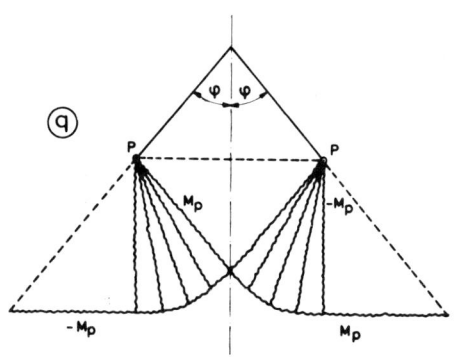

$P = 2 M_p (1 + \varphi)$

$(M_p = M'_p)$

ref. [7-8]

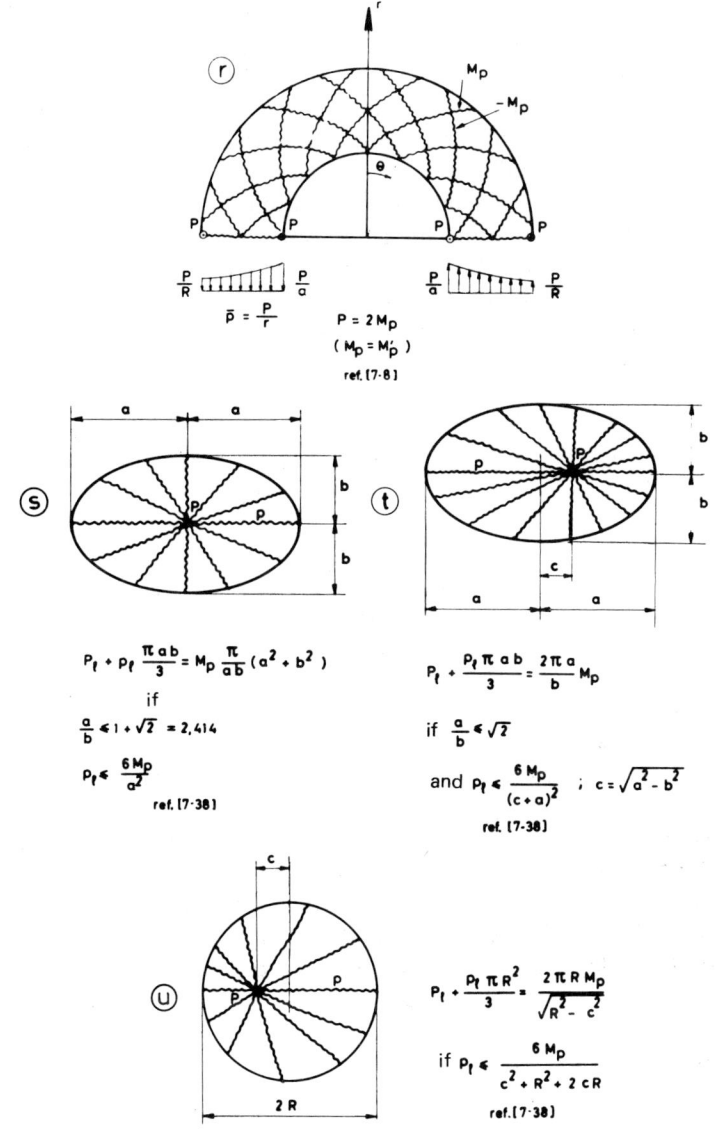

Fig. 7.66.

If the reinforcement is adjusted on the moment field, the beams must be designed with the reactions obtained from *this very field* for the yield condition to be nowhere violated.

In the case of statical boundary conditions, the limit load for a given loading is also well defined but the collapse mechanism strongly depends on the kinematic boundary conditions. Consider once again the square plate of the preceding example. If the supports are rigid, only the diagonal mechanism can form. But if the supporting beams are designed to become fully plastic at the same time as the plate, the mechanism can be the cone of fig. 7.67, (corresponding to a purely radial moment field). This mechanism is only possible when the supports are free to settle. Nevertheless, it furnishes the same load as the diagonal mechanism. This is a general property.

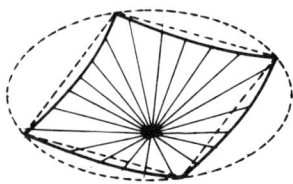

Fig. 7.67.

Indeed, let P_{l_1} and P_{l_2} be the limit loads of two structures differing solely by the kinematic boundary conditions. The stress field of one structure is statically admissible for the other structure. Hence, we simultaneously have $P_{l_1} \leqslant P_{l_2}$ and $P_{l_2} \leqslant P_{l_1}$. Hence, $P_{l_1} = P_{l_2}$.

When different statically admissible stress fields exist for a given structure subjected to given loads, they can be linearily combined to furnish new stress fields. The latter fields are statically admissible for a load equal to the original load multiplied by the sum of the coefficients of the linear combination, provided the yield condition has not been violated [7.38].

7.6. Minimum weight of reinforcement

7.6.1. *Ways of reducing reinforcement*
7.6.1.1. *Circular plates*

Consider, for example, a circular, built-in plate with radius R, that is uniformly loaded. Assume first it is isotropically reinforced, with $M_p \neq M'_p$. Its limit load is

$$p_l = \frac{6(M_p+M'_p)}{R^2}. \tag{7.139}$$

The reinforcement being isotropic, we may place it in any two orthogonal directions. We have

$$M_p = A\sigma_Y h, \qquad M'_p = A'\sigma_Y h, \tag{7.140}$$

where A and A' are the cross-sectional areas per unit of length of upper and lower reinforcement, respectively (in one of the two considered orthogonal directions), σ_Y is the yield stress of the steel, and h the lever arm of internal forces. In a reinforced concrete plate with constant thickness, this arm may be regarded as constant [7.16].

The total volume of a homogeneous (uniform) reinforcement of this kind is

$$V = \pi R^2 (2A+2A'). \tag{7.141}$$

Substitution of the values (7.140) for A and A' in eq. (7.141) yields

$$V = 2\pi R^2 \left(\frac{M_p}{\sigma_Y h} + \frac{M'_p}{\sigma_Y h} \right),$$

or

$$V\sigma_Y h = 2\pi R^2 (M_p+M'_p). \tag{7.142}$$

Following Wood [7.16], introduce the "moment volume"

$$\mathcal{V}_M = \sigma_Y h V,$$

which is proportional to the volume of reinforcement. From eq. (7.142),

$$\mathcal{V}_M = 2\pi R^2 (M_p+M'_p). \tag{7.143}$$

According to eq. (7.139), we may obtain the desired carrying capacity p_l with various ratios

$$c \equiv \frac{M'_p}{M_p}. \tag{7.144}$$

With relations (7.144) and (7.139), eq. (7.143) becomes

$$\mathcal{V}_M = \frac{\pi}{3} R^4 p_l = 1.046 p_l R^4 ,\qquad(7.145)$$

independently of c. This volume can be reduced in two (possibly simultaneous) ways: (a) making the reinforcement nonhomogeneous, (b) making the reinforcement orthotropic. In both cases we must start from a statically admissible moment field. For example,

$$M_\theta = M_p ,$$
$$M_r = M_p - (M_p + M'_p) \frac{r^2}{R^2}. \qquad(7.146)$$

Note that as soon as the reinforcement is made orthotropic we assume it is placed in the principal directions*. Let $M_p = 0$ (no lower reinforcement) and let M'_p follow the variation of the radial moment (isotropic nonhomogeneous upper reinforcement). We have

$$\mathcal{V}_M = 2 \int_0^R M'_p 2\pi r\, dr = 4\pi \int_0^R \left| -p_l \frac{r^2}{6} \right| r\, dr = 0.524 p_l R^4 . \qquad(7.147)$$

As M_θ has been made to vanish in field (7.146), isotropy is obviously not necessary, and the volume of reinforcement can be decreased further by placing *only top radial* reinforcement (and omitting all circumferential steel). In other words, we make $M_p = 0$ in eqs. (7.139) and (7.146), and substitute the expression (7.139) for M'_p in eqs. (7.146); we obtain the statically admissible field

$$M_\theta = 0 ,$$
$$M_r = -p_l \frac{r^2}{6} , \qquad(7.148)$$

---

* Denote these directions by 1 and 2; if we have $M_{p1} \geqslant |M_1|, M_{p2} \geqslant |M_2|$, and reinforcement placed at top or bottom according to sign of principal moment, then, because of the very nature of the yield locus (7.1) and (7.2), the yield condition cannot be violated at the considered point.

for a given load p_l, and we place reinforcement just to suit. Its moment volume is

$$\mathcal{V}_M = \int_0^R |M_r| 2\pi r\, dr = 0.262 p_l R^4 \ . \tag{7.149}$$

The plate resists as adjacent cantilever beams. Comparison of eqs. (7.145), (7.147), and (7.149) shows how $\mathcal{V}_M$ can be reduced.

Alternatively, we might start from field (7.146), place only isotropic lower reinforcement in the center of the plate where $M_r > 0$, and place only radial reinforcement for $M_r < 0$, together with only circumferential bottom reinforcement for $M_\theta > 0$ in the same outer region (fig. 7.68). The moment volume turns out to be

$$\mathcal{V}_M = 0.392 p_l R^4 \ . \tag{7.150}$$

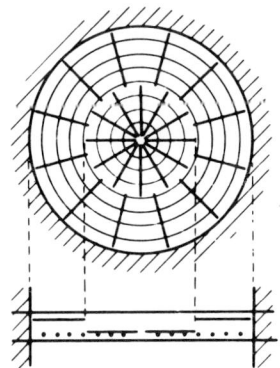

Fig. 7.68.

Consider now the elastic moment field

$$M_r = \frac{pR^2}{16}\left[(1+\nu) - (3+\nu)\frac{r^2}{R^2}\right],$$

$$M_\theta = \frac{pR^2}{16}\left[(1+\nu) - (1+3\nu)\frac{r^2}{R^2}\right], \tag{7.151}$$

where ν is Poisson's ratio. If, for a given p, we reinforce radially and circumferentially *just to match* M_r and M_θ with the (ultimate) *plastic moments*, we clearly have a plate for which the field (7.151) is statically admissible and which is everywhere at yield. Thus,

$$\mathcal{V}_M = \int_0^R (|M_\theta|+|M_r|)2\pi r\, dr = \frac{\pi p R^4}{4} \cdot \frac{1+\nu}{(3+\nu)(1+3\nu)}. \tag{7.152}$$

As compatibility is not required (but statical admissibility is), we choose ν to obtain the minimum $\mathcal{V}_M$, which turns out to be $\mathcal{V}_M = 0.259 p R^4$ for $\nu = 0.0704$. With $\nu = 0$ we obtain

$$\mathcal{V}_M = 0.262 p R^4. \tag{7.153}$$

Obviously, a reinforcement that must be exactly adapted to a given moment field is not likely to be practically feasible. On the other hand homogeneous polar nets of radial and circumferential reinforcing rods can be easily used, whereas economy can be achieved by appropriate orthotropy. This problem has been studied recently by Zawidzki and Sawczuk [7.39]. Because the radial reinforcement per unit central angle is constant, the radial plastic moment per unit length varies with the radius as shown in fig. 7.69.

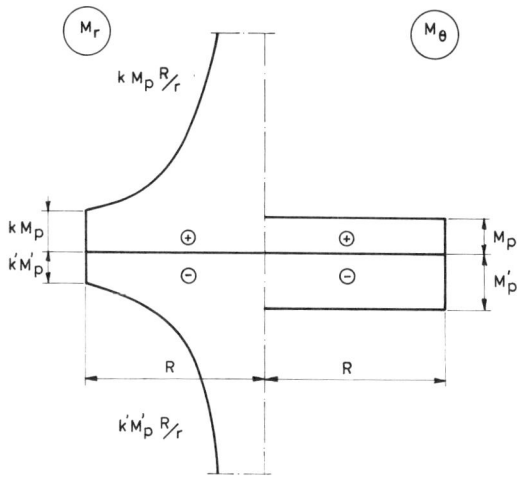

Fig. 7.69.

Complete solutions that depend on the strength parameters, M_p, M'_p, k, and k' are obtained. Comparison with solutions for a square net of bars reveals important difference. Minimum volume of reinforcement is then studied on two examples. For the simply supported uniformly loaded circular plate with $k' = 0$ and $M'_p = M_p$ it turns out that the optimum orthotropy coefficient is $k = 0.385$. This result is almost obvious from fig. 7.70, where the limit load of this plate is given for various k. With $k < 0.385$ the limit load decreases but for $k > 0.385$ the limit load is kept constant.

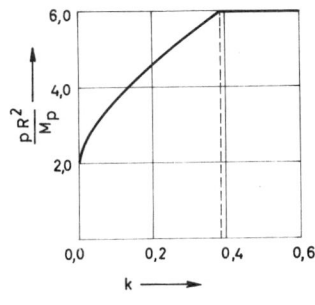

Fig. 7.70.

When k becomes larger than 0.385, the circumferential positive plastic moment is solely responsible for collapse, which occurs in the classical conical mechanism (see Section 6.3.2). For smaller k, a weaker annular region flows plastically, whereas the central part remains rigid. The moment volume obtained with $k = 0.385$ is 1.13 times smaller than that given by eqs. (7.146). It could be further reduced by making $M'_p = 0$ because only positive circumferential resisting moment is needed.

7.6.1.2. *Square and rectangular plates*

Consider a simply supported uniformly loaded square plate. If we place reinforcement just to match the principal moments of the field (7.116) we obtain [7.16]

$$\mathcal{V}_M = 1.45 p \frac{l^4}{24}. \tag{7.154}$$

Vallance ([7.16], p.102) has found another statically admissible moment field for this plate, with plastic yielding along the diagonals only. It is

$$M_x = M_y = \frac{pl^2}{24} \frac{[1-(4x^2/l^2)][1-(4y^2/l^2)]}{\sqrt{[1-(2x^2/l^2)][1-(2y^2/l^2)]}},\qquad(7.155)$$

$$M_{xy} = \frac{pxy}{2} + \frac{pxy}{24}\left\{\frac{[(4x^2/l^2)-3]\sqrt{1-(2y^2/l^2)}}{[1-(2x^2/l^2)]^{3/2}} + \frac{[(4y^2/l^2)-3]\sqrt{1-(2x^2/l^2)}}{[1-(2y^2/l^2)]^{3/2}}\right\}.$$

Because $M_x = M_y$, principal directions are everywhere parallel to the diagonals; this makes the reinforcement easier. When this one is made to match field (7.155), we find

$$\mathcal{V}_M = 1.29 p \frac{l^4}{24}. \qquad(7.156)$$

Comparison with eq. (7.154) shows a saving of 12%, but it must be remarked that field (7.155) gives higher edge reactions, and hence requires stronger supporting beams than field (7.166).

The preceding solutions, as well as those of Section 7.6.1.1, are theoretical solutions that are likely to be difficult to realize in practice, but their main advantage is to form reference solutions with which practical reinforcing could be compared.

Searching for simpler, but still rather economical, reinforcement, Hillerborg [7.40,7.41] suggested the use of a moment field with zero twisting moment. With $M_{xy} = 0$, the plate may be regarded as a fabric of orthogonal strips. Considering a plate with edges parallel to two fixed orthogonal directions x and y, it is assumed that, in regions adjacent to the supports parallel to the y-direction, the load is supported exclusively by the bending moment distribution M_x. The fundamental equilibrium eq. (6.37) then becomes

$$\frac{\partial M_x}{\partial x^2} = -p(x,y),$$

$$\frac{\partial^2 M_y}{\partial y^2} = 0,$$

$$M_{xy} = 0. \qquad(7.157)$$

Similarly, for regions adjacent to supports parallel to the x-direction, we have

$$\frac{\partial^2 M_y}{\partial y^2} = -p(x,y),\qquad \frac{\partial^2 M_x}{\partial x^2} = 0,\qquad M_{xy} = 0. \qquad(7.158)$$

The resulting moment fields have fixed principal directions x and y and meet along straight lines bounding the considered regions. Along these lines, the moment tensor jumps discontinuously from its value in one region to its value in the other, and one must make sure that all equilibrium conditions are satisfied. The appropriate continuity conditions have been studied by Prager [5.15] and others [6.16, 7.16] (see Section 5.6.2). In the particular case of the simply supported square plate uniformly loaded, eq. (7.157) is valid for $|x| < |y|$, and eq. (7.158) for $|y| < |x|$. Integration of eq. (7.157) and use of boundary conditions give

$$M_x = \frac{pl^2}{4}\left(1 - \frac{2|y|}{l}\right)\left(1 - \frac{2x}{l}\right) - \frac{pl^2}{8}\left(1 - \frac{2x}{l}\right)^2,$$

$$M_y = \frac{pl^2}{8}\left(1 - \frac{2x}{l}\right)^2,$$

$$M_{xy} = 0, \qquad (7.159)$$

and symmetry expressions for $|y| < |x|$. In field (7.159) the diagonals are the discontinuity lines, where discontinuity conditions turn out to be satisfied [7.16]. Hence the reinforcement can be placed perpendicular to the edges and curtailed to match the moment field (7.159). In practice, this will be done discontinuously in strips of uniform reinforcement. An average factor k, depending of the width of the strip, must be applied to the maximum moment through the width to furnish the needed reinforcement. This factor k is given by Hillerborg [7.40]. In the preceding example, the moment volume of field (7.159) is

$$\mathcal{V}_M = 1.50 \frac{pl^4}{24}. \qquad (7.160)$$

The strip method is thus less economical than other solutions, eqs. (7.154) and (7.156), but extremely simple and compares favorably with the isotropic reinforcement that imposes

$$\mathcal{V}_M = 2\frac{pl^4}{24}.$$

All preceding methods are based on a statically admissible moment field, and hence furnish a safe design, provided that variable reinforcement is feasible.

When no such moment field is available but only the assumed collapse mechanism, variable reinforcement will be subjected to criticism. Economy can then be pursued through uniform (homogeneous) orthotropic reinforcement, by determining the most economical coefficient of orthotropy for each problem.

Consider, for example, the simply supported, uniformly loaded* rectangular plate. For each value of the ratio $\lambda = L/l$ and each coefficient of orthotropy k (bottom reinforcement only), the yield moment M_p (for given load p) can be obtained by the affinity method (Section 7.4.3).

As the total weight of reinforcement, for given dimensions L and l, is proportional to $2M_p$ in the isotropic plate and to $(1+k)M_p$ in the orthotropic plate, we can trace a diagram of the weight ratio

$$\mathcal{R} \equiv \frac{[(1+k)M_p]_{\text{orthotropic}}}{(2M_p)_{\text{isotropic}}}$$

as a function of $1/k$ and λ (fig. 7.71). The most economical orthotropy coefficient turns out to be given by

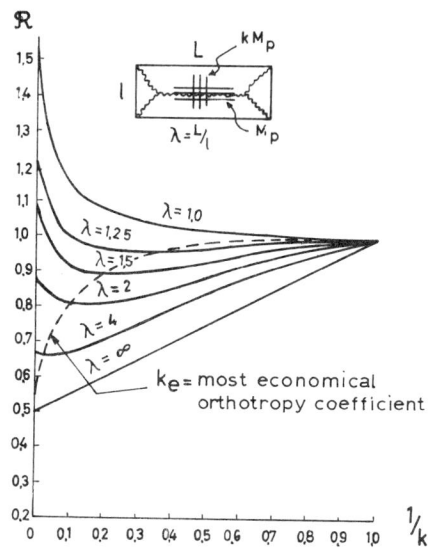

Fig. 7.71. (Reprinted by permission of Thames and Hudson Ltd. from R.H.Wood, *Plastic and Elastic Design of Slabs and Plates*, London, 1961.)

* In this example, a very good static solution also exists (Section 7.5.8).

$$k_e = 3\frac{L^2}{l^2} - 2. \tag{7.161}$$

Note that $\mathcal{R}$ is not very sensitive to variations of k around k_e.

Finally, we emphasize that, as shown by the example of the built-in circular plate (Section 7.6.1.1) an ultimate design based on an elastic moment field gives a very economical amount of reinforcement. Wood [7.16] has studied the uniformly loaded square built-in plate. The theoretical moment volume is $\mathcal{V}_M = 0.021 p l^4$. With a more practical reinforcement pattern, it becomes $\mathcal{V} = 0.028 p l^4$, which remains very good. Contrary to all other designs, these "elastic uniform strength" designs can give us information on serviceability. For sufficiently small loads, deflections can be computed from the elastic moment field and the elastic coefficients E and ν of the plate, which, fortunately, do not depend appreciably on the reinforcement ratio and hence can be regarded as constant throughout the plate.

7.6.2. Approach by constant energy dissipation
7.6.2.1. Introduction

From Section 6.8.2 we know that, for sandwich plates with given midsurface, boundary conditions, and core, and unknown variable sheet thickness, a sufficient condition for minimum volume of the sheets is that a collapse mechanism can be found (with a corresponding moment field) that ensures constant dissipation of energy per unit volume of the sheets.

Because the lever arm h of internal forces may be regarded as constant, reinforced concrete plates with uniform thickness may to some extent be regarded as sandwich plates of core thickness h, the sum of reinforcement areas $(A_x + A_y)$ being the substitute of the thickness of the sheets. But, in reinforced concrete, we may choose independently the directions of the reinforcing bars, the area A_x and the area A_y. In order to achieve absolute minimum-weight of reinforcement, the bars must be placed in the principal directions of the curvature rate tensor [7.42]. The dissipation per unit volume in the two layers of bars are then assigned separately a common constant value*.

Consider, for example, a circular plate under rotationally symmetric load. The dissipation per unit volume of steel is

$$\frac{|M_r||\kappa_r|}{|M_r|/\sigma_Y h},$$

---

* Dissipation in concrete is neglected with respect to dissipation in steel.

for the radial direction, and

$$\frac{|M_\theta||\kappa_\theta|}{|M_\theta|/\sigma_Y h},$$

for the circumferential direction, and the condition for minimum volume is $|\kappa_r| = |\kappa_\theta| = \alpha > 0$, where α is a constant. In regions where reinforcement is absent in at least one of the principal directions, the modulus of the corresponding curvature may not be constant if it is smaller than α, as discussed by Drucker and Shield [6.48] for the general case, and by Morley in a more recent paper [7.42] which discusses minimum reinforcement of concrete plates. Morley [7.42] shows in particular that accounting for the variation of the actual level arm h does not result in significant improvement.

The following examples are borrowed from this paper [7.42].

7.6.2.2. Built-in circular plate, with radius R

We use the conditions $\kappa_r = \kappa_\theta = \alpha$ for $r \leqslant R/2$, and $\kappa_r = \alpha$, $|\kappa_\theta| \leqslant \alpha$ (with $M_\theta = 0$) for $r \geqslant R/2$. The corresponding displacement field satisfying the boundary conditions is

$$w = \alpha \left(\frac{R^2}{4} - \frac{r^2}{2} \right) \quad \text{for} \quad r \leqslant R/2,$$

$$w = \alpha \left(\frac{R^2}{2} - Rr + \frac{r^2}{2} \right) \quad \text{for} \quad r \geqslant R/2. \tag{7.162}$$

It has w and dw/dr continuous throughout, a necessary condition for no local dissipation in hinges. The moment volume is given by [7.42]

$$\mathcal{V}_M = \frac{1}{\alpha} \int_A pw \, dA. \tag{7.163}$$

In the present case, we find

$$\mathcal{V}_M = 0.229 pR^4. \tag{7.164}$$

An elastic uniform-strength design would give [7.16] $\mathcal{V}_M = 0.262 pR^4$. A possible moment field [7.42] is $M_r = M_o[1-(4r^2/R^2)]$, $M_\theta = M_o + [(p/2) - (12M_o/R^2)]r^2$, with $r < R/2$; and $M_\theta = 0$, $M_r = pR^3/48r - pr^2/6$, with $r > R/2$. The central moment M_o is arbitrary, otherwise subjected to $0 \leqslant M_o \leqslant pR^2/16$. Reinforcement cross-sectional areas A_r and A_θ are given by

$$M_r = \sigma_Y h A_r, \qquad M_\theta = \sigma_Y h A_\theta. \tag{7.165}$$

With a circular line load $\bar{p}$ per unit of length, at radius $\rho < R/2$, we obtain [7.42]

$$\mathcal{V}_M = \pi \bar{p} \rho R^2 \left(\frac{1}{2} - \frac{\rho^2}{R^2} \right), \tag{7.166}$$

one possible moment field being

$$M_r = M_\theta = \bar{p}\rho \ln \frac{R}{2\rho}, \qquad \text{for} \qquad 0 < r < \rho,$$

$$M_r = M_\theta = \bar{p}\rho \ln \frac{R}{2r}, \qquad \text{for} \qquad \rho < r < \frac{R}{2},$$

$$M_\theta = 0, \qquad M_r = -\bar{p}\rho \left(1 - \frac{R}{2r}\right), \qquad \text{for} \qquad \frac{R}{2} < r < R.$$

7.6.2.3. *Simply supported square slab*

The slab will be divided into a central region of spherical deflection and corner regions of anticlastic curvatures (fig. 7.72). We take, for the quadrant shown in fig. 7.72, $w = \alpha[xy-(l/2)(x+y)+(l^2/4)]$ for $x+y > l/2$, and $w = \alpha[l^2/8 - \frac{1}{2}(x^2+y^2)]$ in the center. A possible moment field is

$$M_x = M_y = 0,$$

$$M_{xy} = \frac{p}{4}\left(x+y-\frac{l}{2}\right)^2 + \frac{pl}{8}\left(x+y-\frac{l}{2}\right)$$

in the corners $(x+y>l/2)$, and, in coordinates n, t shown in fig. 7.72,

$$M_n = \frac{p}{4}\left(\frac{l^2}{8}-n^2\right), \qquad M_{nt} = 0,$$

$$M_t = \frac{p}{4}\left(\frac{l^2}{8}-t^2\right)$$

in the central region. The moment volume is $\mathcal{V}_M = 0.521 p l^4$. It coincides

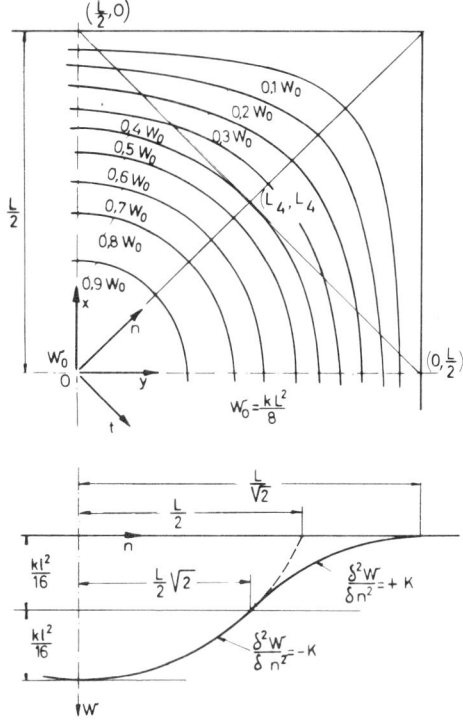

Fig. 7.72.

with that obtained by Wood [7.16] from an elastic uniform strength design with vanishing Poisson's ratio.

7.7. Influence of axial force

7.7.1. *Introduction*

When reinforced concrete plates are loaded up to rupture, their load versus deflection relationship does not exhibit a final flat portion as assumed in the simple plastic theory. The shape of the curve may in fact be very different from that predicted by this theory and the maximum obtainable load may be substantially greater than the theoretical limit load. The discrepancy is strongly dependent on the boundary conditions (and, obviously, on the deflection at which rupture is obtained or the test is ended). These facts are illustrated in figs. 7.73 and 7.74.

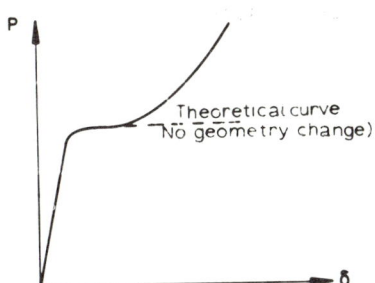

Fig. 7.73. Simply supported plate.

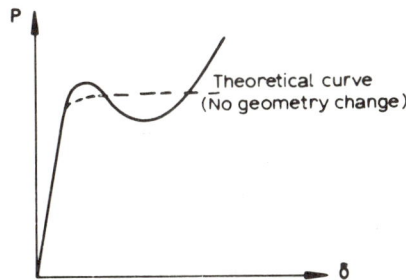

Fig. 7.74. Built-in plate.

The increase in carrying capacity stems primarily from geometry changes, generating axial forces, and secondarily from work hardening of steel, effects that are both disregarded in the simple plastic bending theory. These effects are responsible for the fact that, even when only computed from a good collapse mechanism and thus being upper bounds P_+ to the exact limit load P_l of simple bending theory, most theoretical collapse loads are smaller than the corresponding experimental values (see [6.16,7.3,7.16,7.17,7.18]). The effects of geometry changes will be considered in Sections 7.7.2 to 7.7.4.

7.7.2. *Finite plastic bending with boundaries restrained against sliding over the supports*

Finite plastic bending of reinforced concrete plates has been especially studied by Wood [7.16] and more recently by Sawczuk [6.16] and Janas [7.43]. We consider for illustration a square isotropic plate under a uniform load p; the plate is to consist of a rigid perfectly plastic material, and to be hinged along its supports at the level of the bottom reinforcement layer.

This plane is taken as the reference plane. The (already known) collapse mechanism of fig. 7.75 is not kinematically admissible for nonzero deflection with the present boundary conditions. Indeed, fig. 7.76 shows the deflected position of a cross section normal to a yield line (as cross section *aa* in fig. 7.75). Section *AB* in the yield line undergoes a horizontal displacement (see fig. 7.76)

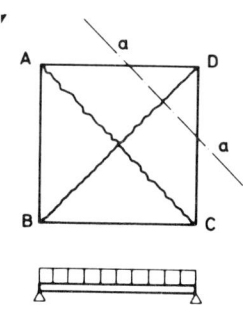

Fig. 7.75.

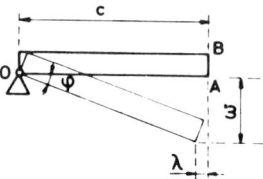

Fig. 7.76.

$$\lambda = c(1-\cos\theta) = \frac{w}{\sin\theta}(1-\cos\theta) \cong w\frac{\theta}{2} \tag{a}$$

(w is small compared to c). The small rotation θ is related to w by

$$\theta = \frac{w}{c}, \tag{b}$$

when

$$\lambda = \frac{w^2}{2c}. \tag{c}$$

7.7] INFLUENCE OF AXIAL FORCE

Differentiating eqs. (b) and (c) with respect to time we obtain

$$\dot{\theta} = \frac{\dot{w}}{c}, \qquad \text{(d)}$$

$$\dot{\lambda} = \frac{w\dot{w}}{c}. \qquad \text{(e)}$$

Use of eq. (d) in eq. (e) yields

$$\dot{\lambda} = w\dot{\theta}. \qquad \text{(f)}$$

Because normals to the reference plane remain normal to the polyhedron into which it transforms, the zero strain rate level is at the distance ξh of the bottom face such that $\dot{\theta}\xi h = \dot{\lambda}$ or, using eq. (f)

$$\xi h = w. \qquad \text{(g)}$$

Consequently, for vanishing w (impending collapse), boundary conditions enforce the zero strain surface to coincide with the restraining plane. The yield lines must therefore be subjected both to a moment M and an axial force N for such a mechanism to develop. This kinematic condition of internal compatibility at zero deflection is discussed by Janas in ref. [7.19]; it is responsible for the increase in the load at incipient collapse.

To obtain the MN interaction curve, we assume that, in a yielding cross section, concrete is yielding in compression at a constant stress $-\sigma_c$ (where σ_c is the crushing strength of the concrete measured by tests on cylinders) above the zero strain rate level, and that, beneath that level, the concrete is cracked and the steel yields in tension at constant stress σ_Y.

If A denotes cross-sectional area of the reinforcement per unit of length of yield line, we have (fig. 7.77)

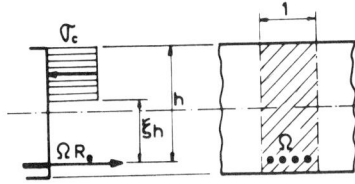

Fig. 7.77.

$$M = \sigma_c \frac{h^2}{2}(1-\xi),$$

$$N = A\sigma_Y - \sigma_c h(1-\xi). \tag{7.167}$$

Denoting

$$M_* \equiv \frac{\sigma_c h^2}{2}, \qquad N_* \equiv \sigma_c h, \tag{7.168}$$

we have

$$M = M_*(1-\xi^2),$$

$$N = N_*(\mu-1+\xi) = \frac{2}{h}M_*(\mu-1+\xi), \tag{7.169}$$

where $\mu = (A/h)/(\sigma_Y/\sigma_c)$ is the "reduced reinforcement ratio".

Elimination of the parameter ξ from eqs. (7.169) gives

$$M = M_*\left[\mu(2-\mu) - \left(\frac{N}{N_*}\right)^2 - 2\frac{N}{N_*}(1-\mu)\right]. \tag{7.170}$$

For vanishing w (initiation of collapse mechanism), eqs. (e) and (g) enforce vanishing λ and ξ. Hence, at this instant,

$$N = -N_*(1-\mu), \tag{7.171}$$

$$M = M_*. \tag{7.172}$$

It can be verified that values (7.171) and (7.172) of N and M correspond to a point on the curve with eq. (7.170) where $\partial N/\partial M = 0$, as is required by the normality law. On the other hand, with no restraint at the boundaries, the plastic moment is

$$M_p = M_*\mu(2-\mu), \tag{7.173}$$

obtained by letting $N = 0$ is eq. (7.170).

The moment in the yield lines is increased by the factor

$$\frac{M_*}{M_p} = \frac{1}{\mu(2-\mu)}, \qquad (7.174)$$

and so is the limit load p_l. With $N_p = \sigma_Y A$, we have $N_* = N_p/\mu$, and we can rewrite relation (7.170) as

$$\frac{M}{M_p} = 1 - \frac{2(1-\mu)}{2-\mu} \frac{N}{N_p} - \frac{\mu}{2-\mu}\left(\frac{N}{N_p}\right)^2. \qquad (7.175)$$

Let

$$\alpha = \frac{2(1-\mu)}{2-\mu}, \qquad \beta = \frac{\mu}{2-\mu}. \qquad (7.176)$$

Eq. (7.175) becomes

$$\frac{M}{M_p} = 1 - \alpha\frac{N}{N_p} - \left(\frac{N}{N_p}\right)^2, \qquad (7.177)$$

and it is readily shown that the enhancement factor [eq. (7.174)] can be written [7.16].

$$\frac{M_*}{M_p} = 1 + \frac{\alpha^2}{4\beta}. \qquad (7.178)$$

For example, with $\sigma_Y = 33{,}000$ psi, $\sigma_c = 3{,}300$ psi, and $A/h = 0.01$, we have $\mu = 0.1$ and $\alpha = 0.946$, $\beta = 0.0526$.

The corresponding M/M_p versus N/N_p interaction curve is given in fig. 7.78. Note that eq. (7.177) is represented by the parabolic part AB, whereas segment BC corresponds to $\xi = 0$ (neutral layer at the level of the reinforcement), a situation where M is kept constant while N may vary from N_p to $-N_p$. With $N = 0$ and $h = 4$ in, we would obtain

$$M_p = \frac{\sigma_c h^2}{2}\mu(2-\mu) = 5.015 \text{ lb in./in.}$$

Knowing the relation between moments and axial forces in the yield line, we can now study the collapse of the considered slab (fig. 7.75). The rate of work of the external loads is the same as for the simple bending theory. For the collapse mode shown in fig. 7.75, it is

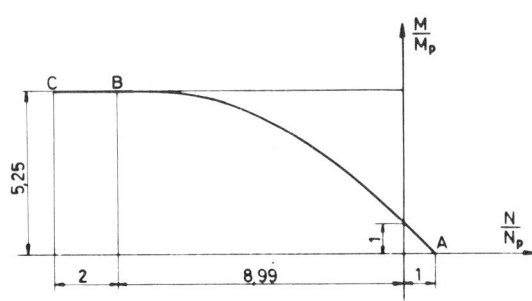

Fig. 7.78.

$$W = \frac{pl^2}{12} \dot{w}_o , \tag{7.179}$$

whereas the rate of dissipation is now expressed by the formula

$$D = \sum_{i=1}^{n} \int_{o}^{l_i} [M_i(l)\dot{\theta}_i + N_i(l)\dot{\lambda}_i] \, dl , \tag{7.180}$$

where summation extends on all n yield lines with the respective lengths l_i and rotation rates $\dot{\theta}_i$.

The internal forces M, N depend on the position of the current neutral axis $\xi = \dot{\lambda}/h\dot{\theta}$ through eqs. (7.169). Note that, because of relation (f), dissipation is now deflection dependent.

For the position of edge restraints shown in fig. 7.76, the rotation axes of the collapse mechanism must lie in the bottom plane of the slab, otherwise horizontal displacements of point O occur. Since the neutral axis in a hinge must be the axis of reciprocal rotations of adjacent parts, initial positions are defined, in agreement with eq. (g), by $\xi = 0$. Internal forces can be found from eqs. (7.169) and the rate of dissipation can be computed. The work equation

$$W = D \tag{7.181}$$

yields the following upper bound for the initial collapse load

$$p_{01} = \frac{24 M_*}{l^2} = p_l \frac{M_*}{M_p} \tag{7.182}$$

whereas the solution derived from the simple bending theory with the limit load given by eq. (7.104) is no longer kinematically admissible.

Since restraints enforce $\xi = 0$, large compressive forces $N = -N_*(1-\mu)$ exist in the hinges, and must be supported by the restraints. It can be shown that, for the reinforcement ratio $\mu < \frac{1}{2}$ and in absence of special reinforcements at supports, plastic deformations must develop there. Since there is a zone of concrete crushed in compression [fig. 7.79 (a)] the collapse mode including rotations around O' can satisfy the zero horizontal displacements condition at the supports, and appears to be admissible. Now, the position of the neutral axis in the hinges is $\xi = \eta$. Internal forces and dissipation change while W remains the same as for the preceding collapse mode. Note, moreover, that the total rate of dissipation must now include a part due to plastic deformation at the supports, where negative rotation rate θ appears. The work equation gives a collapse load dependent on the unknown parameter η which must be found (as in the simple bending theory) from the minimum principle. The minimization procedure is equivalent to satisfying the horizontal equilibrium of rigid portion cut off by the hinges [fig. 7.79 (b)]. Finally, a new bound for the collapse load is found to be

$$p_{02} = p_l \left[1 + \frac{M_*}{M_p} (\tfrac{1}{2} - \mu^2) \right]. \tag{7.183}$$

For plastic flow to continue at finite values of the central deflection w_o, the actual collapse load must not be higher than those derived from the work equation [eq. (7.181)] established for the deformed structure. Assuming that the deflection pattern does not change, the current deformations are defined

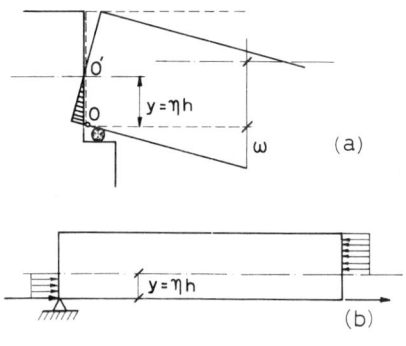

Fig. 7.79.

by one parameter, w_o. Normal strains being dependent on the actual deflections, they are seen to vary along the hinges. If collapse starts with the mode of fig. 7.76, relation (f) can be introduced into eqs. (7.169) and, after the appropriate integration of eq. (7.180), dissipation can be computed. External work rate is still unchanged [eq. (7.179)].

The work equation gives

$$p_1 = p_l \frac{M_*}{M_p} \left[1 - \delta_o(1-\mu) + \frac{\delta_o^2}{3} \right], \tag{7.184}$$

where $\delta_o = w_o/h$. For $\delta_o = 0$ we obtain eq. (7.182), as expected. When initial yielding starts with the mode of fig. 7.79, simultaneous plastic deformations develop both in the inner hinges and at supports. According to fig. 7.79 (a), we now have $\lambda = \theta(y+w)$, $\xi = \eta + w/h$, and the foregoing procedure leads to the solution

$$p_2 = p_l \left\{ 1 + \frac{M_*}{M_p} \left[\frac{(1-\delta_o/2)^2}{2} + \frac{\delta_o^2}{12} - \mu^2 \right] \right\}, \tag{7.185}$$

the unknown parameter η being found in the same way as for initial yielding.

Both solutions, eqs. (7.184) and (7.185), are valid only if the central deflection remains sufficiently small. If the upper face of the slab descends below the plane of the rotation axes, pure membrane response develops in the hinges of a central zone (II of fig. 7.80). Relations (7.169) are not valid there,

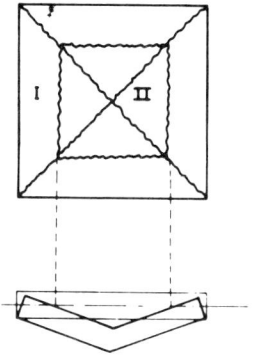

Fig. 7.80.

the internal forces being $M = 0$, $N = N_p$. In zone I however, eq. (7.169) still holds. Accounting for this new situation, we may derive the appropriate formula for the collapse loads for both considered modes. They are, respectively,

$$p_{m1} = p_l \frac{M_*}{M_p}\left(\mu\delta_o + \frac{1}{3\delta_o}\right), \qquad \delta_o \geqslant 1, \tag{7.186}$$

$$p_{m2} = p_l \frac{M_*}{M_p}\left\{(1+\delta_o)^2 - \frac{\delta_o^2}{3} - \frac{2}{3}\sqrt{\delta_o}[\delta_o + 2(1+2\mu)]^{3/2}\right.$$

$$\left. + \mu(4+5\delta_o)\right\}, \qquad \frac{2}{3}(1+2\mu) \leqslant \delta_o \leqslant \frac{1}{4\mu}. \tag{7.187}$$

We see that a continuous transition from the initial compression (even with crushing of the concrete at supports) to membrane tensile response has been obtained. Collapse load is plotted against deflections in fig. 7.81 for the reinforcement ratio $\mu = 0.2$. It is seen that, at the first stage, the collapse load decreases as deflections increase. This unstable situation was also observed experimentally [7.16, 7.44]. Elastic deformations and early crackings transform the theoretical curve of fig. 7.81 as shown in fig. 7.82. The peak A was found experimentally by Park [7.44] to correspond approximately to $w_o = 0.5h$ or $\delta_o = 0.5$. It can be shown (see [7.43]) that the minimum value of p cannot be smaller than p_l (see fig. 7.81). Note also that, if the midplane is restrained instead of the bottom plane, membrane action begins at $w_o = 0.5h$.

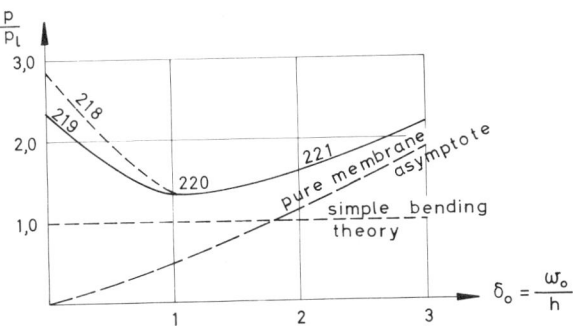

Fig. 7.81.

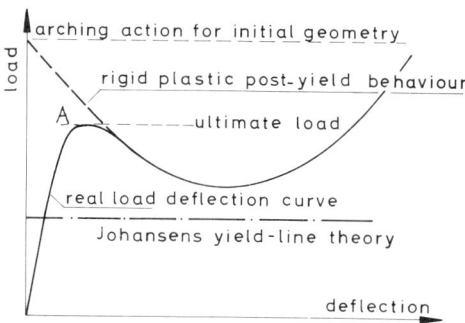

Fig. 7.82.

7.7.3. Finite plastic bending with unrestrained supports

Since the support conditions enable free horizontal displacements of the slab edges, arbitrary positions can be adopted for the rotation axes without inducing plastic deformations at the supports. Consider once again the square slab under a uniformly distributed load. With the yield line pattern of fig. 7.75, the only parameter of the collapse mode is the distance $y = \eta h$ of the plane of the rotation axes (that is the neutral plane) from the (bottom) reference plane. Reinforcement plane and reference plane are assumed to coincide.

At the initiation of collapse, the neutral axes are defined by $\xi = \eta$, and eqs. (7.169) can be applied to determine the rate of dissipation [eq. (7.180)]. Since the rate of external work is always as for simple bending theory [eq. (7.179)], the appropriate expression for collapse load can be derived from the work equation, eq. (7.181). The parameter η can be found both from the minimum principle or from the equilibrium of horizontal forces applied to the rigid portion ABE (fig. 7.75). The last condition imposes zero total stress resultant in each hinge. With

$$\eta = 1 - \frac{N_p}{N_*} = 1 - \mu,$$

there are no normal forces in the hinges [eqs. (7.169)] and the simple bending theory is valid here, giving

$$p_o = p_l = \frac{24 M_p}{l^2}.$$

When deflections increase, the work equation [eq. (7.181)] must be established for the deformed structure, as in Section 7.7.2. The position of the neutral layer varies along the hinge:

$$\xi = \eta + \frac{w}{h},$$

notations being those of Section 7.7.2. Repeating the preceding procedure and after appropriate integration of eq. (7.180), the collapse load is found to be dependent upon w_o and η. The minimum principle yields

$$\eta = 1 - \mu - \frac{\delta_o}{2}, \qquad (7.188)$$

and the current collapse load is expressed by

$$p = p_l \left(1 + \frac{\delta_o}{12} \frac{M_*}{M_p}\right). \qquad (7.189)$$

It can be noticed that since the absolute minimum point occurs at the initial collapse, there is no instability phenomenon, contrary to what was observed in fig. 7.81.

For the central deflection w_o large enough ($\eta + \delta_o > 1$), the membrane response zone appears (see fig. 7.80), with $M = 0$, $N = N_p$, whereas formula (7.169) still holds for the outer region. The limit load is now

$$p_m = p_l \left[1 + \mu \left(\delta_o + \mu - \frac{4}{3}\sqrt{2\mu\delta_o}\right)\frac{M_*}{M_p}\right], \qquad \delta_o > 2\mu. \qquad (7.190)$$

To minimize the collapse load, the position of the rotation axes η must furnish a vanishing total stress resultant in the hinge. A peripherial compressed zone must therefore exist in the slab.

The same problem for the circular slab is studied in detail in ref. [7.16]. Since, however, the internal forces are there related to the finite strains rather than the strain rates, the results can not be accepted within the framework of the flow theory of plasticity.

7.7.4. Practical conclusions

The preceding theoretical analysis, as well as experimental evidence [7.16, 7.44], show that the carrying capacity can exceed the theoretical limit load p_l.

If the slab is strongly restrained at the boundary against extension of its median plane by surrounding beams or slabs, the factor $(1+\alpha^2/4\beta)\,\rho$ can be applied to the theoretical limit load P_l. The factor ρ is a reduction coefficient accounting for the fact that instability occurs before the maximum load at zero deflection can be reached. According to Wood [7.16], the load factor to use with the carrying capacity $P_c = P_l[1+(\alpha^2/4\beta)]\,\rho$ should be at least 4, and ρ should be taken as:

Reduced reinforcement ratio μ	*Reduction coefficient* ρ
$\mu > 0.08$	$\rho = 0.7$
$0.08 > \mu > 0.04$	$\rho = 0.6$
$0.04 > \mu$	$\rho = 0.5$. (7.191)

For rectangular plates with $L/l > 2$, and for plates where a combined beam-plate mechanism can occur, the limit load P_l should be used.

If, for unrestrained plates, one is prepared to accept a maximum deflection w_o, Wood [7.16] suggests the following relations.

$$P_c = P_l(1+0.6w_o/h_t) \qquad \text{for} \qquad \mu \cong 0.02 , \qquad (7.192)$$

and

$$P_c = P_l(1+0.3w_o/h_t) \qquad \text{for} \qquad \mu \cong 0.08 , \qquad (7.193)$$

where h_t is the total height.

Relations (7.192) and (7.193) should not be used for rectangular plates with $L \geqslant 3l$.

7.8. Influence of shear forces. Punching

Stresses due to shear forces are, as a rule, rather small in plates (see [7.16], p.11) and their influence is completely negligible, as has been verified by numerous tests up to collapse. The only case to be considered is the punching of the plate by a high concentrated force or by a column reaction.

The punching fracture under a concentrated force is a complicated phenomenon, the interpretation of which remains open to discussion [7.45]. It is influenced by various parameters, among which we have:

1. The quality of the concrete;
2. The flexural reinforcement (percentage, distribution, adhesion);
3. The distribution of bending moments in the vicinity of the concentrated load;
4. The ratio of the (small) area of application of the load to the effective thickness h of the plate;
5. The presence of special shear reinforcement (inclined bars, etc.);
6. The existence of large compressive forces arising from the constraints at the boundary of the plate in its plane. This last influence is particularly noticeable in the tests described by Guyon [7.46] and Muller [7.47], and related to what was developed in Section 7.7 and explains why the following formula is too conservative.

Indeed, formula (7.194) was empirically deduced by Elstner and Hognestad [7.48] from tests on *simply supported* square slabs of 1.8-m length, 15-cm thickness, loaded by a column with square cross section of 15-cm length. They obtained

$$P_V = \frac{7}{8} A_S \sigma'_{r(\text{cyl})} \left(\frac{23.4}{\sigma'_{r(\text{cyl})}} + \frac{0.046}{\phi_o} \right) \tag{7.194}$$

(units are kg and cm), where P_V is the punching load; A_S is the "sheared area", product of the perimeter of the punch by the effective thickness of the plate; $\sigma'_{r(\text{cyl})}$ is the rupture compressive stress of the concrete on cylinders; $\phi_o = P_V/P_l$ where P_l is the limit load in bending.

Formula (7.194) implies the absence of shear force reinforcement. When this latter exists, formula (7.194) must be replaced by

$$P_V = A_V \sigma_{YV} \sin \theta , \tag{7.195}$$

where A_V is the cross-sectional area of the inclined shear force reinforcing bars; σ_{YV}, the yield limit of the shear force reinforcement; and θ, the angle of the inclined bars with the midplane of the plate.

The safety with respect to P_V must be higher than with respect to P_l because punching is a localized and sudden phenomenon, very sensitive to local imperfections of the concrete and the reinforcement. Despite its limitations, formula (7.194) is one of the most widely used. It gives too large a safefy factor when applied to continuous or built-in slabs.

Numerous papers have been devoted to punching of reinforced concrete plates (see, for example, [7.49,7.50]). Moe [7.51] has suggested the following formula:

$$P_V = \tau A_S = A_S \sqrt{\sigma'_{r(\text{cyl})}} \; \frac{15(1-0.075a/h)}{1 + 5.25(A_S\sqrt{\sigma'_{r(\text{cyl})}})/P_l} , \qquad (7.196)$$

where a is the side of the square area of application of the load. In order to avoid punching failure and obtain the bending collapse mechanism, one must have

$$\tau \equiv \frac{P_V}{A_S} \leqslant \left(9.23 - 1.12\frac{a}{h}\right)\sqrt{\sigma'_{r(\text{cyl})}} \qquad \text{for} \qquad \frac{a}{h} \leqslant 3$$

$$\tau \equiv \frac{P_V}{A_S} \leqslant \left(2.5 + 1.12\frac{a}{h}\right)\sqrt{\sigma'_{r(\text{cyl})}} \qquad \text{for} \qquad \frac{a}{h} \geqslant 3 . \qquad (7.197)$$

Formulas (7.196) and (7.197), like formula (7.194), are deduced from tests on *simply supported* plates.

We finally note with Guerrin [7.52] that punching loads P_V are always very large, especially when the slab is continuous or built-in, even when no special precautions have been taken, and even for very thin plates (4 cm < 2 in.). Hence it is very seldom, except in tests, that multiplication of the actual service loads by the safety factor will result in a load larger than P_V.

7.9. Example of application

We want to design the slab shown in fig. 7.83 [7.13]. It is supported on the boundary *ABCDE* and free along *EF* and *FA*. In the corner *AFE* a staircase will be built. Along the edges *ABCD* the plate is continuous and, hence, can exhibit negative resisting moments. The slab is loaded:

1. On *AF* by the staircase applying a dead weight of 134 lb/ft and a serivce load of 333 lb/ft. With a load factor of $s = 2$, we obtain, at collapse, 800 lb/ft on *AF*;
2. On the double line representing a wall, by a line load of 667 lb/ft at collapse;
3. On its whole area by a uniformly distributed load due to (a) a dead load evaluated at 44.4 lb/ft² (approximately 4-in. thickness); (b) a live load of 66.6 lb/ft² × x = 66.6 × 2 = 133.2 lb/ft².

The total load thus is 177.6 lb/ft².

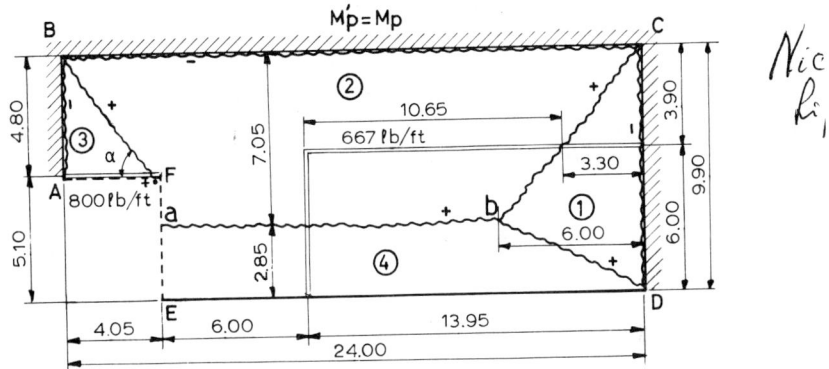

Fig. 7.83.

Solution 1: *isotropic uniform upper and lower reinforcement* ($M_p=M'_p$). We begin with the mechanism shown in fig. 7.83. We shall avoid the differentiation process based on the kinematic theorem by considering the equilibrium of rotation of each portion of the slab about its axis of rotation (see Section 7.4.10 dealing with nodal forces). Note that, at the intersection of the yield line from corner B with the side AF, a twisting moment (see Section 7.4.6.4), $M_{xy} = -M_p \cot \alpha$, must be applied by the staircase to the slab. It will be replaced by two forces $V = M_p \cot \alpha$, one on each side of this intersection and a unit of length apart. Their directions are shown in fig. 7.83; downwards for the force applied to part ③, upwards for that applied to part ②.

Equilibrium of rotation gives, for part ①,

$$9.9(M_p+M_p) = 177.6 \times 9.9 \times \frac{(6.0)^2}{6} + 667 \times \frac{(3.3)^2}{2}.$$

Hence $M_p = 720$ lb. For part ②,

$$24(M_p+M_p) = 177.6 \times 4.05 \frac{(4.8)^2}{6} + 177.6(19.95-6.00)\frac{(7.05)^2}{2} + 177.6$$

$$\times 6.0 \frac{(7.05)^2}{6} + 667 \times 10.65 \times 3.9 + 667(7.05-3.90)$$

$$\left(3.9+ \frac{7.05-3.90}{2}\right) - \frac{4.05}{4.80} M_p \times 4.80.$$

312   REINFORCED CONCRETE PLATES   [Ch. 7

Hence M_p = 2160 lb. For part ③,

$$4.8(M_p+M_p) = 177.6 \frac{(4.05)^2}{6} + 800 \frac{(4.05)^2}{2} + \frac{4.05}{4.80} M_p \times 4.05.$$

Hence M_p = 1,440 lb. For part ④,

$$(24-4.05)M_p = 177.6(19.95-6.0)\frac{(9.9-7.05)^2}{2}$$

$$+ 177.6 \times 6.0 \frac{(9.9-7.05)^2}{6} + 667 \frac{(9.9-7.05)^2}{2}.$$

Hence M_p = 720 lb.

The very different values of M_p reveal that the mechanism is not the right one. To know how it should be modified, we compute an average value of M_p from the work equation:

$$\sum M_i \theta_i = \int pw \, dx \, dy.$$

With a unit displacement for segment ab, we obtain

$$\sum M_i \theta_i = \frac{19.8}{6} M_p + \frac{48}{7.05} M_p + \frac{9.6}{5.04} M_p + \frac{19.95}{1.95} M_p = 19.73 M_p,$$

$$\int pw \, dx \, dy = \frac{14190}{6} + \frac{112380}{7.05} + \frac{8880}{5.04} + \frac{14220}{1.95} = 24790.$$

Whence M_p = 1,256 lb. Consequently, parts ① and ④ should be increased to furnish larger M_p, whereas parts ② and ③ should be decreased. We shall, however, not modify part 3 that exhibits a plastic moment not very different from 1,360 lb, and shall use the mechanism shown in fig. 7.84.

The equilibrium of rotation now gives for part ①,

$$9.9(M_p+M_p) = 177.6 \times 9.9 \frac{(7.95)^2}{6} + 667 \frac{(5.4)^2}{2}.$$

Hence M_p = 1,428 lb. For part ②,

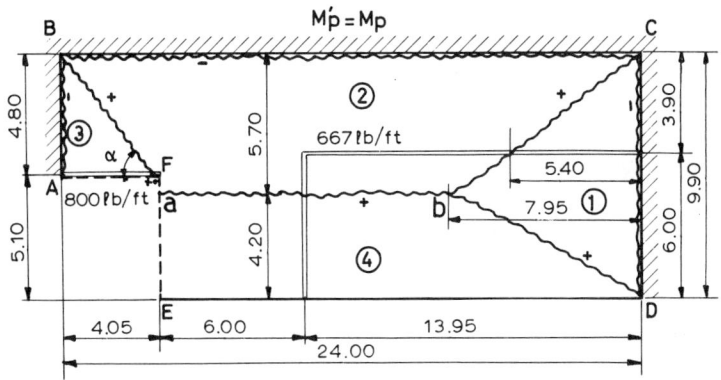

Fig. 7.84.

$$24(M_p + M_p) = 177.6 \times 4.05 \frac{(4.8)^2}{6} + 177.6(24 - 4.05 - 7.95)\frac{(5.7)^2}{2}$$

$$+ 177.6 \times 7.95 \frac{(5.7)^2}{6} + 667 \times 8.55 \times 3.9 + 667(5.7 - 3.9)$$

$$\left(3.9 + \frac{5.7 - 3.9}{2}\right) - \frac{4.05}{4.80} M_p \times 4.80.$$

Hence $M_p = 1.404$ lb. For part ③ (unmodified), $M_p = 1{,}440$ lb. For part ④,

$$19.95 M_p = 177.6(19.95 - 7.95)\frac{(4.2)^2}{2} + 177.6 \times 7.95 \frac{(4.2)^2}{6} + 667 \frac{(4.2)^2}{2}.$$

Hence $M_p = 1{,}448$ lb.

The work equation of the mechanism is

$$\left(\frac{19.8}{7.95} + \frac{4.8}{5.7} + \frac{9.6}{4.8} + \frac{19.95}{4.2}\right) M_p = \frac{28260}{7.95} + \frac{73062}{5.7} + \frac{8892}{4.8} + \frac{28860}{4.2}.$$

When $M_p = 1{,}418$ lb. The variation values of M_p are sufficiently close to accept the mechanism of fig. 7.84 as the best of its family, and take the plastic moment $M_p = 1{,}419$ lb.

Assuming that the right family was chosen (a fact that could not be verified with certainty without using the static theorem) it remains to determine

314   REINFORCED CONCRETE PLATES   [Ch. 7

the reinforcement. With a total thickness $h_t \cong 3$ in., the effective thickness is $h = 2.5$ in. Mild steel reinforcing bars have $\sigma_Y = 34{,}200$ psi and we shall use a concrete with $\sigma'_r = 3{,}420$ psi. The lever of internal forces can be approximated to $0.95h$ [7.16] and we can write

$$M_p = A_S \sigma_Y 0.95h \ . \tag{a}$$

Substituting 1,419 lb for M_p and the values above for σ_Y and h in relation (a), we obtain

$$A_S = 0.0175 \text{ in.}^2/\text{in.} = 0.21 \text{ in.}^2/\text{ft} \ .$$

We shall use two bars of 0.4-in. diameter per foot. The reinforcement ratio A_S/h is thus 0.84%.

The bars are placed in two orthogonal upper layers and two orthogonal identical lower layers. Their total length is, anchorages excluded,

$$8A_{\text{slab}} = 1{,}740 \text{ ft} \ .$$

Solution 2: *exclusively lower isotropic reinforcement* ($M'_p=0$). The work equation of the mechanism shown in fig. 7.84 now gives, with $M'_p = 0$,

$$\left(\frac{9.9}{7.95} + \frac{24}{5.7} + \frac{4.8}{4.8} + \frac{19.95}{5.2} \right) M_p = 25{,}092 \ ,$$

that is, $M_p = 25.092/11.22 = 2{,}240$ lb $> 1{,}448$ lb. Hence, we slightly modify the mechanism to increase the value of M_p in part 4 where M'_p did not enter. With the mechanism of fig. 7.85, we obtain, for part ①,

$$9.9 M_p = 177.6 \times 9.9 \frac{(6.6)^2}{6} + 667 \frac{(5.79)^2}{2}$$

and, hence, $M_p = 2{,}200$ lb. For part ②,

$$24 M_p = 177.6 \times 3.3 \frac{(4.8)^2}{6} + 177.6 \times 0.75 \frac{(4.8)^2}{2} + 800 \times 0.75 \times 4.8$$

$$+ 177.6(24-4.05-6.6) \frac{(4.95)^2}{2} + 177.6 \times 6.6 \frac{(4.95)^2}{6} + 667 \times 8.76$$

$$\times 3.9 + 667(4.95-3.9) \left(3.9 + \frac{4.95-3.9}{2} \right) - \frac{3.3}{4.8} M_p \times 4.8 \ .$$

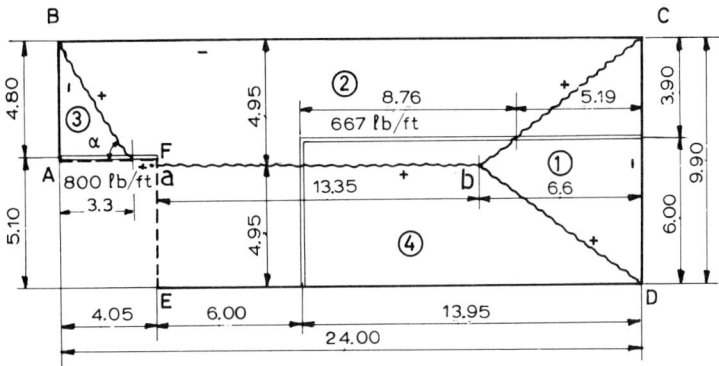

Fig. 7.85.

Hence M_p = 2,384 lb. For part ③,

$$4.8 M_p = 177.6 \times 4.8 \frac{(3.3)^2}{6} + 800 \frac{(3.3)^2}{2} + \frac{3.3}{4.8} M_p \times 3.3.$$

Hence M_p = 2,330 lb. For part ④,

$$19.95 M_p = 177.6 \times 13.35 \frac{(4.95)^2}{2} + 177.6 \times 6.6 \frac{(4.95)^2}{6} + 667 \frac{(4.95)^2}{2}.$$

Hence M_p = 2,120 lb.

The four values of M_p do not differ appreciably from that given by the work equation. On that basis, we accept the mechanism of fig. 7.85 with M_p = 2,380 lb. We then obtain, from relation (a), A_S = 0.352 in.^{2}/ft. We place two orthogonal lower layers of three bars of 0.4-in. diameter per foot. Their total length is $6A_{\text{slab}}$ = 1305 ft.

We see that an economy of 25% on the volume of reinforcement is achieved when the slab is treated as simply supported along $ABCD$. Continuity (that is, $M'_p \neq 0$) proves economic only if upper bars can be placed exclusively where negative moments exist, information that is given by a static approach (or by empirical rules, if reliable). Obviously, with simple supports (M'_p=0) one must be prepared to accept cracks at the upper face in the vicinity of the supports.

7.10. Problems

7.10.1. On the basis of the example treated in Section 7.9, show that built-in edged "push off" the yield lines, whereas simple supports or free edges "attract" them. *Hint*: study the influence of displacing the yield lines on the dissipation and on the work of applied loads.

7.10.2. Determine the approximate limit load p_+ of a square isotropic plate uniformly loaded, simply supported on three edges, and free along the fourth. Show that the mechanism in fig. 7.86 must be rejected. *Answer*: $p_+ = 14.15 M_p/l^2$. *Hint*: start with the mechanism of half a rectangular plate with sides l and $2l$.

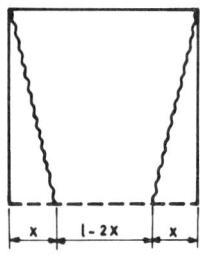

Fig. 7.86.

7.10.3. Determine the approximate limit load p_+ of a square isotropic plate with side l, uniformly loaded, and supported by four corner columns. *Answer*: $p_+ = 8M_p/l^2$.

7.10.4. Determine the approximate limit load p_+ of the triangular, isotropic uniformly loaded plate in fig. 7.87 without accounting for the corner effect.

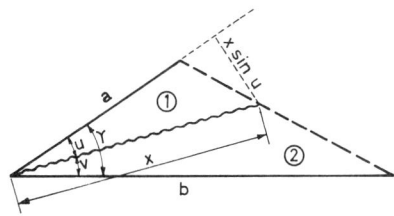

Fig. 7.87.

Obtain the right mechanism by differentiation. *Answer*: (total load)$_+$ = $6M_p/\tan \gamma/2$).

7.10.5. A simply reinforced isotropic rectangular plate is simply supported on three edges and free along the fourth (fig. 7.88). It is subjected, at collapse, to a uniformly distributed load p = 300 lb/ft² and two concentrated loads of 10,000 lb, acting on the free edge as shown in fig. 7.88. Determine the necessary M_p (by mechanisms). *Answer*: M_p = 9,580 lb. *Hint*: apply the superposition method, using the mechanisms shown in fig. 7.88.

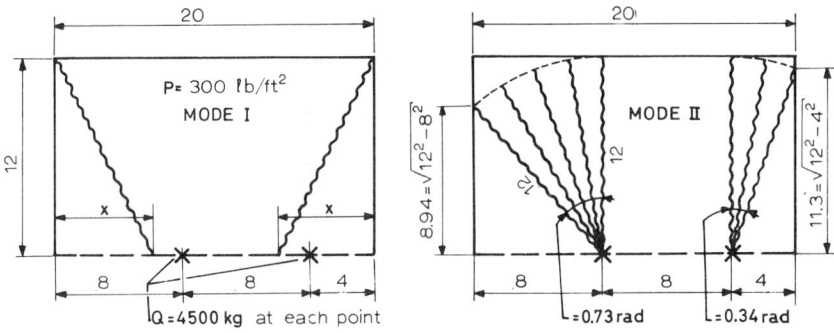

Fig. 7.88. (Reprinted by permission of Thames and Hudson Ltd. from R.H.Wood, *Plastic and Elastic Design of Slabs and Plates*, London, 1961.)

7.10.6. Determine the approximate limit load P_+ of a rectangular isotropic balcony, doubly reinforced, subjected to a concentrated load P at a free corner (fig. 7.89). *Answer*: $P_+ = 2\sqrt{M_p \cdot M'_p} + (M_p + M'_p)(\pi/2 - 2\arctan\sqrt{M_p/M'_p})$. *Hint*: use mechanism of fig. 7.89; differentiate to obtain the best mechanism; note the influence of the plastic moment M_p for positive bending.

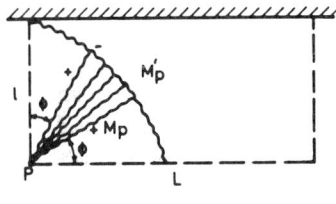

Fig. 7.89.

7.10.7. Determine the approximate limit load p_+ of a square isotropic plate, uniformly loaded, simply supported on one side, and supported by two corner columns on the opposite side (fig. 7.90). *Answer*: $p_+ = 8.5 M_p/l^2$. *Hint*: use mechanism of fig. 7.90. Differentiate with respect to x.

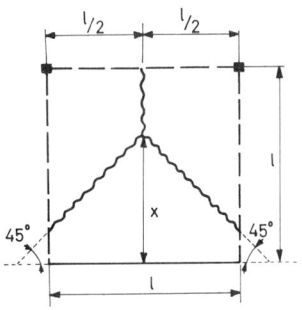

Fig. 7.90.

7.10.8. Same problem as Problem 7.10.7, but with a built-in edge with $M'_p = M_p$ replacing the simply supported edge. *Answer*: $p_+ = 11.2 M_p/l^2$.

7.10.9. Same problem as Problem 7.10.7, but two adjacent sides are simply supported and the remaining corner is column-supported. Take $l = 4$ ft. *Answer*: $p_+ = M_p/1.5$.

7.10.10. Determine the approximate total limit load P_+ of a circular isotropic plate (simple reinforcement), uniformly loaded ($P = p\pi R^2$), simply supported on its edge by four columns angularly apart from $\pi/2$. *Answer*: $P_+ = 14.1 M_p$.

7.10.11. Obtain the formula for the dissipation in a circular fan when the reinforcement is orthotropic (fig. 7.91). *Answer*: $D = (M_p + M'_p)\{[(1-k)/2] \cos(\beta+\alpha) \sin\phi + [(1+k)/2]\}$.

With the obtained formula, determine the limit load P_+ of a circular orthotropic plate subjected to a concentrated load at the center. *Answer*: $P_+ = \pi(1+k)(M_p + M'_p)$.

7.10.12. A very long, isotropic, doubly reinforced rectangular plate is supported on its two long sides and subjected to a concentrated load P at midspan. Its plastic moments M_p and M'_p are different.

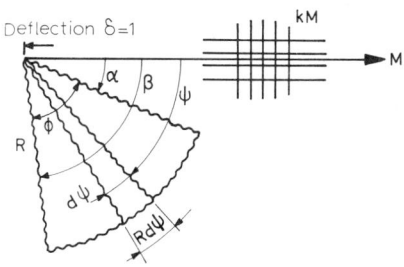

Fig. 7.91.

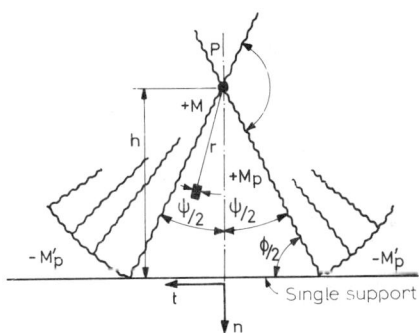

Fig. 7.92. (Reprinted by permission of Thames and Hudson Ltd. from R.H.Wood, *Plastic and Elastic Design of Slabs and Plates*, London, 1961.)

1. Show that the angle $\phi/2$, fig. 7.92, of the fans with the side must be larger than or equal to $\text{arccot}\sqrt{M'_p/M_p}$.

2. With the following equilibrium equations in polar coordinates, applicable to a field without polar symmetry,

$$\frac{1}{r^2}\frac{\partial}{\partial r}\left(r^2\frac{\partial M_r}{\partial r}\right) - \frac{1}{r}\frac{\partial M_\theta}{\partial r} + \frac{1}{r^2}\frac{\partial^2 M_\theta}{\partial \theta^2} - \frac{2}{r^2}\frac{\partial}{\partial r}\left(r\frac{\partial M_{r\theta}}{\partial \theta}\right) = -p,$$

$$V_r = \frac{M_r}{r} + \frac{\partial M_r}{\partial r} - \frac{M_\theta}{r} - \frac{1}{r}\frac{\partial M_{r\theta}}{\partial \theta},$$

$$V_\theta = -2\frac{M_{r\theta}}{r} - \frac{\partial M_{r\theta}}{\partial r} + \frac{1}{r}\frac{\partial M_\theta}{\partial \theta},$$

show that the moment field

$$M_\theta = +M_p,$$

$$M_r = -M_p \tan^2 \theta,$$

$$M_{r\theta} = 0,$$

is statically admissible (assuming the rigid part ourside the fans to be strong enough).

Determine the approximate limit load P_-, prescribing that ϕ be such to give plasticity on the edge. *Answer*: $P_- = 4\sqrt{M_p \cdot M'_p} + 2(M_p + M'_p)\phi$.

Compare with eq. (7.66), after having used eq. (7.67).

3. Determine the edge reactions between the fans.

7.10.13. Determine the plastic moment of the isotropic plate shown in fig. 7.93, with $M_p = M'_p$. *Answer*: $M_p = 1194$ lb. *Hint*: use the mechanism shown in fig. 7.93, and obtain the values of x_1, x_2, x_3, and x_4 by a "cut and try" process, for largest M_p.

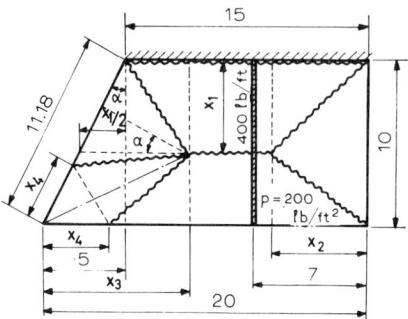

Fig. 7.93. (Reprinted by permission of Thames and Hudson Ltd. from R.H.Wood, *Plastic and Elastic Design of Slabs and Plates*, London, 1961.)

7.10.14. Determine the plastic moment of the orthotropic rectangular plate in fig. 7.94, built-in on one short side and column-supported on the two opposite corners. The loads are

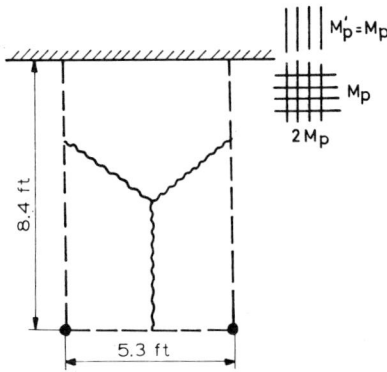

Fig. 7.94.

	lb/ft²
Dead weight (h_t=10 in.)	120
Covering	20
Live load at collapse	300
Total p	440

Answer: M_p = 12300 lb. *Hint*: use the affinity method.

References

[7.1] "Flexural Mechanics of Reinforced Concrete", *Proc. A.S.C.E.-A.C.I. Int. Symp.*, Miami, Nov. 1964, publ. by A.S.C.E., 1965.
[7.2] C.E.MASSONNET, "Limit Design Applied to Steel and Concrete Structures" (lecture notes), Univ. of California at Berkeley, Civil Eng. Dept., 1965.
[7.3] K.W.JOHANSEN, "Yield-line Theory" (translated from the Danish), *Cement and Concrete Association*, London, 1962.
[7.4] Comité Européen du Béton: "Recommandations pratiques unifiées pour le calcul et l'exécution des ouvrages en béton armé" (European Committee for Concrete: unified practical recommendations), 1964.
[7.5] W.PRAGER, "The General Theory of Limit Design", *Proc. 8th Int. Cong. Appl. Mech.*, Istanbul, 1952, **2**: 65, 1956.
[7.6] M.A.SAVE, "A Consistent Limit-Analysis Theory for Reinforced Concrete Slabs", *Magazine of Concrete Research*, **19**: 58, 3, March 1967, and discussion in **19**: 61, 252, December 1967.

[7.7] R.BAUS, S.TOLACCIA, "Calcul à la rupture des dalles en béton armé et étude expérimentale du critère de rupture en flexion pure", (Yield-line Theory and Experimental Investigation of the Yield Criterion of Reinforced Concrete Slabs in Pure Bending), *Ann. Inst. Tech. Bat. Trav. Pub.*, Paris, June 1963.

[7.8] M.P.NIELSEN, "Limit Analysis of Reinforced Concrete Slabs", *Acta Polytechnica Scandinavia Ci* 26, Copenhagen, 1964.

[7.9] K.O.KEMP, "The Yield Criterion for Orthotropically Reinforced Concrete Slabs", *Int. J. Mech. Sci.* 7: 11, November 1965.

[7.10] C.T.MORLEY, "On the Yield Criterion of an Orthogonally Reinforced Concrete Slab Element", *J. Mech. Phys. Solids*, **14**: 1, 33, January 1966.

[7.11] C.E.MASSONNET, M.A.SAVE, *Calcul plastique des constructions* (Plastic Analysis of Structures), Vol. 2, Centre belgo-luxembourgeois d'information de l'acier, Brussels, 1963.

[7.12] R.WOLFENSBERGER, "Traglast und optimale Bemessung von Platten", *Technische Forschungs- und Beratungsstelle der Schweizerischen Zement Industrie*, Wildegg, 1964.

[7.13] G.A.STEINMANN, "La théorie des lignes de rupture" (Yield-line Theory), Comite Europeen du Beton, *Bull. d'Inf.*, 27, September 1960.

[7.14] R.J.LENSCHOW, M.A.SOZEN, "A Yield Criterion for Reinforced Slabs", *A.C.I. Journal*, May 1967. Discussion of this paper is in *A.C.I. Journal*, November 1967.

[7.15] P.LENKEI, "On the Yield Condition for Reinforced Concrete Slabs", *Archiwum Inzynierii Ladovej*, **XIII**: 1, 5, Warszawa, 1967.

[7.16] R.H.WOOD, *Plastic and Elastic Design of Slabs and Plates*, Thames and Hudson, London, 1961. See also, by L.L.JONES, R.H.WOOD, *Yield-line Analysis of Slabs*, Thames and Hudson, Chatto and Windus, London, 1967.

[7.17] Comité Européen du béton (C.E.B.) (European Committee for Concrete), Bulletins 27, 43, 45.

[7.18] A.R.RJANITSYN, *Calcul à la rupture et plasticité des constructions* (Limit Analysis and Plasticity of Structures), Eyrolles, Paris, 1959.

[7.19] M.JANAS, "Kinematical Compatibility Problems in Yield-line Theory", *Mag. Conc. Research*, **19**: 58, 33, March 1967.

[7.20] E.H.MANSFIELD, "Studies in Collapse Analysis of Rigid-plastic Plates with a Square Yield Diagram", *Proc. Roy. Soc., Series A*, **241**: 311, 1957.

[7.21] A.SAWCZUK, "Grenztragfähigkeit der Platten", *Bauplanung-Bautechnik*, **11**: 7, 8, 1957.

[7.22] M.P.NIELSEN, "On the Calculation of Yield-line Patterns with Curved Yield Lines", *R.I.L.E.M. Bull.* 19, 67, June 1963.

[7.23] L.L.JONES, "Recent British Advances in Yield-line Analysis by the Equilibrium Method", in "Flexural Mechanics of Reinforced Concrete", *Proc. Int. Symp.*, Miami, Nov. 1964, A.S.C.E., 1965.

[7.24] R.H.WOOD, "Plastic Design of Slabs using Equilibrium Methods", in "Flexural Mechanics of Reinforced Concrete", *Proc. Int. Symp.*, Miami, Nov. 1964, A.S.C.E., 1965.

[7.25] K.O.KEMP, "The Evaluation of Nodal and Edge Forces in Yield-line Theory", *Mag. Conc. Research Special Pub.*, London, May 1965.

[7.26] C.T.MORLEY, "Equilibrium Methods for Least Upper Bounds of Rigid-plastic Plates", *Mag. Conc. Research Special Pub.*, London, May 1965.

REFERENCES

[7.27] R.H.WOOD, "New Techniques in Nodal-force Theory for Slabs", *Mag. Conc. Research Special Pub.*, London, May 1965.

[7.28] L.L.JONES, "The Use of Nodal Forces in Yield-line Analysis", *Mag. Conc. Research Special Pub.*, London, May 1965.

[7.29] L.L.JONES, *Ultimate Load Analysis of Reinforced and Prestressed Concrete Structures*, Chatto and Windus, London, 1962.

[7.30] M.HOLMES, K.A.STEEL, "Upper and Lower Bound Solutions to the Collapse of a Continuous Slab under Uniform Load", *Mag. Conc. Research*, **16**: 47, 83, June 1964.

[7.31] Z.SOBOTKA, "La limite supérieure et inférieure de la capacité portante des rectangulaires dalles encastrées" (Upper and Lower Bounds to the Collapse Load of Rectangular Built-in Slabs), Comité Européen du Béton, *Report of the 10th General Meeting*, London, October 1965.

[7.32] Z.SOBOTKA, "La capacité portante plastique des dalles rectangulaires encastrées avec la charge uniforme" (Plastic Carrying Capacity of Rectangular Plates with Uniform Load), *Acta Technica Csaw*, 6, 676, 1966.

[7.33] A.SAWCZUK, "Grenztragfähigkeit der Platten", *Bauplanung-Bautechnik*, **11**: 7, 315, July 1957; 8, 359, August 1957.

[7.34] K.O.KEMP, "A Lower-Bound Solution to the Collapse of an Orthotropically Reinforced Slab on Simple Supports", *Mag. Conc. Research*, **14**: 41, 79, July 1962.

[7.35] G.CERADINI, C.GAVARINI, "Calcolo a rottura e programmazione lineare", *Giornale del Genio Civile*, Jan.-Feb. 1965,

[7.36] C.GAVARINI, "I theoremi fondamentali del calcolo a rottura e la dualita in programmazione lineare", *Ingegneria civile*, 18, 1966.

[7.37] G.SACCHI, "Contribution à l'analyse limite des plaques minces en béton armé" (these), 1966, Fac. Polytechnique de Mons, Mons, Belgium.

[7.38] C.E.MASSONNET, "Complete Solutions Describing the Limit State of Reinforced Concrete Slabs", *Mag. Conc. Research*, **19**: 58, 13, March 1967.

[7.39] J.ZAWIDZKI, A.SAWCZUK, "Plastic Analysis of Fiber-Reinforced Plates under Rotationally Symmetrie Conditions", *Int. J. Solids and Struct.*, **3**: 3, 413, May 1967.

[7.40] A.HILLERBORG, *Strimlemethoden för Platter pa Pelare, Vinkelplattor M.M.*, Svenska Riksbyggen, Stockholm, 1959.

[7.41] R.E.CRAWFORD, "Limit Design of Reinforced Concrete Slabs", *Proc. A.S.C.E., J. Eng. Mech. Div.*, **90**: EM5, 321, October 1964.

[7.42] C.T.MORLEY, "The Minimum Reinforcement of Concrete Slabs", *Int. J. Mech. Sci.*, **8**: 4, 305, April 1966.

[7.43] M.JANAS, "Large Plastic Deflections of Reinforced Concrete Slabs", *Int. J. Solids and Struct.*, **3**: 4, November 1967.

[7.44] R.PARK, "Ultimate Strength of Rectangular Concrete Slabs under Short-term Uniform Loading with Edge Restrained Against Lateral Movement", *Proc. Inst. Civ. Eng.*, **28**: June 1964.

[7.45] "Effort tranchant", Colloquium de Wiesbaden, 1963. Comité Européen du Béton, Bull. d'Inf. 40, 41, 42, Paris, 1964 ("Shear strength", European Committee for Concrete).

[7.46] Y.GUYON, *Béton précontraint* (Prestressed Concrete), Vol. 2, Collection de l'ITBTP, Eyrolles, Paris, 1958.

[7.47] M.J.MULLER, "Quelques aspects du comportement des dalles et des poutres précontraintes en phase élastique et à la rupture" (Some Aspects of the Behavior of Prestressed Slabs and Beams in Elastic and Ultimate Ranges), Groupement belge de la precontrainte, publication A.B.E.M. 15, Brussels, March 1959.

[7.48] R.C.ELSTNER, E.HOGNESTAD, "Shearing of Reinforced Concrete Slabs", *J. Amer. Concrete Inst.*, July, 1956.

[7.49] C.FORSELL, A.HOLMBERG, "Stampellast pa plattor av betong" (Concentrated Loads on Concrete Slabs), *Betong* **31**: 2, Stockholm, 1946.

[7.50] G.D.BASE, "Some Tests on the Punching Shear Strength of Reinforced Concrete Slabs", *Technical Report*, Cement and Concrete Association, July 1959.

[7.51] J.MOE, "Shearing Strength of Reinforced Concrete Slabs and Footings under Concentrated Loads", Portland Cement Association, Research and Development Laboratories, Stokie, Ill., U.S.A., April 1961.

[7.52] A.GUERRIN, *Traite de Béton armé*, t. IV, Dunod, Paris, 1960.

8

Metal Shells

8.1. Introduction

Because the median surface of a shell exhibits at least one nonvanishing principal curvature, applied forces can be balanced exclusively, from the very beginning of the loading process, by forces acting at every point of the median surface in the corresponding tangent plane. The shell is then said to act as a membrane. This situation occurs when the shell has a very small flexural rigidity, supports with reactions in the tangent planes to the shell median surface at the boundary, and discontinuities in, neither geometry, (curvature, thickness) nor loading (concentrated loads). When the preceding conditions are not satisfied, flexural stresses arise in addition to membrane stresses.

Because shells frequently have a large slenderness ratio (span-to-thickness), the conservation of material normals to the median surface is well verified and transverse shear forces may therefore be classified as "reactions". On the other hand, large slenderness ratios may result in appreciable elastic and elastic-plastic deflections and particular attention must therefore be given to the danger of buckling. This problem, however, is beyond the scope of this book. The reader is referred to both theoretical ([8.1,8.2,8.3]) and experimental ([8.4 to 8.8]) papers on the subject. The dangers of fatigue and brittle fracture ([8.9,8.10,8.11]), are also very important, especially in welded shells with holes and nozzles. If, however, ductility is preserved by suitable precautions, like stress-relieving or annealing, plastic flow will level out stress concentrations due to various discontinuities. These stress concentrations in the elastic range can be calculated rather easily in many cases [8.12,8.13,8.14] but are not taken into account by building codes that implicitly rely on the ductile behavior of the material. Further development of limit analysis should provide a stronger basis for the specifications of the building codes [8.15].

8.2. Experiments on metal shells

8.2.1. *Conical shells*

Figs. 3.6 and 6.53 show typical load versus deflection diagrams of the conical shells with slenderness ratios of $\mu = 10/0.5 = 20$ tested by Onat [3.9]. We recall that P_{lo} is the limit load for the plate, regarded as a cone with vanishing height δ, and that the theoretical limit load for any cone corresponds to the intersection of the dashed curve (representing eq. (6.54)) with the parallel to the load axis at the distance δ/t from the origin. By inspection of these figures we conclude that:

1. For shells subjected to tension (fig. 6.53) the load versus deflection curve tends to exhibit a sharp bent in the neighborhood of the limit load, especially for the cone with the largest δ/t ($\delta/t=2$). The limit state is nevertheless stable and changes of geometry require the load for continued plastic flow to increase with the deflection.
2. For shells in compression (fig. 3.6) at the limit load (or even at a lower load as for shell with $\delta/t = 3$), large deflections occur at constant (or decreasing) load. The limit load then is a true failure load.

8.2.2. *Cylindrical shells*

Cylindrical shells of mild steel with external annular stiffeners were subjected to external hydrostatic pressure by M.E.Lunchick [8.16]. Without referring to load versus deflection diagrams, he gives experimental values of a "collapse pressure". Because the annular stiffeners did not participate in the collapse, the shell can be regarded as built-in at the successive annuli, and subjected to external pressure and axial force due to the pressure on the ends. In table 8.1 the theoretical limit pressure given by eq. (8.27) is compared with the experimental collapse pressure.

We see that simple plastic theory, which neglects geometry changes, is quite satisfactory.

An experiment similar to test in table 8.1, but with a cylinder machined from a solid piece rather than welded, gave an experimental collapse load of 0.97 times the limit load [8.17]. It is worth noting that the slenderness $\mu = 2R/t$ of the test cylinders varied from 113 (test 1) to 210 (test 6).

Table 8.1.

Test number	Shell parameter ω	Spacing of stiffeners $l'/2R$	Experimental collapse pressure / Theoretical limit pressure
1	0.932	0.114	0.914
2	0.857	0.095	0.905
3	0.857	0.095	0.975
4	1.388	0.110	1.029
5	1.951	0.202	0.971
6	1.290	0.109	1.026
7	2.077	0.168	1.063

Note: The notations used in this table are as follows: l', distance between median planes of adjacent stiffeners; R, mean radius of shell; l, clear distance between stiffeners; t, shell thickness, $\omega^2 = l^2/2tR$.

Aluminium cylinders with internal annular stiffeners were tested by Dehart and Basdekas [8.18] also under hydrostatic pressure. Though the main purpose of the experiments was to determine spacing and cross section of the stiffeners to avoid buckling of the shell and collapse of the stiffeners, the collapse pressure can be used for comparison with the limit pressure [eq. (8.27)]. We obtain table 8.2. The slenderness was $\mu = 17.1$, except for test 3G where $\mu = 18.3$.

Table 8.2.

Test number	Shell parameter ω	Spacing of stiffeners $l'/2R$	Experimental collapse pressure / Theoretical limit pressure
3F	4.36	1.060	0.905
3H	1.64	0.398	0.985
3G	3.40	0.790	0.87
3J	3.275	0.795	0.92

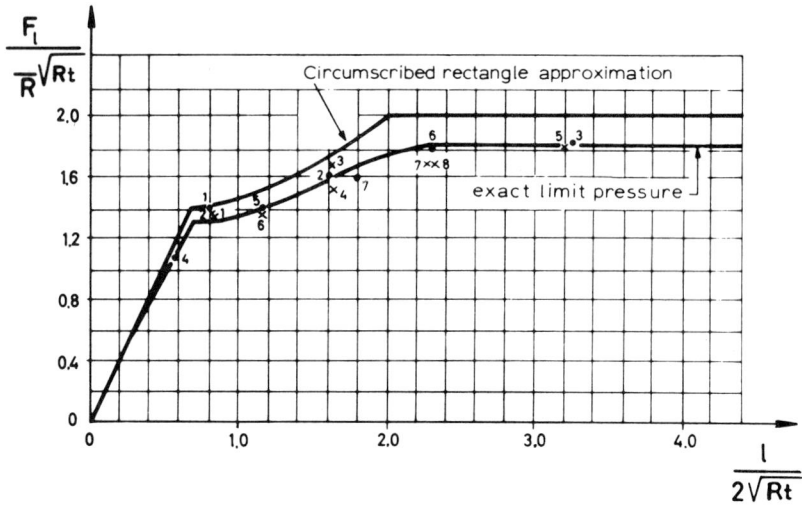

Fig. 8.1. Steel shells: ×. Aluminium shells: ●.

Fig. 8.2.

8.2] EXPERIMENTS ON METAL SHELLS 329

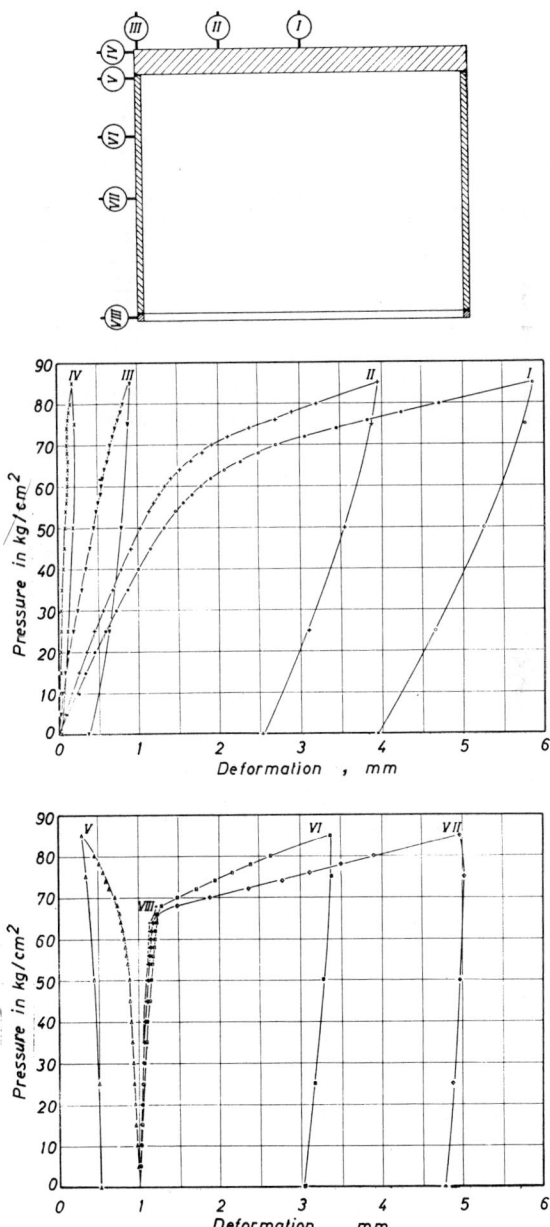

Fig. 8.3. Background plate (flat bottom).

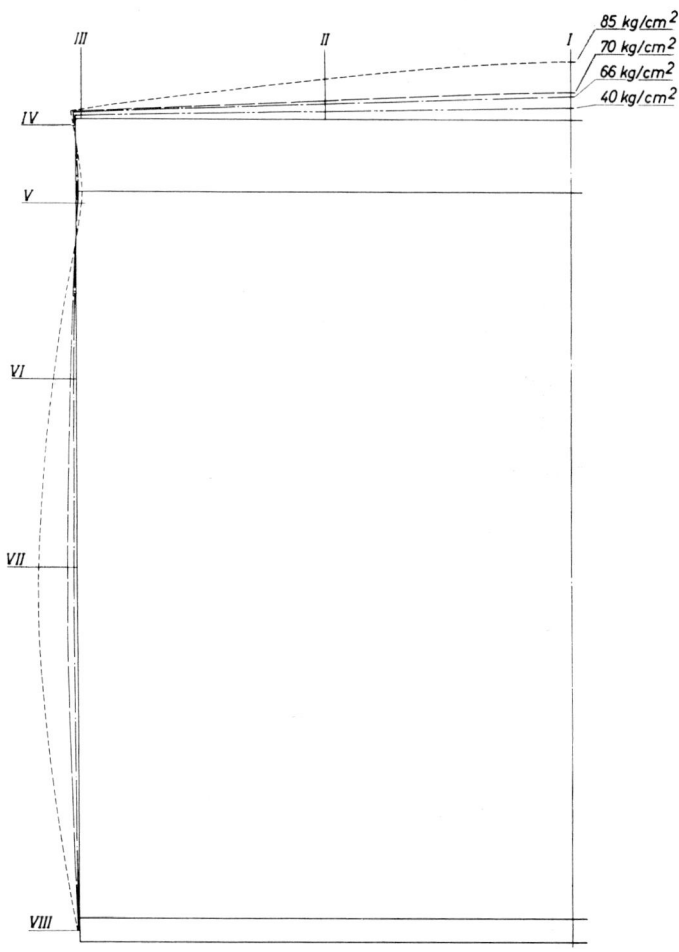

Fig. 8.4. Plastic deformation pattern: cylindrical vessel with flat head.

8.2] EXPERIMENTS ON METAL SHELLS

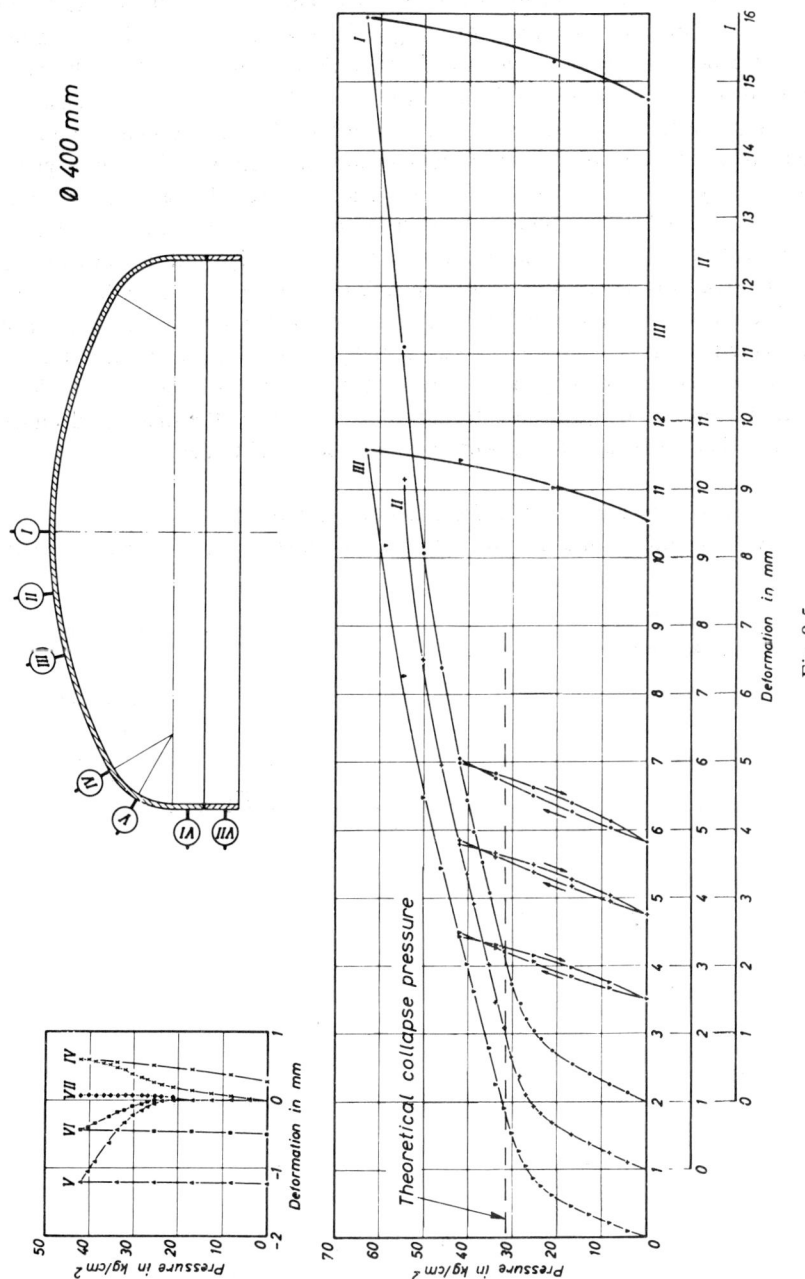

Fig. 8.5.

Very careful experiments by Demir and Drucker [8.19] on steel and aluminium cylinders subjected to a ring of force with total magnitude F show a sharp bent in the load versus deflection diagram, followed by a flat part. Hence changes of geometry do not interfere with the collapse mechanism by an appreciable amount. Experimental limit loads, corresponding approximately to the end of the sharp bent in the load versus deflection diagram, agree very well with theoretical prediction, as can be seen in fig. 8.1.

Twelve mild-steel cylinders, built-in at the bottom end and closed by a welded flat plate at the upper end (fig. 8.2) were subjected by Save [8.20] to internal pressure up to very large plastic deformations. Typical load versus deflection diagrams and deformed shapes are given in figs. 8.3 and 8.4. The predicted collapse mechanisms did actually occur and experimental limit pressures (corresponding to the end of the elastic-plastic bent in the load versus deflection diagrams, points marked △) differ from the theoretical values by less than 7%, (but are all lower than the theoretical values).

8.2.3. *Torispherical and toriconical pressure vessel heads*

Two toriconical and nine torispherical heads were each welded to a rigid

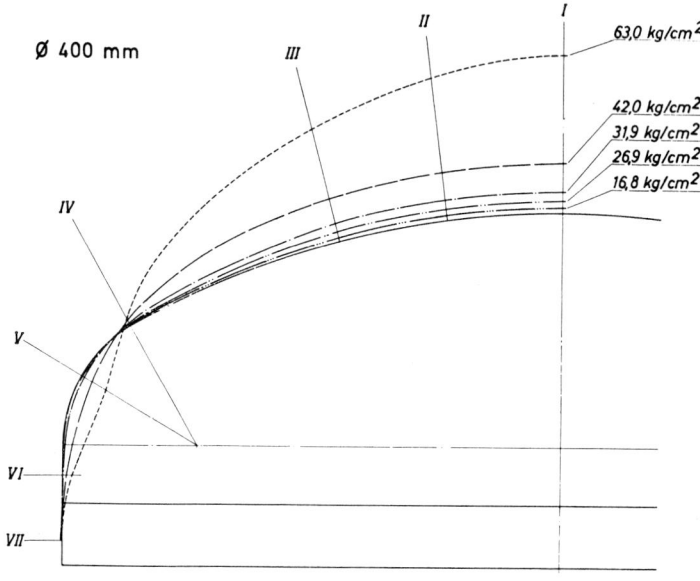

Fig. 8.6. Torispherical head deformations.

annulus and subjected to internal pressure by Save [8.20]. These test shells were annealed mild-steel industrial heads made by deep-drawing. Typical load versus deflection diagrams and deformed shapes are shown in fig. 8.5 ands 8.6. The predicted collapse mechanisms actually occur without being appreciably influenced by changes of geometry. Deflection at the theoretical limit

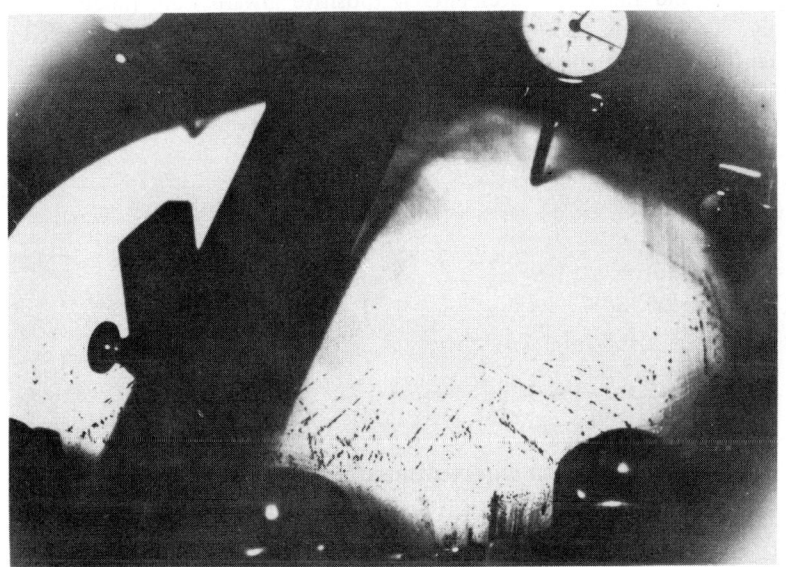

Fig. 8.7.

pressure remains less than 1% of the diameter. Experimental values of the limit pressure are larger than theoretical values by 0 to approximately 10%. The expected slip lines were clearly visible (fig. 8.7).

8.2.4. *Conclusions*

From the preceding review of experimental results, it can be concluded that shells are far less "degenerate" problems (that is, exhibiting a lesser influence of geometry change) than plates, and that limit loads actually correspond to strong modifications of the shell behavior, resulting in very large permanent deflections. Hence, limit loads can be regarded as real failure loads when unrestrained plastic flow is to be avoided.

8.3. Circular cylindrical shells axisymmetrically loaded

8.3.1. *Introduction*

We refer to the cylindrical coordinate system x, θ, r of fig. 8.8, with the origin at one end of the shell with length l. Because of the symmetry of revolution, circumferential displacements v vanish, whereas longitudinal displacements u and radial displacements w (positive inwards) are functions of x only. As noted in Section 5.1.4, the curvature rate $\dot{k}_\theta$ vanishes and, consequently, the generalized stresses are M_x, N_x, and N_θ (fig. 8.8). The corresponding generalized strain rates are

$$\dot{k}_x = -\frac{d^2 \dot{w}}{dx^2},$$

$$\dot{\epsilon}_x = \frac{d\dot{u}}{dx},$$

$$\dot{\epsilon}_\theta = -\frac{\dot{w}}{R}. \tag{8.1}$$

The rate of dissipation per unit median surface is

$$D = -M_x \frac{d^2 \dot{w}}{dx^2} + N_x \frac{d\dot{u}}{dx} - N_\theta \frac{\dot{w}}{R}. \tag{8.2}$$

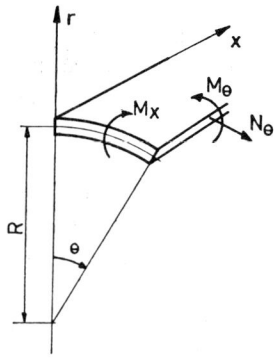

Fig. 8.8.

8.3] CIRCULAR CYLINDRICAL SHELLS AXISYMMETRICALLY LOADED

Using the "reduced stresses"

$$n_x = \frac{N_x}{N_p},$$

$$m_x = \frac{M_x}{M_p},$$

$$n_\theta = \frac{N_\theta}{N_p}, \qquad (8.3)$$

where $N_p = \sigma_Y t$ and $M_p = \sigma_Y t^2/4$, relation (8.2) becomes

$$D = \sigma_Y \left[m_x \left(-\frac{t^2}{4} \frac{d^2 \dot{w}}{dx^2} \right) + n_x t \frac{d\dot{u}}{dx} + n_\theta \left(-t \frac{\dot{w}}{R} \right) \right]. \qquad (8.4)$$

Hence, the generalized strain rates corresponding to the reduced stresses are

$$\dot{\Phi}_x = -\frac{t^2}{4} \frac{d^2 \dot{w}}{dx^2},$$

$$\dot{\lambda}_x = t \frac{d\dot{u}}{dx},$$

$$\dot{\lambda}_\theta = -t \frac{\dot{w}}{R}, \qquad (8.5)$$

and relation (8.4) is written as

$$D = \sigma_Y (m_x \dot{\Phi}_x + n_x \dot{\lambda}_x + n_\theta \dot{\lambda}_\theta). \qquad (8.6)$$

When the shell is subjected to an external radial pressure p (that depends on x only), the equilibrium equations of a shell element are (fig. 8.8)

(a) $R\dfrac{dV}{dx} + N_\theta + Rp = 0$,

(b) $\dfrac{dM_x}{dx} = V$. (8.7)

Elimination of V from these two equations yields

$$\dfrac{d^2 M_x}{dx^2} + \dfrac{N_\theta}{R} + p = 0.$$ (8.8)

Defining a "reduced pressure"

$$p^* \equiv \dfrac{pR}{\sigma_Y t},$$ (8.9)

Eq. (8.8) can be rewritten

$$\dfrac{tR}{4}\dfrac{d^2 m_x}{dx^2} + n_\theta + p^* = 0.$$ (8.10)

Except where otherwise stated, we shall use in the following the linearized Tresca yield surface represented in fig. 5.30, with eqs. (5.89) for the various planes.

8.3.2. Built-in cylindrical shell under external uniform pressure p

As noted in Section 8.2, the present case represents fairly accurately the behavior of a part of a shell between rigid stiffening rings. The part of the shell is built-in at the ring to the extent that strains must vanish there, but the shell is subjected to an axial compressive force

$$2\pi R N_x = p\pi R^2.$$ (8.11)

Hence, there is a constant longitudinal normal stress

$$n_x = \dfrac{-p\pi R^2}{2\pi R \sigma_Y t} = -\dfrac{p^*}{2}.$$ (8.12)

Considering first very short shells, we can reasonably assume that they will collapse by axial compression, with $n_x = -1$. For sufficiently short shells, we

8.3] CIRCULAR CYLINDRICAL SHELLS AXISYMMETRICALLY LOADED 337

thus may expect $-1 \leq n_x \leq -\frac{1}{2}$. Because the shell will certainly contract circumferentially, we have $n_\theta < 0$ and $\dot{\lambda}_\theta < 0$. Hence, the stress point is likely to lie on a "stress profile" such as the segment ab in fig. 5.30, but located on face I' symmetrical of face I with respect to the origin. The equation of face I' is

$$n_\theta = -1 . \tag{8.13}$$

The normality law gives:

$$\dot{\lambda}_\theta \equiv -t\frac{\dot{w}}{R} = -1 ,$$

$$\dot{\lambda}_x \equiv t\frac{d\dot{u}}{dx} = 0 ,$$

$$\dot{\Phi}_x \equiv -\frac{t^2}{4}\frac{d^2\dot{w}}{dx^2} = 0 ,$$

wherewith we obtain

$$\dot{w} = C_1 x + C_2 ,$$

$$\dot{u} = C_3 . \tag{8.14}$$

The shell tends to deform into two conical parts (fig. 8.9) with "hinge circles" at $x = 0, x = \frac{1}{2}$ and $x = 1$. At these hinge circles, which are similar to the plastic hinges in beam theory, the rate of slope, $d\dot{w}/dx$, in the meridional plane varies discontinuously. This situation can occur only if the stress points are on edges of the yield surface. For $x = 0$, the stress point is on the intersection of

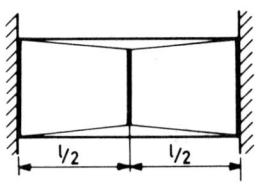

Fig. 8.9.

faces I′ and III′. Equation of face III′ is $-n_x - m_x = 1$, obtained by substituting $-m_x$ for m_x in eq. (5.89) of face III. On this face, from the normality law and eqs. (8.5) we have:

(a) $\quad \dfrac{t^2}{4} \dfrac{d^2 \dot{w}}{dx^2} = 1$,

(b) $\quad \dot{w} = 0$,

(c) $\quad t \dfrac{d\dot{u}}{dx} = -1$. \hfill (8.15)

From eqs. (8.15a) and (8.15c) we obtain

$$\frac{d^2 \dot{w}}{dx^2} = -\frac{4}{t} \frac{d\dot{u}}{dx}. \tag{8.16}$$

Relation (8.16) shows that the slope $d\dot{w}/dx$ is discontinuous by an amount equal to $-(4/t)\dot{u}$.

For $x = l/2$, the stress point is on the intersection of faces I′ and III. The equation of face III is $m_x - n_x = 1$.

For $0 < x < l/2$, the stress point is in the face I′ with the equation $n_\theta = -1$. Therefore, taking into account eq. (8.12), we have, for $x = 0$,

$$m_x = -1 - n_x = -1 + \frac{p^*}{2}, \tag{8.17}$$

and for $x = l/2$,

$$m_x = 1 + n_x = 1 - \frac{p^*}{2}, \qquad \frac{dm_x}{dx} = 0. \tag{8.18}$$

Integration of eq. (8.10) with $n_\theta = -1$, and use of boundary conditions (8.17) and (8.18) yields

8.3]   CIRCULAR CYLINDRICAL SHELLS AXISYMMETRICALLY LOADED   339

(a) $m_x = \dfrac{2l^2}{tR}(1-p^*)\left(\dfrac{x^2}{l^2} - \dfrac{x}{l}\right) - 1 + \dfrac{p^*}{2}$,

(b) $n_x = -\dfrac{p^*}{2}$,

(c) $n_\theta = -1$, (8.19)

and the limit value

$$p^* = 1 + \dfrac{1}{1 + l^2/2tR},\qquad(8.20)$$

obtained by substituting $l/2$ for x and $1 - p^*/2$ for m_x in eq. (8.19a).

The collapse mechanism corresponding to the stress field (8.19) is given by eqs. (8.14) and the discontinuity conditions at $x = 0$ and $x = l/2$. Assuming vanishing rates of displacement at the left built-in end ($x=0$), we have,

$$\dot{w} = \dot{w}_o x \qquad \text{for} \qquad 0 < x < \dfrac{l}{2}, \qquad (8.21)$$

where $\dot{w}_o$ is a positive parameter which is the product of l by half the rate of deflection in the median cross section. From eq. (8.21) we see that $d\dot{w}/dx = \dot{w}_o$. Hence, the jump of $\dot{u}$ at $x = 0$ is $\dot{u}] = -(t/4)\dot{w}_o$ and, from eq. (8.14),

$$\dot{u} = -\dfrac{t}{4}\dot{w}_o \qquad \text{for} \qquad 0 < x < \dfrac{l}{2}. \qquad (8.22)$$

In the interval $l/2 < x < l$, we have

$$\dot{w} = \dot{w}_o(l-x), \qquad (8.23)$$

$$\dfrac{d\dot{w}}{dx} = -\dot{w}_o, \qquad (8.24)$$

$$\dot{u} = \dfrac{t}{4}\dot{w}_o. \qquad (8.25)$$

The discontinuity in slope at $x = l/2$ is $2\dot{w}_o$. Having found a statically admissible stress field, eqs. (8.19), and a corresponding kinematically admissible mechanism, eqs. (8.21) to (8.25), the limiting value (8.20) of the pressure will be exact if the considered stress field and mechanism remain admissible for any possible value of the shell parameter $l^2/2tR$. Inspection of eqs. (8.20) and (8.19b) immediately shows that n_x is always bounded by $-\frac{1}{2}$ and -1 regardless of the value of $l^2/2tR$, and that m_x is monotonically increasing as x varies from 0 to $l/2$. Hence, the stress profile is always adequate and, if we let

$$\omega^2 = \frac{l^2}{2tR}, \tag{8.26}$$

the limit pressure can be written, using eqs. (8.20) and (8.9), as

$$p_l = \frac{\sigma_Y t}{R}\left(1 + \frac{1}{1+\omega^2}\right). \tag{8.27}$$

B.Paul [8.21] has obtained the exact limit pressure for the shell considered above when it is strengthened by a *ring stiffener at midspan*. His results are summarized in fig. 8.10. The yield surface for the shell is that of fig. 5.30. The yield condition of the annulus was given by B.Paul [8.22]. It is greatly simplified in the present case where, on account of symmetry, the ring can collapse exclusively from compression. If the ring remains rigid, the reduced limit pressure is,

$$p_l^* = 1 + \frac{1}{1+\omega^2/4} \quad \text{if} \quad \frac{4\omega^2}{4+\omega^2} \leq \gamma, \tag{8.28}$$

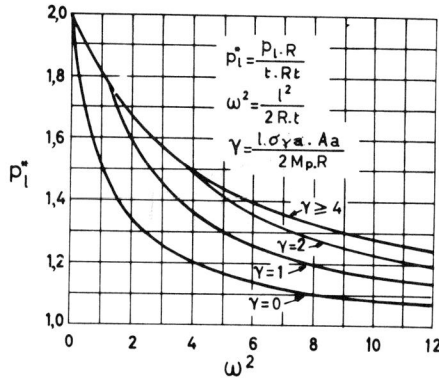

Fig. 8.10.

8.3] CIRCULAR CYLINDRICAL SHELLS AXISYMMETRICALLY LOADED   341

where $\gamma = (l/2M_p R)\sigma_{Ya} A_a$, and σ_{Ya} is the yield stress of the annulus, A_a the cross-sectional area of the annulus, l the total span of the cylinder, and R the mean radius of the cylinder. When $\gamma \leqslant 4\omega^2/(4+\omega^2)$, the ring collapses with the shell.

8.3.3. Cylindrical shell with axial load. Various other cases
8.3.3.1. Cylindrical shell simply supported at both ends and subjected to uniform pressure

The solution of this problem is due to Hodge and Paul [8.23] who found the following exact limit pressure:

$$p_l = \frac{\sigma_Y t}{R} \left(1 + \frac{1}{1+2\omega^2}\right). \tag{8.29}$$

8.3.3.2. Cylindrical shell built-in at one end, free at the other end, and subjected to uniform radial pressure and independent axial load

Let p be the pressure and $2\pi R N$ the total axial load. If we replace the section of the yield surface of fig. 5.12 with the plane $n_1 = -|N|/N_p$ by the circumscribed rectangle, the following interaction formula is obtained:

$$p = \frac{N_p}{R}\left(1 + \frac{N}{N_p}\right) + \frac{2}{l^2} M_p \left[1 - \left(\frac{N}{N_p}\right)^2\right], \tag{8.30}$$

where l is the length of the shell (Onat [8.24]).

The same problem was treated in a slightly different manner by Sankaranarayanan [8.25].

8.3.3.3. Cylindrical shell joined a both ends to rigid plates, and subjected to radial pressure and independent axial load

A detailed solution can be found in a paper by Hodge and Panarelli [8.26] or in the recent book by Hodge [8.27]. Close bounds on the interaction curve p versus t are given in figs. 8.11 and 8.12 [8.26] for a Tresca and a von Mises material, respectively. The dimensionless pressure p^* and axial load n^* are defined by

$$p^* = \frac{pR}{\sigma_Y t}, \qquad n^* = \frac{N}{2\pi R \sigma_Y t},$$

where N is the total axial load, the other symbols having the same meaning as in Section 8.3.2.

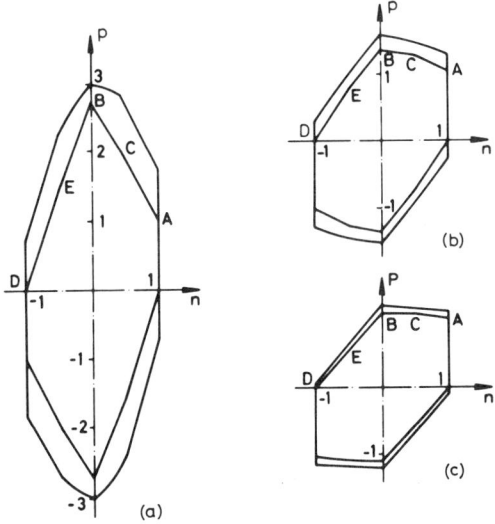

Fig. 8.11. Best bounds on interaction curves for Tresca shell. (a) $\omega = 1$. (b) $\omega = 2$. (c) $\omega = 4$.

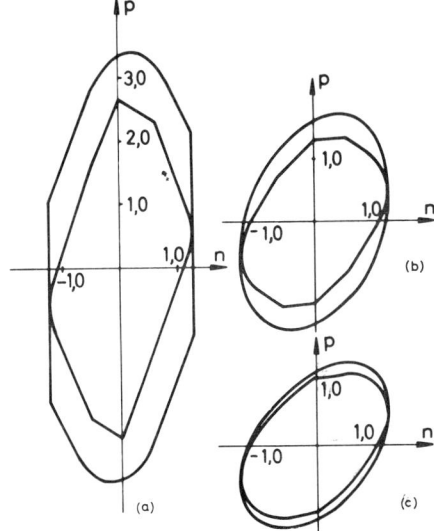

Fig. 8.12. Best bounds on interaction curve for Mises shell. (a) $\omega = 1$. (b) $\omega = 2$. (c) $\omega = 4$.

8.3.3.4. *Closed cylindrical vessel subjected to internal pressure*

This problem of important practical interest has been treated by Hodge [8.28] on the basis of his limited-interaction yield condition consisting of two independent Tresca hexagons for bending moments and axial forces. The solution is very simple and yet, when compared with more refined and laborious analysis [8.29,8.30], retains all essential results with sufficient accuracy for all vessels with current proportions (that is, with $L/R > \sqrt{0.5P/\sigma_Y}$, $t' > t$, fig. 8.13). It can be further simplified if we note that, for a simply supported circular plate obeying the yield condition of Tresca and subjected to transversal uniformly distributed load p and edge moments M_o, the limiting value of p is given by [8.31]

$$p_l \cong \sigma_Y \left[\frac{3}{2}\left(\frac{t'}{R}\right)^2 - 5.64 \frac{M_o}{\sigma_Y R^2} \right], \qquad (8.31)$$

as long as M_o does not exceed half the plastic moment of the plate. The limit pressure of the cylinder is related to its thickness by

$$\frac{t}{R} = \frac{p_l}{\sigma_Y} \frac{2}{1 + (1 + 4p_l R^2/L^2 \sigma_Y)^{1/2}} . \qquad (8.32)$$

Eq. (8.32) is a modified version of Hodge's formula, more suitable for accurate numerical computations.

Note that, because the end plates never collapse by in-plane deformation

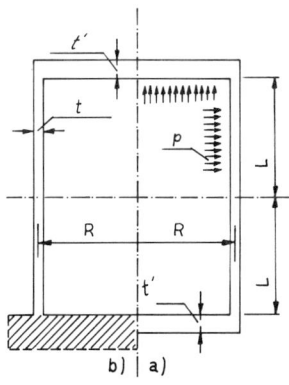

Fig. 8.13.

and because the yield conditions for axial forces and bending moments are independent, the solution is valid for both types of vessel shown in fig. 8.13, either with two end plates or with one end plate and one built-in end.

8.3.3.5. *Rib-reinforced cylindrical shell*

Yield loci for rib-reinforced shells consisting of ribs and symmetric sheeting have been derived by Nemirovsky and Rabotnov [8.32]. Cylindrical shells

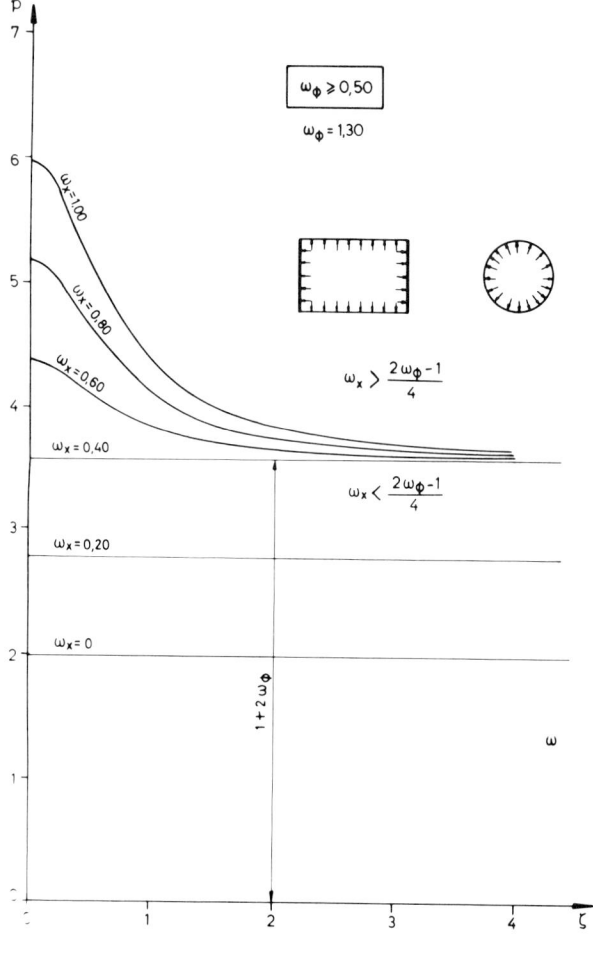

Fig. 8.14.

8.3] CIRCULAR CYLINDRICAL SHELLS AXISYMMETRICALLY LOADED

without axial load, with purely longitudinal ribs on one side, have been treated by Biron and Sawczuk [8.33] who derived the yield curves and gave some illustrative examples. Recently, Capurso and Gandolfi [8.34] have given the yield locus of an axisymmetrically loaded shell of revolution with I reinforced ribs placed along meridians and parallels. Ribs were considered as ideal sandwich beams. These authors have treated examples of cylindrical shells with various end conditions [8.35]. The curves of figs. 8.14 and 8.15 give the reduced limit pressure p^* that depends on

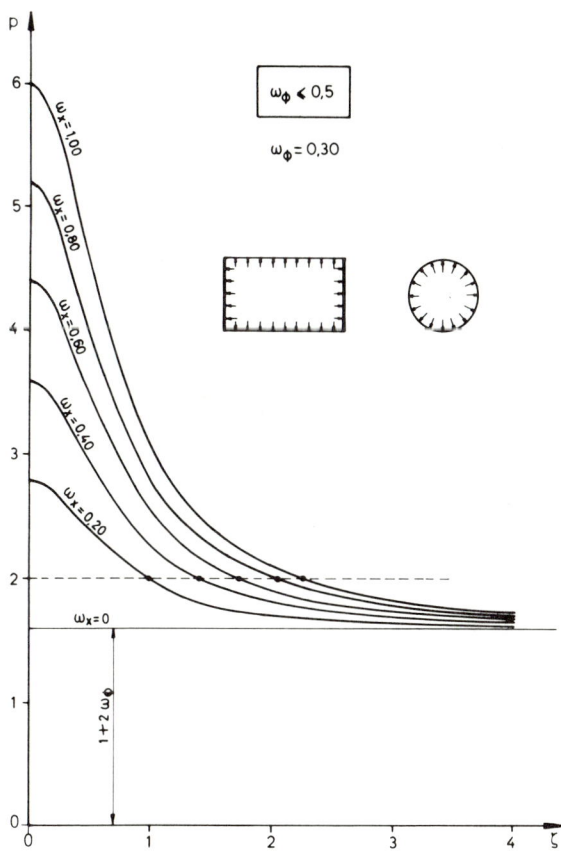

Fig. 8.15.

$$\omega_x = \frac{N_{px}}{2\sigma_Y B_x t}, \qquad \omega_\theta = \frac{N_{p\theta}}{2\sigma_Y B_\theta t},$$

and on $\zeta = L^2/2RH_x^*$ for the particularly interesting case of a closed vessel subjected to uniform pressure.

In the preceding relations, R is the radius of the shell sheet, N_p the plastic axial force of the sheet, ω_{px} and $\omega_{p\theta}$ the plastic axial forces of a longitudinal and a circumferential rib, respectively, B_x and B_θ the spacings of these ribs, t the sheet thickness, and $2L$ the length of the cylinder. H_x^* is the ratio of the plastic moment of a longitudinal rib (with a part of the cover sheet included) to its plastic axial force ω_{px}.

8.3.4. Cylinder without axial load

8.3.4.1. *Cylindrical shell built-in at both ends and subjected solely to radial uniform pressure*

In the absence of axial loads, and with the symmetry of revolution, the yield surface reduces to the M_x versus N_θ interaction curve and the problem is accordingly simplified. Using the linearized Tresca yield condition of fig. 5.24, Hodge [3.8] has obtained the diagrams of fig. 8.16 where the reduced limit pressure $p^* = p_l R/\sigma_Y t$ and the extent $\eta = x/(l/2)$ of the shell in regime

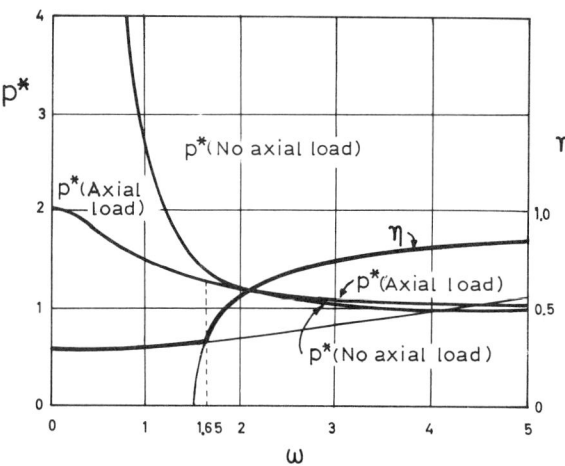

Fig. 8.16. (From *Plastic Analysis of Structures* by P.G.Hodge Jr. Copyright 1959, McGraw-Hill Book Company. Used by permission of McGraw-Hill Book Company, Inc.)

8.3] CIRCULAR CYLINDRICAL SHELLS AXISYMMETRICALLY LOADED

ED are given as functions of the shell parameter ω. We note that there are two types of solution, depending on the value of ω with respect to 1.65.

The limit pressure given by eq. (8.20) is also represented in fig. 8.16 for comparisons.

8.3.4.2. Cylindrical shell subjected to a ring of pressure

When the shell is infinitely long and subjected to a ring of pressure with magnitude $2F$, we have seen in Section 5.5.4 that the reduced load $f_{ol} = 2F_{ol}/\sqrt{M_p N_p/R}$ depends on the yield condition as shown in table 5.4. Shells with finite length were treated by Eason and Shield [8.36] with a yield rectangle circumscribed to the Tresca yield curve of fig. 5.24, and by Eason [8.37] and Demir [8.38] with the exact Tresca curve of the same figure. When the ring of load is located in the transversal plane of symmetry of the shell limit loads are given in fig. 8.1. Loads not located in the plane of symmetry and comparison with elastic analysis are treated in refs. [8.37] and [8.38], respectively.

8.3.4.3. Cylindrical tank subjected to hydrostatic pressure

The following yield condition is used:

$$n_\theta = \pm 1,$$

$$m_x = \pm 1, \tag{8.33}$$

with

$$n_\theta = \frac{N_\theta}{N_{p\theta}},$$

$$m_x = \frac{M_x}{M_{px}}. \tag{8.34}$$

Condition (8.33) represents a rectangle circumscribed to the Tresca hexagon of fig. 5.24. Possible orthotropy of the shell can be taken into account regarding $N_{p\theta}$ and M_{px} as independent. *For a liquid-filled tank shell that is vertical, free at the upper edge, and built-in at the bottom edge*, the stress distributions and the collapse mechanism are shown in fig. 8.17.

We have [8.39]

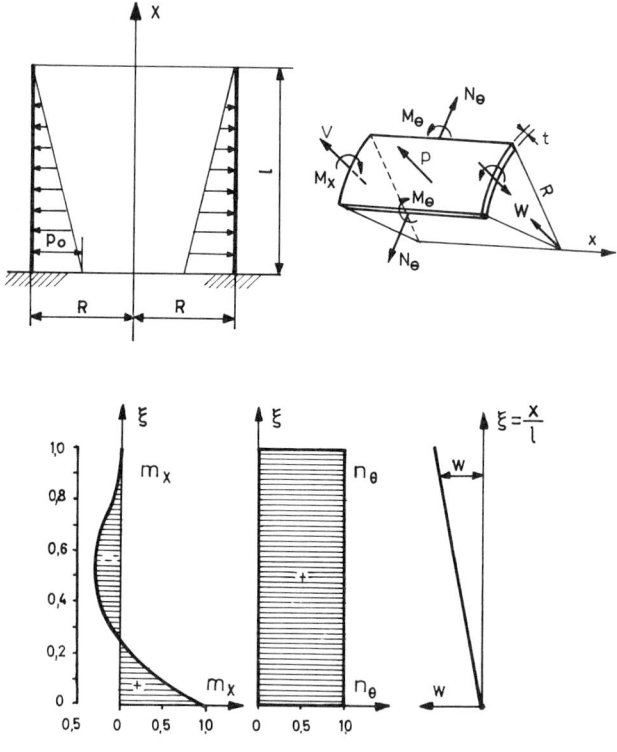

Fig. 8.17.

$$m_x = \frac{C^2}{2}(p_o^{**}-1)\xi^2 - \frac{C^2}{6}p_o^{**}\xi^3 + (3C^2-3-p_o^{**}C^2)\xi + 1,$$

where $\xi = x/l$, $C = l\sqrt{N_{p\theta}/RM_{px}}$, $p_o^{**} = p_o N_{p\theta}/R$, and $p = p_o(1-\xi)$. The limit pressure is

$$p_{ol}^{**} = 3\left(1 + \frac{2}{C^2}\right). \tag{8.35}$$

It is given by the curve a in fig. 8.18. The solution (8.35) is valid only for "short" shells, for which $C^2 < 17.1$, because the maximum absolute value of the bending moment for $x > 0$ must remain smaller than (or at most equal to) M_{px}.

8.3] CIRCULAR CYLINDRICAL SHELLS AXISYMMETRICALLY LOADED 349

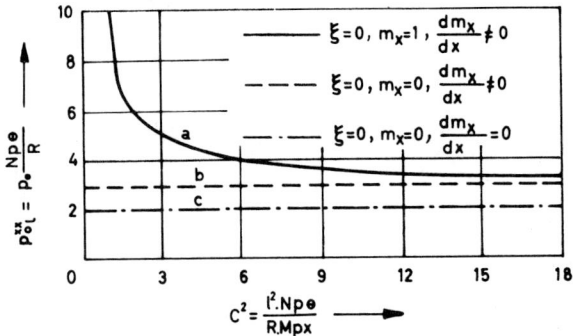

Fig. 8.18.

When the bottom edge is simply supported or free, curve *a* must be replaced by curve *b* or *c*, respectively.

For shells that we simply supported at both edges, the various possible mechanisms, moment distributions, limit loads, and hinge locations are given in figs. 8.19 to 8.21 [8.39]. For all cases, $n_x = \pm 1$.

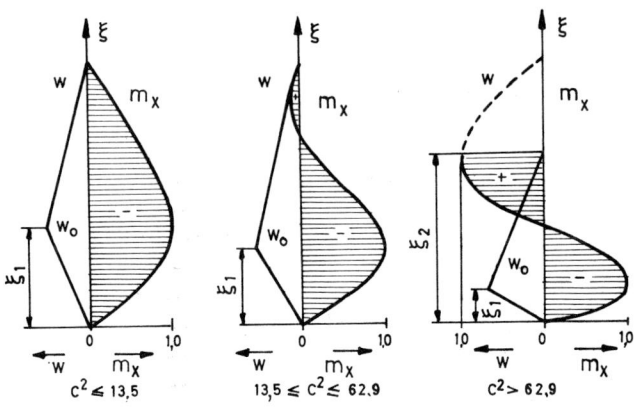

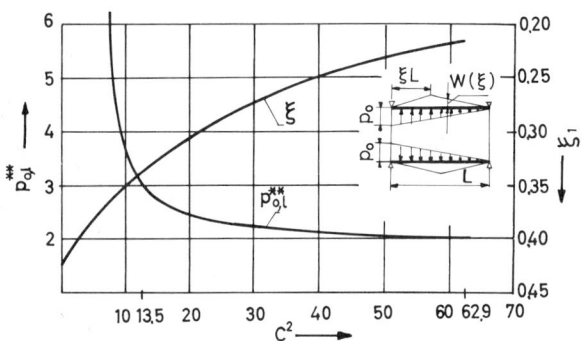

Fig. 8.20.

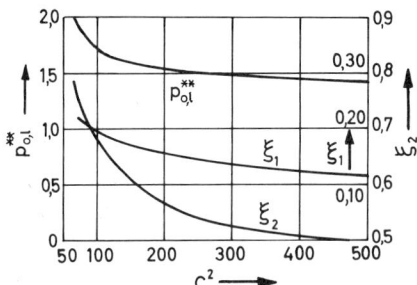

Fig. 8.21.

8.3.5. *Second-order effects in cylindrical shells*

Limit loads obtained in Sections 8.3.1 to 8.3.4 are based on the rigid perfectly plastic idealization, but real shells are neither rigid nor perfectly plastic but elastic work hardening. When cylindrical shells are subjected to both transversal and axial loads, inward radial deflections provide lever arms for the compressive axial forces, introducing supplementary bending moments. The situation is similar to the divergence of equilibrium state in beams. The considered second order effects occur in elastic-plastic ranges and are likely to decrease the load corresponding to unrestrained plastic flow, if we neglect the counteracting influence of work hardening. We hereafter follow B.Paul [8.40] in studying a closed cylindrical shell subjected to uniform external pressure. Equilibrium must now be formulated in the deflected situation (fig. 8.22). Let $w(x)$ be the radial displacements. Moment equilib-

8.3] CIRCULAR CYLINDRICAL SHELLS AXISYMMETRICALLY LOADED 351

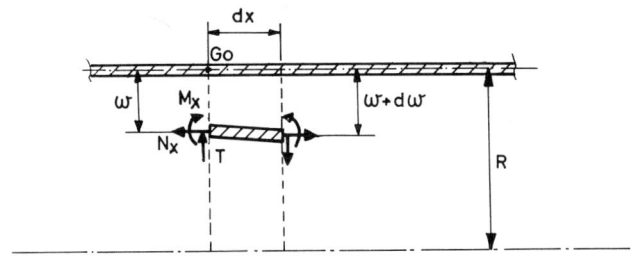

Fig. 8.22.

rium about the centroid, in the indeformed state, of an elementary cross section with central angle $d\theta$ is

$$dM_x + N_x\, dw + dN_x w = V\, dx\ .$$

Because N_x is a constant, the preceding equation can be rewritten as

$$\frac{dM_x}{dx} + N_x \frac{dw}{dx} = V\ . \tag{8.36}$$

Eq. (8.7a) remains unaltered. Eliminating V from eqs. (8.7a) and (8.36), and using the already defined reduced variables, we obtain

$$\frac{tR}{4}\frac{d^2 m_x}{dx^2} + Rn_x \frac{d^2 w}{dx^2} + n_\theta + p^* = 0\ , \tag{8.37}$$

to couple with

$$n_x = -\frac{p^*}{2}\ . \tag{8.38}$$

We try a stress profile of the type used in Section 8.3.2, a profile on face I' and its intersection with faces III and III', fig. 5.30. We have

$$-1 \leqslant n_x \leqslant -\tfrac{1}{2}\ ,$$

$$-(1+n_x) \leqslant m_x \leqslant 1 + n_x\ . \tag{8.39}$$

Except for the end points of the stress profile, the plastic curvature κ_x^p

vanishes so that the curvature is purely elastic and hence given by Hooke's law. Using reduced variables and taking account of $\kappa_\theta = 0$, we may write

$$\Phi_x \equiv -\frac{t^2}{4}\frac{d^2w}{dx^2} = \frac{3}{4}m_x(1-\nu^2)\frac{t\sigma_Y}{E}, \qquad (8.40)$$

where E is Young's modulus and ν is Poisson's ratio for the shell material. With the origin of the abscissa x at midspan, we have

(a) $\left(\dfrac{dm_x}{dx}\right)_{x=0} = 0$.

Substituting the expression (8.40) for d^2m_x/dx^2 and -1 for n_θ (stress profile on face I′) in eq. (8.37), we obtain a second order differential equation with constant coefficients in the unknown m_x. Integration of this equation, and use of relation (a), yields

$$m_x = C\cos\frac{2kx}{l} + 2\frac{\omega^2}{k^2}(1-p^*), \qquad (8.41)$$

where

$$k^2 = 3(1-\nu^2)\beta^2\omega^4 p^*$$

and

$$\beta^2 = \frac{2\sigma_Y R^2}{El^2}.$$

The integration constant C and the limit value p^* are obtained from the conditions that the extreme points of the stress profile correspond to the central and the end cross sections (sections of hinge circles). We thus obtain [8.40]:

1. For the shell built-in at both ends:

$$C = 1 - \frac{p^*}{2} - 2\frac{\omega^2}{k^2}(1-p^*),$$

$$p^* = \frac{1-\cos k + \dfrac{k^2}{2\omega^2}(1+\cos k)}{1-\cos k + \dfrac{k^2}{4\omega^2}(1+\cos k)}. \qquad (8.42)$$

8.3] CIRCULAR CYLINDRICAL SHELLS AXISYMMETRICALLY LOADED

Note that eq. (8.42) is transcendental because k depends on p^*. If E tends to infinity, k^2 tends to zero and eq. (8.42) reduces to eq. (8.27), valid for the rigid-plastic shell. Let p_o^* be the corresponding limit value: $p_o^* = 1 + (1/1+\omega^2)$. The value (8.42) of p^* decreases monotonically from p_o^* to unity when k varies from zero to π. For $k > \pi$, relation (8.39) ceases to be satisfied and the stress profile must be modified (except for $k = \pi + 2n$ with $n = 1,2,3,...$, corresponding to $p^* = 1$). It is shown in ref. [8.40] that, for all $k > \pi$, several statically admissible stress fields can be found that all give $p^* = 1$. Moreover, corresponding mechanisms can be obtained for all these stress fields (either for $k < \pi$ or $k > \pi$). We define a new shell parameter γ by the following relation

$$\gamma^2 \equiv \frac{k^2}{p^*} = 3(1-\nu^2)\beta^2\omega^4 = \frac{3}{2}(1-\nu^2)\frac{\sigma_Y l^2}{Et^2}. \tag{8.43}$$

It is easily seen that,

$$\gamma^2 < \pi^2 \quad \text{when} \quad k^2 < \pi^2 \quad \text{because} \quad p^* > 1,$$

and

$$\gamma^2 > \pi^2 \quad \text{when} \quad k^2 > \pi^2 \quad \text{because} \quad p^* = 1.$$

A shell will therefore be called "short" if $\gamma^2 \leq \pi^2$, and eq. (8.42) will hold, whereas it will be called "long" when $\gamma^2 \geq \pi^2$, with $p^* = 1$.

2. Shell simply supported at both ends. It is found [8.40] that

$$p^* = \frac{[(k^2/2\omega^2)-1]\cos k + 1}{[(k^2/4\omega^2)-1]\cos k + 1}, \tag{8.44}$$

for short shells with $\gamma^2 \leq (\pi/2)^2$, and that $p^* = 1$ for long shells with $\gamma^2 \geq (\pi/2)^2$.

3. Shell simply supported at one end and built-in at the other.

$$p^* = (2+2\omega^2\psi)(1+2\omega^2\psi), \tag{8.45}$$

with $\psi = 2/k^2 (\sec ku - 1)$ and $2 \cos ku = 1 + \cos k(2-u)$ for short shells with $\gamma^2 \leq (\frac{3}{4}\pi)^2$. For long shells with $\gamma^2 \geq (\frac{3}{4}\pi)^2$, $p^* = 1$.

Closed cylindrical shells subjected to external uniform pressure can thus be classified into "short shells" or "long shells" when

$$\gamma^2 = \frac{3}{2}(1-\nu^2)\frac{\sigma_Y l^2}{Et^2} \qquad (8.46)$$

is smaller or larger than $(n\pi/4)^2$, respectively, with n = 2, 3, or 4 for the shell simply supported at both ends, simply supported at one end and built-in at the other, or built-in at both ends, respectively. It must be emphasized that the values of the pressure obtained above should be regarded, from a rigorous point of view, exclusively as upper bounds to the actual carrying capacity, because the theorems of limit analysis do not apply when equilibrium equations are referred to the deformed state.

The "carrying capacities" p^* obtained above are given in figs. 8.23, 8.24, and 8.25. In fig. 8.23, the lower parts of the curves give the buckling pressures according to Batdorf [8.3], that have been used when these pressures are smaller than the carrying capacities. It turns out that, in the whole considered range of β, the carrying capacity obtained in Section 8.3.5 is smaller than the buckling pressure for *all* short shells. Moreover, the range of long shells in the same situation is very narrow. Accordingly, this simple rule may be accepted: long shells tend to collapse by buckling and short shells by unrestrained

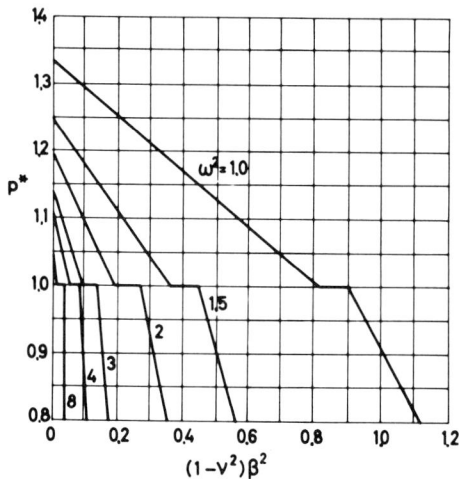

Fig. 8.23.

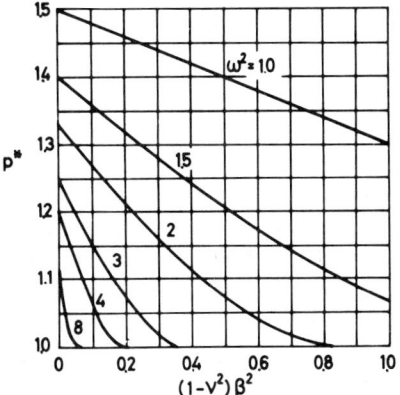

Fig. 8.24.

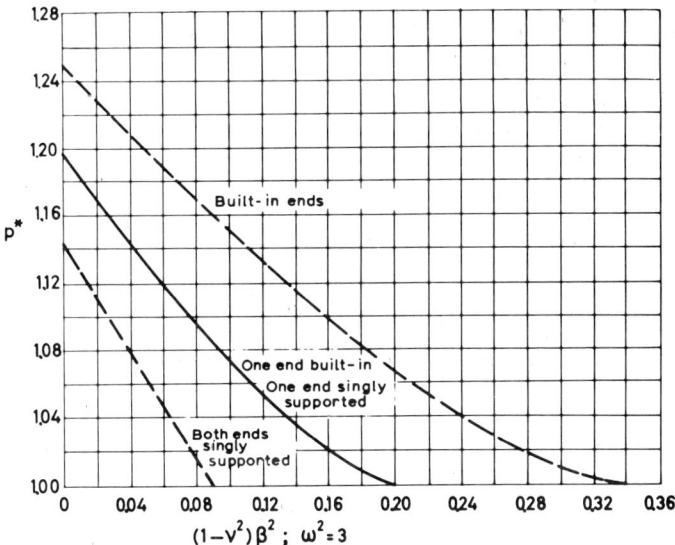

Fig. 8.25.

plastic flow. The terms *short* and *long* are defined unambiguously by means of the parameter γ. It is also worth noting that, as for all long shells $p^* = 1$, annular rigid stiffeners will influence the value of the pressure for unrestrained plastic flow only if they transform the shell into "short" subshells, that is, if their spacing is smaller than

$$l = \pi \frac{t}{2} \sqrt{8E(1-\nu^2)/3\sigma_Y} \ . \tag{8.47}$$

Regarding the deflections prior to unrestrained plastic flow, Hodge [8.41], and Paul and Hodge [8.23] have shown in examples that a load as high as 98% of the limit load could be attained without the maximum elastic-plastic deflection exceeding five times the maximum elastic deflection.

Work hardening seems to have a very small effect on the load-carrying capacity [8.42]; it should in any case increase it with respect to theoretical predictions based on perfect plasticity. Also, the use of a linearized yield condition "inscribed" to the exact Tresca condition, which in turn is "inscribed" to the more realistic von Mises condition, should also result in extra safety.

8.4. Rotationally symmetric shells

8.4.1. Introduction

Generalized stresses M_ϕ, M_θ, N_ϕ, N_θ are shown in fig. 5.8. The yield condition of von Mises for a shell that has rotational symmetry in shape and loading was derived independently by Ilyushin [1.2] and Hodge [5.14]. The Tresca condition is given in Section 5.4.2, tables 5.2 and 5.3. Both conditions are nonlinear; thus their use is difficult. For this reason, approximations to the Tresca condition have been used to obtain better bounds or exact solutions. One possible approximation is the Tresca condition for a sandwich shell, as discussed in Section 5.5. Hodge [8.43] has suggested the use of a limited interaction yield locus, assuming that either bending moments or axial forces but not both at the same time are important. Hence, interaction of generalized stresses of the same nature is fully taken into account, whereas generalized stresses of different nature (moments and axial forces) are uncoupled. The corresponding yield hypersurface for a Tresca material is formed of two independent hexagonal hyperprisms, one for the M_ϕ, M_θ interaction, one for the N_ϕ, N_θ interaction. The equations of the twelve hyperplanes, together with the components of the corresponding strain-rate vectors, are given in table 8.3, in which $m_i = M_i/M_p$, $n_i = N_i/N_p$, $\dot{\lambda}_i \equiv \dot{\epsilon}_i$, $\dot{\Phi}_i \equiv m_i \dot{k}_i$, $(i=\phi,\theta)$. Hodge [3.8] has described a way of visualizing this four-dimensional hypersurface and has shown that, except for the plate, only part of the hypersurface can

Table 8.3.

Face nr	Equation	$\dot{\lambda}_\theta$	$\dot{\lambda}_\phi$	$\dot{\Phi}_\theta$	$\dot{\Phi}_\phi$
1	$n_\theta = 1$	1	0	0	0
2	$n_\phi = 1$	0	1	0	0
3	$-n_\theta + n_\phi = 1$	-1	1	0	0
4	$-n_\theta = 1$	-1	0	0	0
5	$-n_\phi = 1$	0	-1	0	0
6	$n_\theta - n_\phi = 1$	1	-1	0	0
7	$m_\theta = 1$	0	0	1	0
8	$m_\phi = 1$	0	0	0	1
9	$-m_\theta + m_\phi = 1$	0	0	-1	1
10	$-m_\theta = 1$	0	0	-1	0
11	$-m_\phi = 1$	0	0	0	-1
12	$m_\theta - m_\phi = 1$	0	0	1	-1

furnish a collapse mechanism. Because the limited interaction yield hypersurface is completely circumscribed to the Tresca hypersurface and must be shrunk by a factor of 0.618 to become completely inscribed, an exact limit load P_o with the limited interaction condition will, according to Section 5.5.2, bound the limit load with the exact Tresca conditions as follows:

$$0.618 P_o \leqslant P_l \leqslant P_o . \tag{8.48}$$

Recently, Flügge and Nakamura [8.44] have proposed the use of the six parabolic hypercylinders, the equations of which are given in table 5.3, and reproduced in table 8.4, together with the corresponding strain rates. This set

Table 8.4. Hypersurfaces of parabolic hypercylinders

Label of hypersurface	Yield condition	$\dot{\epsilon}_\phi$	$:\dot{\epsilon}_\theta$	$:\dot{\Phi}_\phi$	$:\dot{\Phi}_\theta$
H_ϕ^+	$m_\phi + n_\phi^2 = 1$	$2n_\phi$	$: 0$	$: 1$	$: 0$
H_ϕ^-	$-m_\phi + n_\phi^2 = 1$	$2n_\phi$	$: 0$	$:-1$	$: 0$
$H_{\phi\theta}^+$	$m_\phi - m_\theta + (n_\phi - n_\theta)^2 = 1$	$2(n_\phi - n_\theta)$	$:-2(n_\phi - n_\theta)$	$: 1$	$:-1$
$H_{\phi\theta}^-$	$-m_\phi + m_\theta + (n_\phi - n_\theta)^2 = 1$	$2(n_\phi - n_\theta)$	$:-2(n_\phi - n_\theta)$	$:-1$	$: 1$
H_θ^+	$m_\theta + n_\theta^2 = 1$	0	$: 2n_\theta$	$: 0$	$: 1$
H_θ^-	$-m_\theta + n_\theta^2 = 1$	0	$: 2n_\theta$	$: 0$	$:-1$

of hypercylinders, when limited to their mutual intersections, coincides partly with the exact Tresca hypersurface and completely circumscribe it. An exact limit load P^* for this approximate yield hypersurface gives the following bounds

$$0.851 P^* \leqslant P_l \leqslant P^* . \tag{8.49}$$

Comparison of relations (8.48) and (8.49) shows a noticeable improvement. It is shown in ref. [8.44] that the collapse mechanism obtainable with the hypercylinders H_ϕ^+ and H_ϕ^- is merely a rigid-body translation and does not represent any plastic deformation. General expressions of the velocity fields and stress fields are then obtained for the two other pairs of hypercylinders. In these expressions, only quadratures remain to be performed with the integration constants to be determined from boundary and continuity conditions. The main difficulty lies in the choice of appropriate hypercylinders for the particular problem at hand.

As will be seen in Section 8.5, the Tresca condition for a cylindrical shell can also be used successfully for some other shells of revolution, as suggested by Drucker and Shield [8.45, 8.46].

8.4.2. Spherical cap subjected to radial uniformly distributed load
8.4.2.1. Introduction

Let R be the radius, 2α the cap angle and t its thickness (fig. 8.26). To begin with, assume that a certain stress profile is appropriate. With this stress profile, the equilibrium equations can often be integrated to yield a statically admissible stress field, from which a mechanism is obtained using the normality law. It then remains to be checked whether the stress field and the mechanism remain admissible for the whole range of shell parameters.

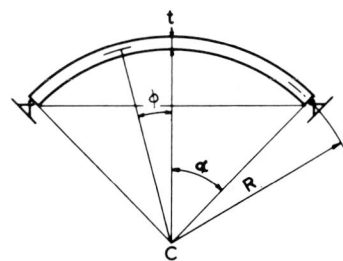

Fig. 8.26.

8.4.2.2. *Spherical cap hinged at the edge*

With the limited interaction yield condition, a complete solution can be obtained [8.43]. It yields

$$p_l^* \equiv \left(\frac{pR}{N_p}\right)_l = 2 + 2k \sin\alpha \left[\ln \frac{1+\sin\alpha}{\cos\alpha} - \sin\alpha\right]^{-1}, \qquad (8.50)$$

with the following stress field (see fig. 8.26 for definition of angle ϕ)

$$n_\theta = -\tfrac{1}{2}[p^* - (p^*-2)\sec^2\phi],$$

$$n_\phi = -1,$$

$$m_\theta = 1,$$

$$m_\phi = 1 - \frac{p^*-2}{2k}\left[\frac{1}{\sin\phi}\ln\frac{1+\sin\phi}{\cos\phi} - 1\right]. \qquad (8.51)$$

In formulas (8.50) and (8.51), $k = M_p/RN_p$. The shear force is

$$V = -N_p \frac{p^*-2}{2} \tan\phi. \qquad (8.52)$$

For a shell with a centered cutout, yield pressures have been obtained by Hodge and Lakshmikantham [8.47] and are given in the book by Hodge [8.27].

With the Tresca yield condition for a sandwich shell, the following good bounds are given in Hodge's book [8.27]:

$$p_-^* = \frac{2k}{(1+k)(1-\alpha\cot\alpha)}, \qquad \text{if} \quad p_-^* \geq 2, \qquad (8.53)$$

or

$$p_-^* = 2. \qquad (8.54)$$

$$p_+^* = \frac{2k}{1 - \alpha\cot\alpha}, \qquad \text{if} \quad \cos\alpha \geq 1 - k, \qquad (8.55)$$

$$p_+^* = 2\left(\frac{\sin\alpha - \phi_1\cos\alpha}{\sin\alpha - \alpha\cos\alpha} - \frac{\cos\alpha}{\cos\phi_1}\frac{\sin\alpha - \sin\phi_1}{\sin\alpha - \alpha\cos\alpha}\right), \tag{8.56}$$

with $\cos\phi_1 = \cos\alpha/(1-k)$, if $\cos\alpha \leq 1-k$.

8.4.2.3. Spherical cap built-in at the edge

With the limited interaction yield condition, the stress field (8.51) is valid for $\phi \leq \phi_1$, the particular value ϕ_1 corresponding to $m_\phi = 0$. For $\phi_1 \leq \phi \leq \alpha$, n_θ and n_ϕ continue to be given by relation (8.51), whereas

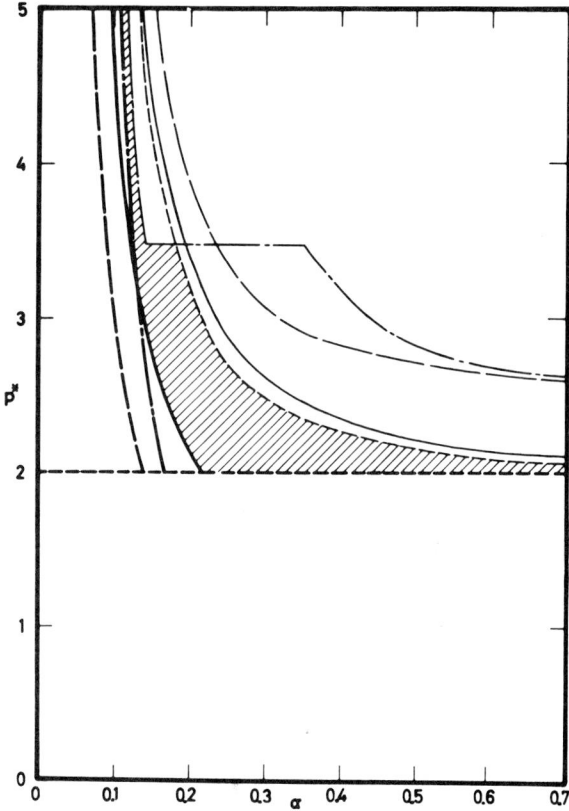

Fig. 8.27. - - - Bounds by Onat and Prager [5.8]; – – – Bounds based on a sandwich structure [3.9]; – · – Improved bounds based on a sandwich structure; —— Bounds based on the limited interaction yield surface. Light lines: upper bounds, heavy lines lower bounds $k = M_p/RN_p = t/4R = 0.005$.

$$m_\phi = -1 + \frac{p^* - 2}{2k} \ln \frac{\cos \phi}{\cos \alpha} - \ln \frac{\sin \alpha}{\sin \phi},$$

$$m_\theta = m_\phi + 1. \tag{8.57}$$

The limit load p^* and the value of ϕ_1 are given by the following system of equations:

$$p^* = 2 + 2k \sin \phi_1 \left[\ln \frac{1+\sin \phi_1}{\cos \phi_1} - \sin \phi_1 \right]^{-1},$$

$$p^* = 2 + 2k \frac{1 + \ln (\sin \alpha / \sin \phi_1)}{\ln (\cos \phi_1 / \cos \alpha)}. \tag{8.58}$$

Various bounds for p^* are compared in fig. 8.27. The exact values for a solid shell obeying the Tresca yield condition fall in the shaded region.

8.4.3. Conical shell loaded by a finite boss

When the boss is hinged to the shell (fig. 8.28) and if the *limited interaction yield condition* is used, with the following dimensionless variables

$$a = \frac{r_a}{R}, \tag{8.59}$$

$$q = Q \frac{1}{2\pi R N_p \sin \alpha} - j, \tag{8.60}$$

$$j = \frac{M_p \cos^2 \alpha}{R N_p \sin \alpha}, \tag{8.61}$$

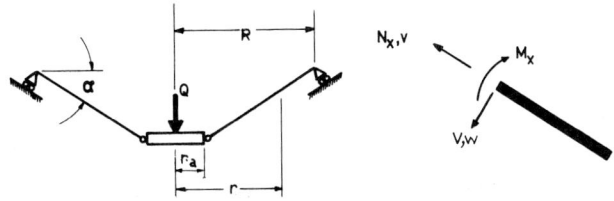

Fig. 8.28.

the limiting value q_l is given, together with a subsidiary unknown b, by [8.48]

$$q_l = \frac{1 - b^2 + 2b^2 \ln b}{4(1-b)} = \frac{b(1-\ln b) - (a+1/e)}{\ln (b/a)}. \tag{8.62}$$

In relation (8.62), e is the basis of the natural logarithms. Relation (8.62) gives the exact limit load for

$$j \geqslant 0.134, \tag{8.63}$$

and must be regarded as an upper bound for smaller j.

Because $M_p/N_p = t/4$ (uniform shell), relation (8.61) can be rewritten as

$$j = \frac{1}{\mu} \frac{\cos^2 \alpha}{2 \sin \alpha},$$

where μ is the slenderness ratio $2R/t$. Condition (8.63) then becomes

$$\frac{1}{\sin \alpha} - \sin \alpha \geqslant 0.268\mu. \tag{8.64}$$

The curve in fig. 8.29 gives α_{max} versus μ for eq. (8.62) to give the exact limit

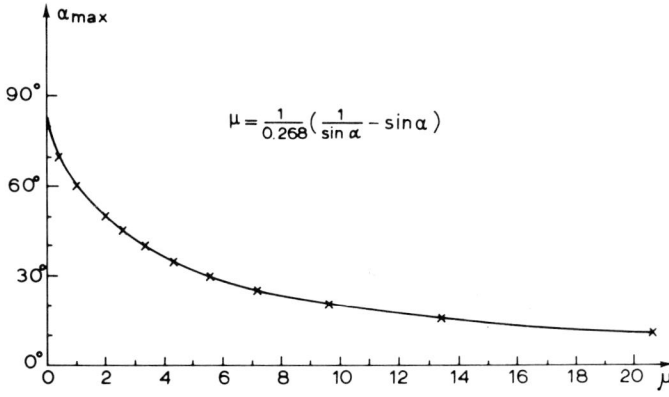

Fig. 8.29.

load. If we now consider vanishingly small boss size ($r_a \to 0$), we obtain a concentrated load at the vertex. The solution is found to be $q = 0$, that is

$$Q = 2\pi M_p \cos^2 \alpha .\tag{8.65}$$

It must be emphasized that the limit load, eq. (8.65), is valid for arbitrary connection at the boss and for arbitrary edge conditions, for all shell geometry such that $\mu \geqslant \frac{1}{2} \sin \alpha \cos^2 \alpha$. This condition is obviously always satisfied for what can be regarded as a shell. When the edge is built-in, the limit value (8.65) is also valid with the exact Tresca yield condition and with the linearized Tresca condition, because the stress field is $n_x = n_\theta = m_x = 0, m_\theta = 1$ in the whole shell and the corresponding stress point falls on the three yield surfaces.

The *exact Tresca yield condition* has been used by Onat [3.10] and by Lance and Onat [8.49] to analyze the simply supported conical shell loaded by a central boss rigidly connected to the shell. The problem is solved numerically, in the limited range of the shell parameter $\beta = 1/4j$ given in table 8.5.

Table 8.5.

$\beta = \dfrac{1}{4j} = \dfrac{R \sin \alpha}{t \cos^2 \alpha}$	$q = \dfrac{Q}{2\pi M_p \cos^2 \alpha}$
0	1.1111
0.1	1.1145
0.2	1.1251
0.3	1.1427
0.4	1.1673
0.418	1.1725
0.5	1.1985
0.6	1.2348
0.7	1.2760
0.8	1.3219
0.9	1.3725
1.0	1.4278

8.4.4. *Other cases of conical shells*

The simply supported conical shell loaded with a uniformly distributed pressure is analyzed in the paper by Kuech and Lee [8.50] on the basis of the limited interaction yield condition.

The approximate yield hypersurface of Flügge and Nakamura [8.44] was used by them to study a conical shell with a centered cut-out along the edge of which acts a uniformly distributed shear force. The outer edge of the shell is built-in.

The reader is referred to the original papers for more information.

8.5. Torispherical and toriconical thin pressure vessel heads

8.5.1. *Introduction*

The correct design of metal pressure vessels and tanks, and especially of their heads, is a very important engineering problem. We therefore devote a whole section to it.

Following Drucker and Shield [8.51], consider a cylindrical vessel subjected to uniform internal pressure. The studied head is made of a torus with radii r and R (see fig. 8.30). Results will also apply when the spherical cap is replaced by the tangent cone drawn with a dashed line in fig. 8.30. The torispherical shape is used to keep the height H of the head reasonably small with the same thickness as the cylinder. An end plate would be much thicker, whereas achieving some kind of optimum (with respect to stress or weight; see Section 8.8.2) with a thickness not larger than that of the cylinder results in values of H ranging from $0.255D$ to $0.3938D$ [8.52, 8.53].

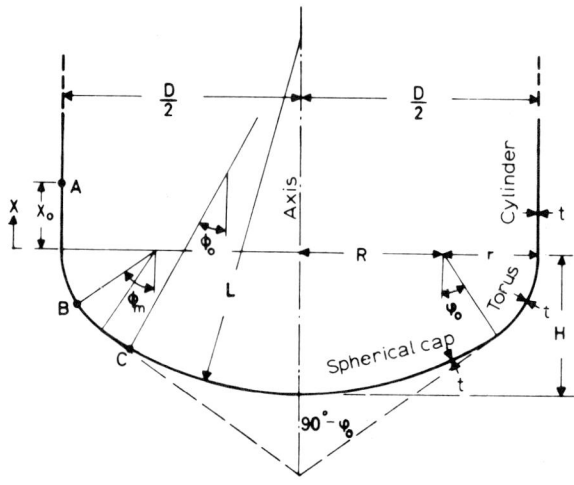

Fig. 8.30.

8.5] TORISPHERICAL AND TORICONICAL THIN PRESSURE VESSEL HEADS

At present, the torispherical heads are designed according to (semi-empirical) specifications of various codes, the most widely used of which is probably the A.S.M.E. Code for Unfired Pressure Vessels. Collapse of a large vessel with torispherical head during hydrostatic pressure tests in the United States some years ago has caused critical examination of the code formulas, which turn out to be unsafe for large $\mu = D/t$, that is, for large vessels subjected to relatively low pressure. Consider a torispherical head with a slenderness ratio $\mu = 440$. Galletly [8.54] has shown that the stresses predicted by the formulas of the A.S.M.E. Code were less than half those given by a correct elastic analysis. Compressive circumferential stresses as high as 30,000 psi under 1.5 times the service pressure, were shown to exist, though completely neglected in the code. Hence, a head of the type studied made of a steel with $\sigma_Y = 29,000$ psi, would have been regarded as "correctly designed" according to the A.S.M.E. Code, and would have exhibited, under hydrostatic testing, large plastic deformations and possibly cracks and rupture.

It is therefore of great practical interest to evaluate the limiting value of the pressure that produces a plastic collapse mechanism. As already noted in Section 8.2, the corresponding theoretical predictions are fairly well supported by experiments [8.20].

8.5.2. *Limit pressure*

Referring to fig. 8.31, the equilibrium equations of a shell element are

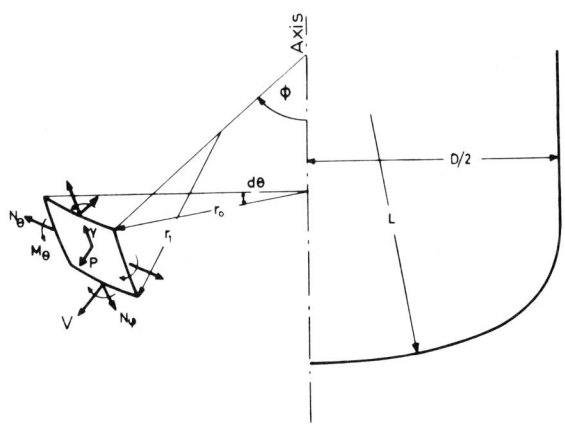

Fig. 8.31.

$$r_0 \frac{dN_\phi}{d\phi} + (N_\phi - N_\theta)r_1 \cos\phi - Vr_0 + Yr_1 r_0 = 0,\quad (8.66)$$

$$r_0 N_\phi + r_1 \sin\phi\, N_\theta + \frac{d}{d\phi}(Vr_0) - pr_0 r_1 = 0,\quad (8.67)$$

$$r_0 \frac{dM_\phi}{d\phi} + (M_\phi - M_\theta)r_1 \cos\phi - Vr_1 r_0 = 0.\quad (8.68)$$

These formulas may be simplified in accordance with the following arguments [8.45]. For the bending state to contribute in an appreciable manner to balancing the pressure p, the shear force V must be of the order of $N_p = \sigma_Y t$ (or at least not much smaller) as can be seen from eqs. (8.66) and (8.67). But because $M_\phi - M_\theta$ cannot exceed $2M_p = \sigma_Y t^2/2$, the term $Vr_0 r_1$ in eq. (8.68) is large with respect to $(M_\phi - M_\theta)r_1 \cos\phi$ provided t/r_0 remains small, that is, at all points of the shell far enough from the axis. Eq. (8.68) can then be simplified into

$$\frac{dM_\phi}{d\phi} - Vr_1 = 0.\quad (8.69)$$

If it is now assumed that the rate of curvature $\dot{\kappa}_\theta$ remains negligible, M_θ is a reaction, and the yield surface derived from the Tresca condition is shown in fig. 8.32 (a) (see Section 5.4.2). To achieve further simplification, we shall use

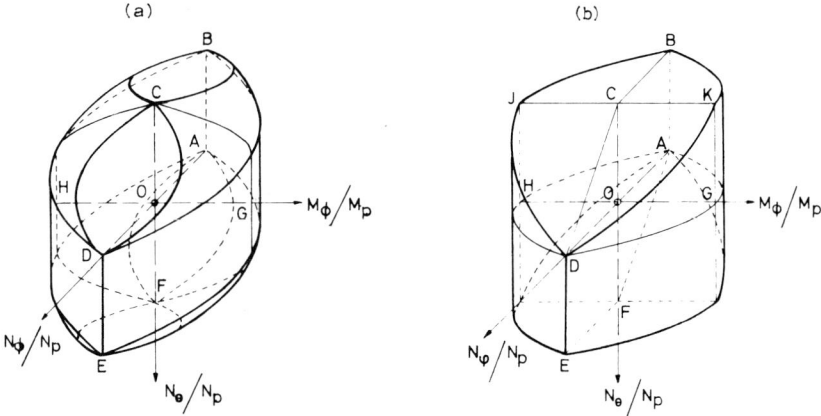

Fig. 8.32.

8.5] TORISPHERICAL AND TORICONICAL THIN PRESSURE VESSEL HEADS

the simpler circumscribed yield surface of fig. 8.32 (b), which consists of a parabolic cylinder and four planes.

In the momentless state of stress ($M_\theta = M_\phi = V = 0$, pure membrane response) we only have eqs. (8.66) and (8.67). If the first is integrated with respect to ϕ, taking into account that $r_0/\sin\phi$ is equal to the value r_2 of the principal radius of curvature in the plane normal to the meridian, we obtain

(a) $\quad N_\phi = \dfrac{pr_2}{2}$,

(b) $\quad \dfrac{N_\phi}{r_1} + \dfrac{N_\theta}{r_2} = p$. $\hfill (8.70)$

A membrane reaches its limit load as soon as the most stressed point attains the yield limit. Indeed, as the state of stress is statically determinate, there is no possible stress redistribution and, hence, no increase in load in the absence of a change of geometry.

The dangerous point is at the junction of the torus with the spherical cap, where $N_\phi = pL/2$. With this value of N_ϕ, eq. (8.70b) gives $N_\theta = -pL(L/2r-1)$ because $r_1 = r_0$. The limit pressure for the membrane is reached when $N_\phi - N_\theta - \sigma_Y t$, that is, for

$$p^M = \dfrac{2t\sigma_Y r}{L(L-r)}. \qquad (8.71)$$

Table 8.6 displays the values of

$$\dfrac{p^M D}{2t\sigma_Y} = \dfrac{rD}{L(L-r)}.$$

Actually, the torus is subjected to significant bending stresses. Under increasing pressure, the meridional axial forces N_ϕ applied by the cylinder and the sphere to the torus tend to depress its curvature $1/r$, that is, to push the region B in fig. 8.30 inward. If the tore thickness, though small with respect to its radius r, is large enough to avoid buckling under the compressive axial forces N_θ, a hinge circle will occur at B (fig. 8.30) enabling the central region to contract circumferentially and deflect inward. A hinge circle with the rate of rotation of the opposite sign will form in the spherical cap at C, and a similar hinge circle will occur at A, usually in the cylinder, but sometimes in the torus. The whole region ABC must be plastic to enable inward displacement. In this region, because $N_\phi > 0$ and $N_\theta < 0$, the yield condition will be

Table 8.6.

$\frac{L}{D}$	$\frac{r}{D}$	ϕ_0 deg	$\frac{H}{D}$	$\frac{pM_D}{2t\sigma_Y}$	$\frac{L}{D}$	$\frac{r}{D}$	ϕ_0 deg	$\frac{H}{D}$	$\frac{pM_D}{2t\sigma_Y}$
1.0	0.06	27.91	0.1694	0.064	0.9	0.06	31.59	0.1844	0.079
	0.08	27.16	0.1815	0.087		0.08	30.81	0.1957	0.108
	0.10	26.39	0.1937	0.111		0.10	30.00	0.2072	0.139
	0.12	25.58	0.2063	0.136		0.12	29.16	0.2188	0.171
	0.14	24.75	0.2190	0.163		0.14	28.27	0.2306	0.205
	0.16	23.88	0.2319	0.190		0.16	27.35	0.2427	0.240
0.8	0.06	36.48	0.2050	0.101	0.7	0.06	43.43	0.2353	0.134
	0.08	35.69	0.2152	0.139		0.08	42.64	0.2440	0.184
	0.10	34.85	0.2256	0.179		0.10	41.81	0.2528	0.238
	0.12	33.97	0.2360	0.221		0.12	40.93	0.2619	0.296
	0.14	33.06	0.2468	0.265		0.14	40.01	0.2710	0.357
	0.16	32.09	0.2577	0.312		0.16	39.02	0.2804	0.423
0.6	0.06	54.57	0.2869	0.185					
	0.08	53.87	0.2934	0.256					
	0.10	53.13	0.3000	0.333					
	0.12	52.34	0.3068	0.417					
	0.14	51.50	0.3136	0.507					
	0.16	50.60	0.3207	0.606					

$$H = L - (L-r)\cos\phi_0$$

$$\sin\phi_0 = \frac{1}{2} - \frac{r}{D} \bigg/ \frac{L}{D} - \frac{r}{D}$$

$$\frac{pM_D}{2t\sigma_Y} = \frac{rD}{L(L-r)}$$

$$N_\phi - N_\theta = \sigma_Y t,$$

$$|M_\phi| \leq \frac{t^2}{4}\sigma_Y \left[1 - \left(\frac{N_\phi}{t\sigma_Y}\right)^2\right], \tag{8.72}$$

according to fig. 8.32 (b).

Assuming that M_ϕ exhibits analytical extremums in A, B, and C, where consequently $V = 0$, integration of the equilibrium equations, eqs. (8.66), (8.67), and (8.69) gives the stress field in the plastic region ABC and the corresponding value of the pressure. The results are:

1. In the cylinder,

$$M_\phi = -\frac{t^2}{4}\sigma_Y\left[1 - \left(\frac{pD}{4t\sigma_Y}\right)^2\right] + \frac{t\sigma_Y}{2D}\left(2 + \frac{pD}{2t\sigma_Y}\right)(x_o - x)^2, \tag{8.73}$$

8.5] TORISPHERICAL AND TORICONICAL THIN PRESSURE VESSEL HEADS

$$V = -\frac{t\sigma_Y}{D}\left(2+\frac{pD}{2t\sigma_Y}\right)(x_o-x), \qquad (8.74)$$

where x is the abscissa from the junction with the torus, and x_o the abscissa of the hinge circle A.

2. In the torus,

$$\frac{M_\phi}{rt\sigma_Y} = \frac{t}{4r}\left[1-\left(\frac{pD}{2t\sigma_Y}\right)^2\frac{(R+r\sin\phi_m)^2}{D^2\sin^2\phi_m}\right] - \frac{pR}{2t\sigma_Y}\frac{[1-\cos(\phi-\phi_m)]}{\sin\phi_m}$$

$$+ \frac{r}{R}\cos\phi[k(\phi_m)-k(\phi)] + \ln\frac{R+r\sin\phi}{R+r\sin\phi_m}, \qquad (8.75)$$

$$\frac{V}{t\sigma_Y} = \frac{pR}{2t\sigma_Y}\frac{\sin(\phi_m-\phi)}{\sin\phi_m} + \frac{r}{R}\sin\phi[k(\phi)-k(\phi_m)], \qquad (8.76)$$

$$k(\phi) = \int_{\phi_o}^{\phi}\frac{R\,d\phi}{R+r\sin\phi} = \frac{2R}{(R^2-r^2)^{1/2}}\left[\tan^{-1}\left\{\frac{r+R\tan\phi/2}{(R^2-r^2)^{1/2}}\right\}\right]_{\phi_o}^{\phi}. \qquad (8.77)$$

ϕ is the angle of the normal to the meridian curve with the axis of rotation; ϕ_m is the angle ϕ corresponding to the hinge circle B. In the sphere (regarding $\phi - \phi_o$ as a small quantity)

$$\frac{M_\phi}{Lt\sigma_Y} = -\frac{t}{4L}\left[1-\left(\frac{pL}{2t\sigma_Y}\right)^2\right] + \frac{1}{2}(\phi-\phi_s)^2, \qquad (8.78)$$

$$\frac{V}{t\sigma_Y} = \phi - \phi_s, \qquad (8.79)$$

where ϕ_s corresponds to the hinge circle C.

The four unknowns p, ϕ_m, ϕ_s, and x_o are obtained from the continuity conditions on M_ϕ and V at the junction of the cylinder and the tore ($x=0$, $\phi=\pi/2$) and at the junction of the tore and the spherical cap ($\phi=\phi_o$). The four equations were solved by successive approximations for these values of the shell parameters:

$$\frac{t}{D} = 0.002; 0.004; 0.006; 0.008; 0.010; 0.012; 0.014;$$

$$\frac{L}{D} = 1.0; 0.9; 0.8; 0.6;$$

$$\frac{r}{D} = 0.06; 0.08; 0.10; 0.12; 0.14; 0.16.$$

The limit loads obtained are exact for the yield surface of fig. 8.32 (b) because collapse mechanisms corresponding to the stress fields are shown to exist [8.45]. They are upper bounds p_+ to exact limit load p_l for the yield surface of fig. 8.32 (a), completely inscribed to that of fig. 8.32 (b). Lower

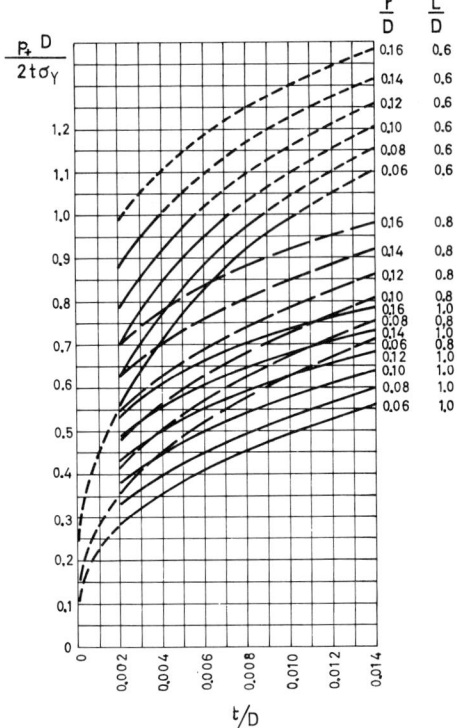

Fig. 8.33.

8.5] TORISPHERICAL AND TORICONICAL THIN PRESSURE VESSEL HEADS 371

bounds $p_- = \alpha p_+$ are obtained with a factor α such that the stress point αN_ϕ, αN_θ, αM_ϕ is on or within the yield surface of fig. 8.32 (a), when applied to the stress field above. It turns out that

$$\alpha = \frac{T^2 - 4T + 12}{2(T^2 - 4T + 8)}, \tag{8.80}$$

with $T = (p+D)/2t\sigma_Y$, the critical point being A. The factor α varies from 0.82 to 0.90. The values of p_+ are given in figs. 8.33 and 8.34. The values of p_- are given in figs. 8.35 and 8.36. If p_s is the service pressure and λ the load factor (or safety factor) it will be sufficient for practical applications to use the mean value

$$p_l \equiv \lambda p_s = \frac{p_- + p_+}{2}. \tag{8.81}$$

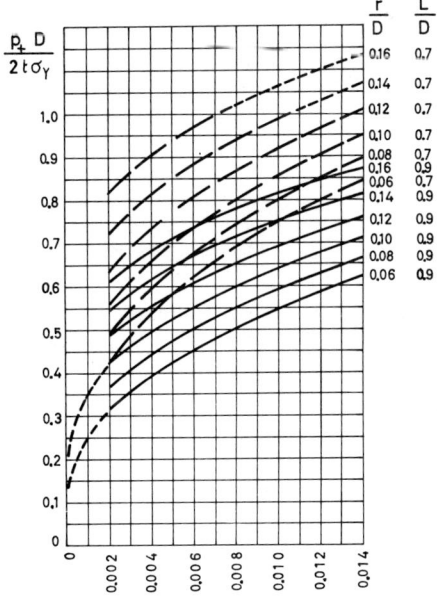

Fig. 8.34.

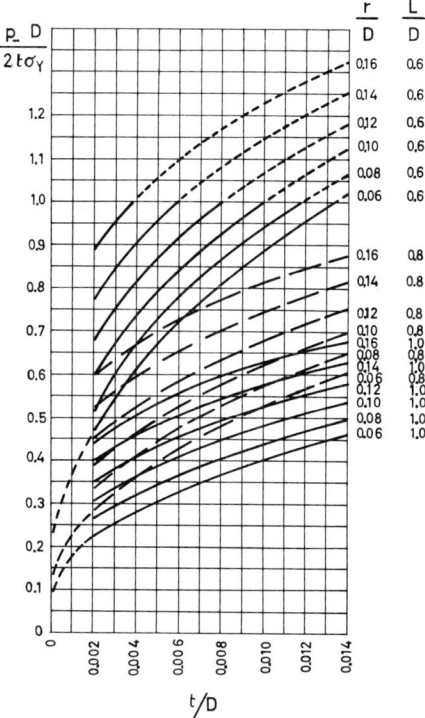

Fig. 8.35.

For a given t/D, the limit pressure p_l increases with increasing r/D and with decreasing L/D. The ratio H/D varies in a similar manner with r/D and L/D (see table 8.6). In fig. 8.37 the straight lines represent approximately the relations t/D versus H/D for fixed p_l/σ_Y, in the range $0.17 < H/D < 0.28$ corresponding to r/D varying from 0.06 to 0.16 and L/D from 0.7 to 1.0. The exact points are indicated to show the accuracy of the approximation.

It also turns out that, for given r/D, the variation of p_l/σ_Y with t/L is nearly independent of L/D. For $L/D = 0.7$ and 0.8, the formula

$$\frac{p_l}{\sigma_Y} = \left(0.33 + 5.5\,\frac{r}{D}\right)\frac{t}{L} + 28\left(1 - 2.2\,\frac{r}{D}\right)\left(\frac{t}{L}\right)^2 - 0.0006 \qquad (8.82)$$

8.5] TORISPHERICAL AND TORICONICAL THIN PRESSURE VESSEL HEADS 373

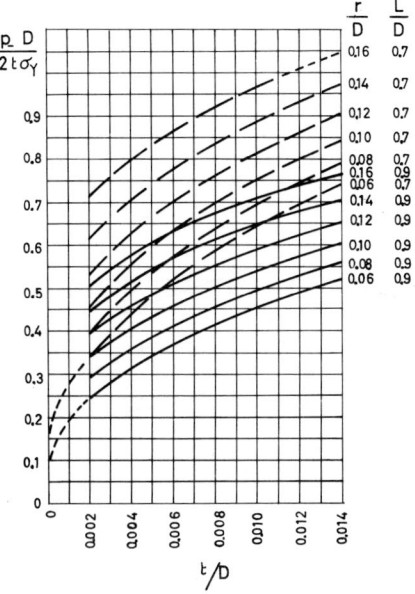

Fig. 8.36.

gives a value of p_l/σ_Y very near of that given by eq. (8.81) (e.g., with less than 3% discrepancy for $t/L = 0.01$). Relation (8.82) can also be used for $L/D = 1.0, 0.9$, and 0.6 as shown in fig. 8.38. It is represented in fig. 8.39.

8.5.3. Discussion

The analysis in Section 8.5.2 implies that the connection between the cylinder and the spherical cap is the weak part of the vessel. This situation occurs when this torical connection is thin and of large curvature. Large storage tanks or vessels for relatively low internal pressure (1 to 5 kg/cm², for example), when designed according to the A.S.M.E. code for Unfired Pressure Vessels, belong to this class. As shown in fig. 8.39 by the dashed ray, the A.S.M.E. code formulas can then be unsafe, especially in the region of small t/L.

On the other hand, for high-pressure vessels, the relative thickness of the torus, as given by the code, increases. The connection between the cylinder and the sphere becomes rigid and behaves like a reinforcing ring. Because

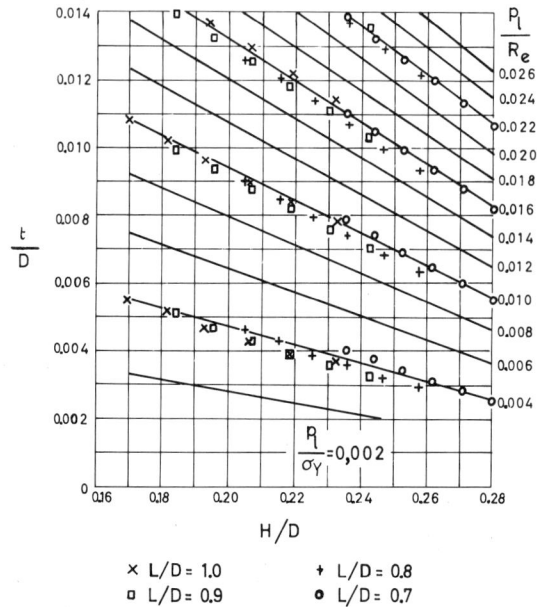

Fig. 8.37.

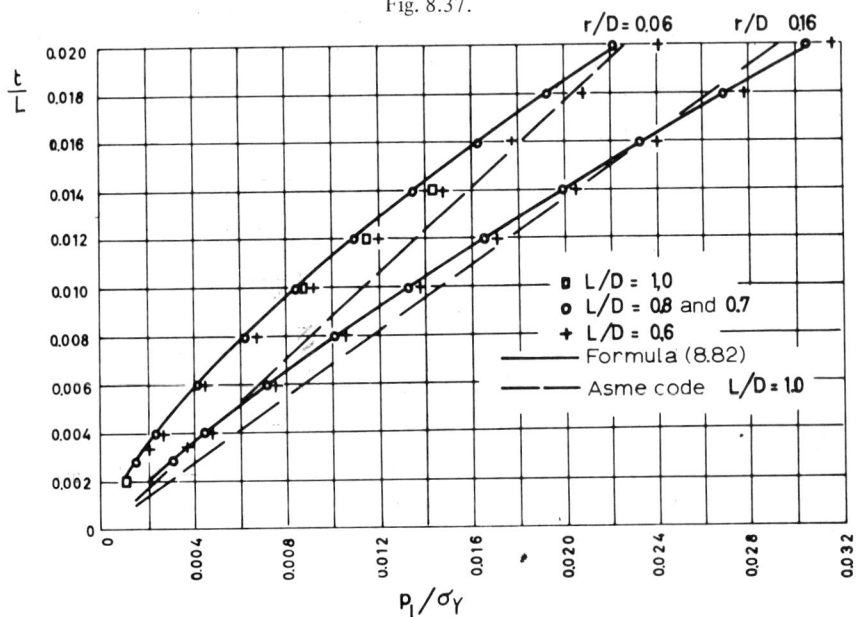

Fig. 8.38.

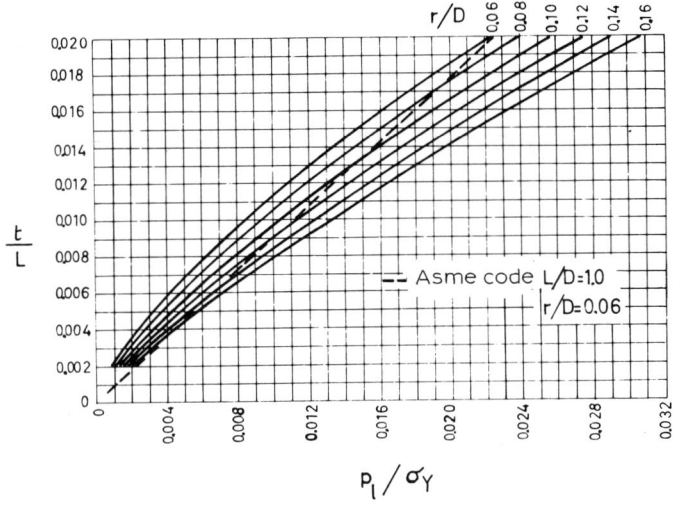

Fig. 8.39.

L/D is usually greater than 0.5, the cylinder now becomes the weakest part of the vessel. The curves giving the limit pressure are thus valid only for $pD/2t\sigma_Y < 1$. This is the reason why they have been drawn in dashed lines for $pD/2t\sigma_Y \geqslant 1$.

8.6. Thin-wall beam with circular axis

8.6.1. Introduction

It is well known that the flanges of I beams (or channel beams) with curved axis are subjected to transversal bending because the axial resultant forces in a flange in neighboring sections forming the angle $d\phi$ (fig. 8.40) are statically equivalent to a radial force. Hence, when the dimensions of the flanges are such that this effect becomes noticeable, the collapse mechanism does not retain the shape of the cross section but includes transversal bending of the flanges [8.55, 8.56].

8.6.2. Wide-flange I beam

Consider a wide flange I beam with circular axis, subjected to a uniform bending moment diagram with magnitude M (fig. 8.40). Due to the constancy of M with ϕ, the deformed axis will remain circular.

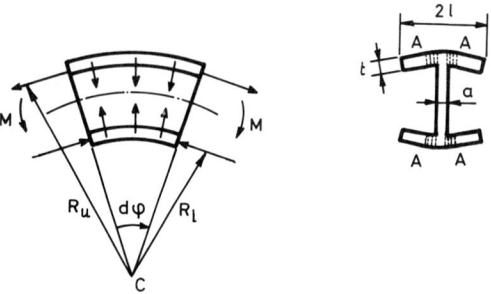

Fig. 8.40.

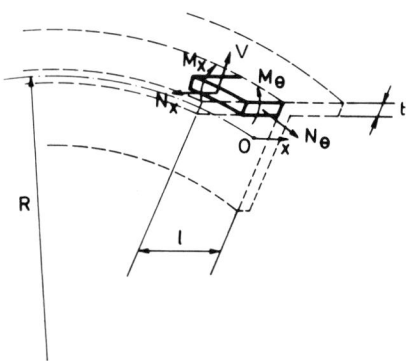

Fig. 8.41.

The flanges will behave like cylindrical, axisymmetrically loaded, shells that are built-in at the web. An element of the upper flange is shown in fig. 8.41. In the present situation, $N_x = 0$ whereas V_x and M_θ are reactions. The generalized stresses M_x and N_θ are related by the equilibrium equation

$$\frac{d^2 M_x}{dx^2} + \frac{N_\theta}{R} = 0 , \qquad (8.83)$$

obtained by eliminating V_x from the equilibrium equations

8.6] THIN-WALL BEAM WITH CIRCULAR AXIS

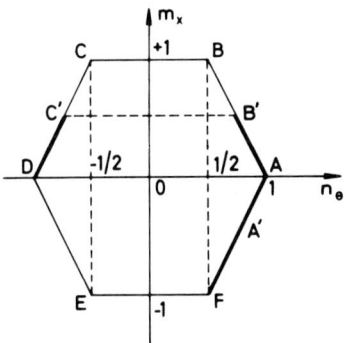

Fig. 8.42.

(a) $\dfrac{dV_x}{dx} + \dfrac{N_\theta}{R} = 0 ,$

(b) $\dfrac{dM_x}{dx} = V_x .$ (8.84)

We use the hexagonal yield condition of fig. 8.42 for the flange, with $m_x = M_x/M_{pf}$ and $n_\theta = N_\theta/N_{pf}$, where $M_{pf} = \sigma_Y t^2/4$ and $N_{pf} = \sigma_Y t$. This is an inscribed approximation to the exact curve derived from the yield criterion of Tresca (see Section 5.4.4, fig. 5.24). Eq. (8.83) can be rewritten as

$$\frac{d^2 m_x}{dx^2} + \frac{4}{tR} n_\theta = 0 .$$ (8.85)

For a flange with a not-too-large span l (fig. 8.41), the bending moment M_x at the connection with the web will remain smaller than the plastic moment and the stress profile will lie entirely on side AF (fig. 8.42) with the equation

$$m_x - 2n_\theta = -2 .$$ (8.86)

The boundary conditions at $x = 0$ are

$$m_x = 0 , \qquad \frac{dm_x}{dx} = 0 .$$ (8.87)

Integration of eq. (8.85), taking account of eqs. (8.86) and (8.87), furnishes $n_\theta = \cos(\sqrt{2x/tR})$. If we let

$$\alpha = l\sqrt{\frac{2}{tR}}, \tag{8.88}$$

the expression for n_θ becomes

$$n_\theta = \cos\frac{\alpha x}{l}, \tag{8.89}$$

valid for $1 \geqslant n_\theta \geqslant \frac{1}{2}$, that is, for $0 \leqslant x \leqslant \sqrt{0.548tR}$. As long as

$$l \leqslant \sqrt{0.548tR}, \tag{8.90}$$

we can use eq. (8.89) to calculate the *efficiency* ρ_f of the flange, defined as

$$\rho_f = \frac{\int_0^l n_\theta \, dx}{l} \equiv \frac{\int_0^l N_\theta \, dx}{N_{pf} l}. \tag{8.91}$$

When condition (8.90) is not satisfied, the stress profile must be modified. Since transversal bending deflections produce circumferential strains ϵ_θ with a sign opposite to those produced by the direct bending moment M on the I beam, we are led to choose the stress profile $DC'B'AF$. In this profile, the extremity of the flange is in regime DC', with $\dot{\epsilon}_\theta < 0$, though the considered flange is, as an average, subjected to traction. The jump from C' to B' (discontinuity in n_θ with x) is admissible, and continuity of m_x (and dm_x/dx) is preserved. We now have the following situation:

1. Regime DC' over $0 \leqslant x \leqslant d_1$. The yield condition is

$$m_x = 2 + 2n_\theta. \tag{8.92}$$

The boundary conditions are

$$m_x = 0, \qquad \frac{dm_x}{dx} = 0, \qquad \text{at } x = 0. \tag{8.93}$$

The resulting stress distribution is

$$n_\theta = -\cos\frac{\alpha x}{l},$$

$$m_x = 2 - 2\cos\frac{\alpha x}{l}. \tag{8.94}$$

2. Regime $B'A$ over $d_1 \leqslant x \leqslant d_2$. The yield condition is

$$m_x + 2n_\theta = 2. \tag{8.95}$$

Elimination of m_x from eqs. (8.85) and (8.95), and integration, gives

$$n_\theta = Ae^{\alpha x/l} + Be^{-\alpha x/l}. \tag{8.96}$$

3. Regime AF over $l - (d_1 + d_2)$. The yield condition is

$$m_x = -2 + 2n_\theta. \tag{8.97}$$

Substitution of eq. (8.97) into eq. (8.85), and integration yields

$$n_\theta = C\sin\frac{\alpha x}{l} + D\cos\frac{\alpha x}{l}. \tag{8.98}$$

The six unknowns A, B, C, D, d_1, and d_2 are obtained from the following six conditions: at

$x = d_1$	discontinuity of n_θ by mere sign change,
$x = d_1$	continuity of V_x, that is, of $\dfrac{dm_x}{dx}$,
$x = d_1 + d_2$,	$n_\theta = 1$,
$x = d_1 + d_2$	continuity of n_θ,
$x = d_1 + d_2$	continuity of V_x, that is, of $\dfrac{dm_x}{dx}$,
$x = 1$,	$n_\theta = \frac{1}{2}$.

If we let $a_1 \equiv d_1/l$, $a_2 \equiv d_2/l$, the conditions just stated are

$$\cos \alpha a_1 = Ae^{\alpha a_1} + Be^{-\alpha a_1},$$

$$\sin \alpha a_1 = -Ae^{\alpha a_1} + Be^{-\alpha a_1},$$

$$1 = Ae^{\alpha(a_1+a_2)} + Be^{-\alpha(a_1+a_2)},$$

$$1 = C \sin \alpha(a_1+a_2) + D \cos \alpha(a_1+a_2),$$

$$C \cos \alpha(a_1+a_2) - D \sin \alpha(a_1+a_2) = -Ae^{\alpha(a_1+a_2)} + Be^{-\alpha(a_1+a_2)},$$

$$C \sin \alpha + D \cos \alpha = \tfrac{1}{2}. \tag{8.99}$$

The system of eqs. (8.99) was solved on the Bull computer of Liege University, for α varying from $\alpha = 1.045$ (corresponding to $\cos \alpha = \tfrac{1}{2}$) to $\alpha = 5.0$. With the curves of A, B, C, D, a_1, and a_2 versus α, the efficiency of the flange can be calculated. It is found to be

$$\rho_f = -\frac{\sin \alpha a_1}{a_1} + \frac{A}{\alpha}\left[e^{\alpha(a_1+a_2)} - e^{\alpha a_1}\right] - \frac{B}{\alpha}\left[e^{-\alpha(a_1+a_2)} - e^{-\alpha a_1}\right]$$

$$- \frac{C}{\alpha}\left[\cos \alpha - \cos \alpha(a_1+a_2)\right] + \frac{D}{\alpha}\left[\sin \alpha - \sin \alpha(a_1+a_2)\right]. \tag{8.100}$$

The curve ρ_f versus α is given in fig. 8.43.

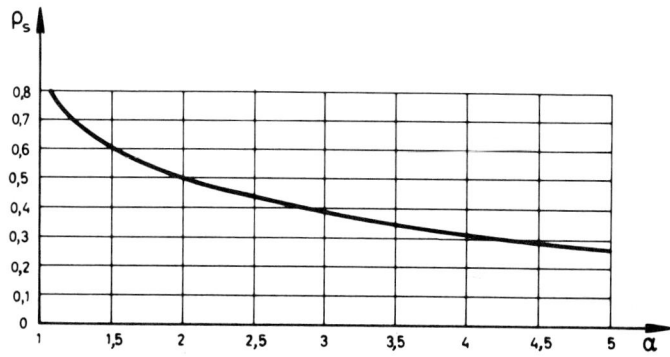

Fig. 8.43.

We now turn to the problem of the *efficiency of the web*. The total axial forces N_u and N_l in the upper and lower flange with radii R_u and R_l, respectively, can be determined from the preceding analysis. The radial stresses applied by the flanges to the web with thickness a are (see figs. 8.40 and 8.44)

$$\sigma_{ou} = -\frac{N_u}{aR_u} < 0 ,$$

$$\sigma_{ol} = \frac{N_l}{aR_l} < 0 . \tag{8.101}$$

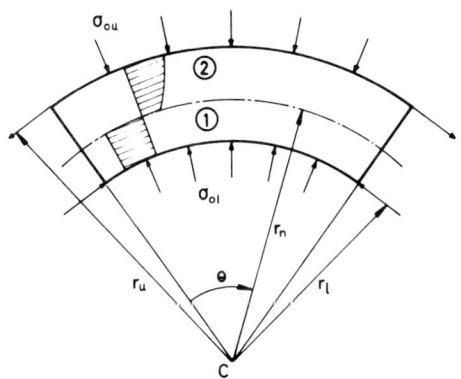

Fig. 8.44.

The web is also subjected to bending stresses σ_θ. In region 1, fig. 8.44, we have $\sigma_\theta < 0$ and $\sigma_r \leq 0$. Hence, the yield condition of Tresca is simply

$$\sigma_\theta = -\sigma_Y \quad \text{with} \quad -\sigma_Y \leq \sigma_r \leq \sigma_Y . \tag{8.102}$$

In region 2, $\sigma_\theta > 0$ and $\sigma_r \leq 0$. The Tresca yield condition is

$$\sigma_\theta - \sigma_r = \sigma_Y . \tag{8.103}$$

The equilibrium equation is (see Section 10.6, eq. (10.83))

$$\sigma_\theta - \sigma_r - r\frac{d\sigma_r}{dr} = 0 . \tag{8.104}$$

382 METAL SHELLS [Ch. 8

With the considered yield conditions and the boundary conditions (fig. 8.44)

$$\sigma_r = \sigma_{ou} \quad \text{at} \quad r = r_u ,$$

$$\sigma_r = \sigma_{ol} \quad \text{at} \quad r = r_l , \tag{8.105}$$

integration of eq. (8.104) yields the following relations:

$$\sigma_r = \frac{1}{r}[-\sigma_Y r + r_l(\sigma_{ol} + \sigma_Y)] , \quad \sigma_\theta = -\sigma_Y \tag{8.106}$$

in region 1, and

$$\sigma_r = \sigma_Y \ln \frac{r}{r_u} + \sigma_{ou} , \quad \sigma_\theta = \sigma_Y\left(1+\ln \frac{r}{r_u}\right) + \sigma_{ou} \tag{8.107}$$

in region 2. The boundary radius r_n between regions 1 and 2 is obtained from the condition that the net resultant force over the cross section vanishes. We obtain in this manner

$$r_n \ln \frac{r_u}{r_n} - (r_n - r_l) + \frac{2lt}{a}\left[\rho_u\left(1 - \frac{r_u - r_n}{R_u}\right) - \rho_l\right] = 0 , \tag{8.108}$$

where ρ_u and ρ_l are the efficiencies of the upper and lower flanges, respectively.

Finally, the plastic moment of the I beam is evaluated as the total moment of the internal forces with respect to the center of curvature.

$$M_p = \sigma_Y\left[2tl(\rho_u R_u - \rho_l R_l) - \frac{a}{2}(r_n^2 - r_l^2)\right.$$

$$\left. + \left(\frac{a}{4} - \frac{tl\rho_u}{R_u}\right)(r_u^2 - r_n^2) + \frac{a}{2}r_n^2 \ln \frac{r_u}{r_n}\right]. \tag{8.109}$$

The plastic moment M_{ps} of the same I beam with straight axis is

$$M_{ps} = 2lth\sigma_Y + a\frac{h_a^2}{4}\sigma_Y , \tag{8.110}$$

with $h = R_u - R_l + t$ and $h_a = r_u - r_l$.

Fig. 8.45.

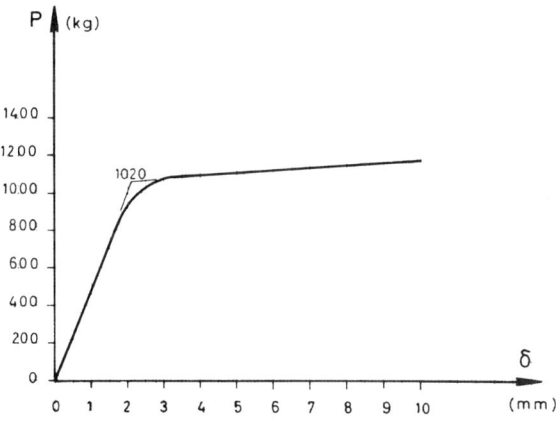

Fig. 8.46.

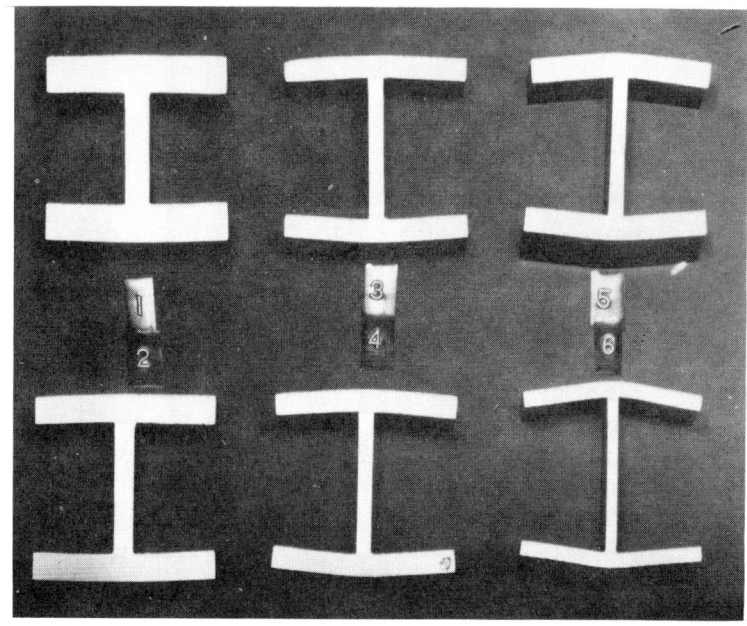

Fig. 8.47.

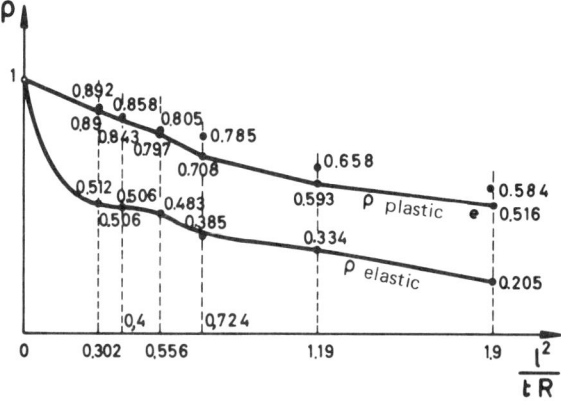

Fig. 8.48.

The overall efficiency ρ of the cross section is thus obtained as $\rho = M_p/M_{ps}$ from eqs. (8.109) and (8.110).

The theoretical predictions given (eqs. (8.91), (8.100), (8.109)) have been tested experimentally on six small-scale models of mild steel [8.55]. The testing device is shown in fig. 8.45. Fig. 8.46 shows a typical moment versus curvature diagram, with the experimental value of M_p. In fig. 8.47, plastically deformed cross sections can be seen, whereas the diagram of fig. 8.48 enables comparison of theoretical efficiencies with experimental values. Experimental values are larger than theoretical ones because the yield condition used is wholly internal to the more realistic von Mises condition.

8.6.3. Circular beam with box shape cross section

The circular beam with the cross section shown in fig. 8.49, subjected to uniform bending moment, can be analyzed in a manner similar to that used in Section 8.6.2 for the I beam. It is assumed that the webs are not bent transversally. To achieve this goal, one must have $l \geqslant \sqrt{0.548 tR}$. The stress field is then continued in a statically admissible manner in the central part of the flange. The stress profile is either FA', FAB' or even FAB. If we let

$$\alpha_c = L\sqrt{2/tR}\,, \tag{8.111}$$

the efficiency ρ_L of the central part is given in fig. 8.50 (a) as a function of α_c [8.56]. The efficiency ρ_f of a flange is then given by

$$\rho_f = \frac{l\rho_l + L\rho_L}{l + L} \tag{8.112}$$

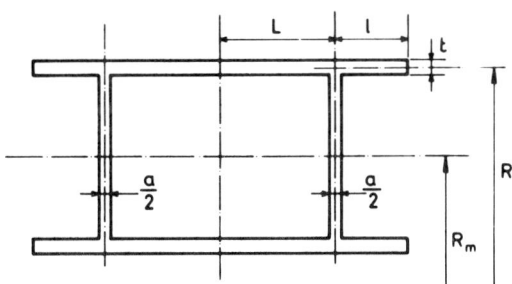

Fig. 8.49.

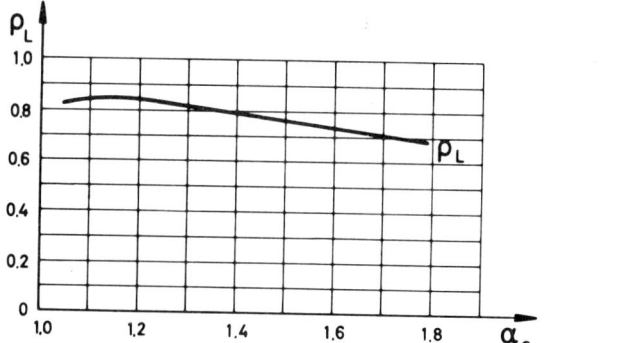

(a)

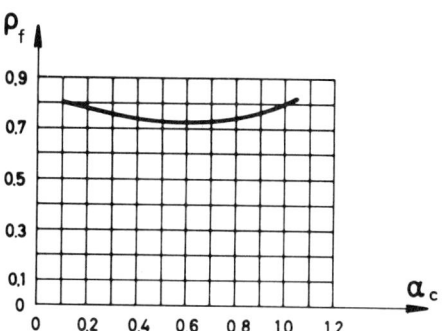

(b)

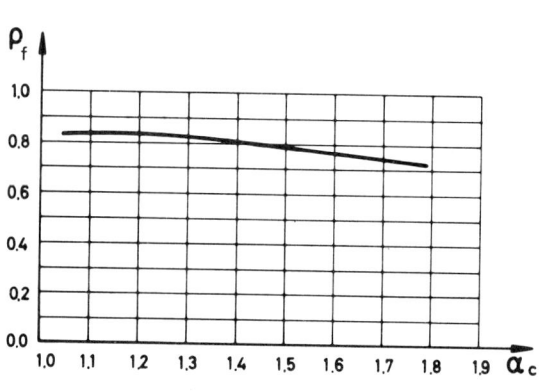

Fig. 8.50.

Table 8.7.

No.	$2L$	R_m	ρ_u	ρ_l	ρ_{th}	ρ_{exp}	Differences
1	30	120	0.802	0.758	0.787	0.844	6.1
2	48	120	0.746	0.652	0.713	0.777	8.2
3	44	170	0.811	0.754	0.791	0.858	7.8
4	40	170	0.826	0.772	0.806	0.858	6.1
5	36	170	0.821	0.785	0.810	0.951	14.8
6	30	170	0.756	0.750	0.769	0.887	13.3
7	20	70	0.737	0.672	0.729	0.850	14.2

where ρ_l is the efficiency of the cantilever part of the flange. For the particular case where $l = \sqrt{0.548tR}$, the ρ_f versus α_c curves are given in fig. 8.50 (b). An experimental verification of the theory has been conducted on seven small-scale models. Dimensions and results are summarized in table 8.7. The yield stress of the material was 35,150 psi (mild-steel) except for a region 2 in. wide near the welds where its average value was 43,200 psi. In fig. 8.51, the deformed cross sections can be seen. In model 1, buckling of the webs occurred as soon as a plastic collapse mechanism was formed.

Fig. 8.51.

8.7. Indications on other problems

Cylindrical shells with circular cut-outs, either with an annular reinforcing ring or without reinforcement have been studied by Hodge [8.57] and by Coon, Gill, and Kitching [8.58], respectively. Ring-reinforced radial branches in cylindrical and spherical vessels have also been considered [8.59].

The limit analysis of a pressure vessel consisting of the junction of a cylindrical shell and a spherical shell has been thoroughly treated by Dinno and Gill [8.60,8.61]. Theoretical predictions have found good experimental support in six collapse tests [8.62].

Shallow shells are treated in the book [8.46] by Hodge.

The reader is referred to the original papers for more information and to the recent survey book by Olszak and Sawczuk [8.63].

8.8. Minimum-weight design

8.8.1. *The general plastic approach*

As in the case of plastic plates, minimum-weight designs were first based on the concept that the minimum-weight structure had to collapse into a mechanism with a number of degrees of freedom equal to the number of design unknowns. On this basis, Onat and Prager [6.62] have studied a cylindrical shell with radius R, subjected to uniform radial internal pressure p and an axial force $Q \leqslant pR$ per unit of length of the cross-sectional perimeter. The shell is simply supported at both ends on an inextensible annulus with negligible torsional rigidity (fig. 8.52). The design unknown is the thickness function $t(x)$. The collapse mechanism obviously is rotationally symmetric. It must have a simple infinity of degrees of freedom. The generalized rates $\dot{\epsilon}_x, \dot{\epsilon}_\theta$, and $\dot{\kappa}_x$ must derive from two functions $\dot{u}$ and $\dot{v}$ (rates of displacement in the longitudinal and circumferential directions) and hence must satisfy a compatibility condition. For these two conditions to be satisfied, the stress point must lie on an edge of the yield surface. With the Tresca condition, fig. 5.12, it is easily seen that the stress point must be on the edge of face I ($N_\theta > N_x$ because $Q < pR$) where $M_x < 0$ (internal pressure). With due account for the symmetry with respect to the median cross section and for the boundary conditions, integration of eq. (8.8) gives

$$t = t^* \left[1 - \left(1 - \frac{Q}{pR} \right) \frac{\cosh \alpha x}{\cosh \alpha l/2} \right]. \tag{8.113}$$

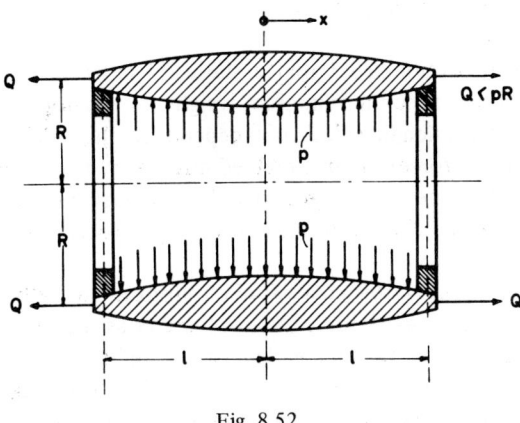

Fig. 8.52.

In relation (8.113), $t^* = pR/\sigma_Y$ is the uniform-strength membrane thickness and $\alpha^2 = 2\sigma_Y/QR$. The ratio $\mathcal{R}$ of the weight of the membrane design to the minimum-weight is

$$\mathcal{R} = \frac{\alpha l/2}{\alpha l/2 - [1-Q/pR]\tanh \alpha l/2}, \qquad (8.114)$$

that tends to unity when l tends to infinity.

Solution (8.113) has been extended by Freiberger [6.63] to the case where p is an arbitrary function of x, whereas Q can take any value. Numerical calculations are rather laborious.

The "constant dissipation" approach described in Section 6.8.2 has been applied by Shield [8.64] to a cylindrical sandwich shell subjected to internal pressure. In the case where either the internal pressure is uniform or the shell is loaded by a ring of force in its midspan cross section, it is shown that economy in weight with respect to a membrane design disappears when $l > 2.634\sqrt{RH}$ for the simply supported shell and $l > 3.382\sqrt{RH}$ for the built-in shell. In the inequalities above, l is the length of the shell, R its radius and H the core thickness. It must be noted that Shield's theoretical solutions contain unrealizable features (stiffening flanges with vanishing thickness but finite area, at the junction of the central part, subjected to membrane stresses, with the external parts subjected to bending). Nevertheless, the minimum-weight designs can be used as references for more practical designs.

8.8.2. *Uniform strength metal shells*
8.8.2.1. *Membrane of revolution*

It was seen in Section 6.8.3 that a statically determinate membrane of uniform strength has absolute minimum weight. It was implicitly assumed in that section that the loading does not depend on the thickness (negligible dead load). Internal forces can be obtained directly by using only the equilibrium equations. Because the stresses are uniformly distributed over the thickness, this latter can be chosen to satisfy $\sigma_R = \sigma_Y$ at each point of the median surface.

Olszak and Sawczuk [8.65] have studied uniform strength membranes of revolution. The plastic axial force N_p is regarded as the "plastic nonhomogeneity function" of the membrane. The considered nonhomogeneity can be obtained with constant thickness and variable yield stress, or, more often, with constant yield stress and variable thickness. In the latter case, the nonhomogeneity function is $N_p(\phi) = t(\phi)\sigma_Y$. Consider a spherical membrane subjected to a load uniformly distributed over the projection of the membrane on a plan normal to the axis of rotation (fig. 8.53). Equilibrium equations of the axisymmetrically loaded membrane of revolution are [5.1] (fig. 8.54):

$$N_\phi = - \frac{1}{r_2 \sin^2 \phi} \ [\int r_1 r_2 (Y \sin\phi + Z \cos\phi) \sin\phi \ d\phi + C] \ , \qquad (8.115)$$

$$N_\theta = -r_2 \left(Z + \frac{N_\phi}{r_1} \right) , \qquad (8.116)$$

where C is an integration constant. For the spherical membrane, where $r_1 = r_2 = R$, we obtain

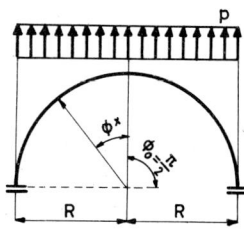

Fig. 8.53.

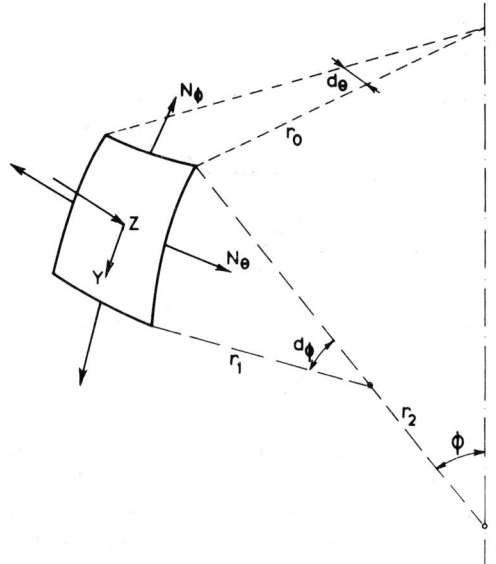

Fig. 8.54.

$$N_\phi = p\frac{R}{2},$$

$$N_\theta = pR\cos 2\phi, \quad \text{for} \quad 0 \leqslant \phi \leqslant \pi/2. \tag{8.117}$$

Satisfying eq. (8.117) and the yield condition of Tresca at every point of the membrane results in

$$N_p(\phi) = \frac{pR}{2} = C^t \quad \text{for} \quad 0 \leqslant \phi \leqslant \frac{\pi}{4}, \tag{8.118}$$

$$N_p(\phi) = pR(1-\cos^2\phi) \quad \text{for} \quad \frac{\pi}{4} \leqslant \phi \leqslant \frac{\pi}{2}. \tag{8.119}$$

With the yield condition of von Mises, we obtain

$$N_p(\phi) = \frac{pR}{2}(1-\cos 2\phi - \cos^2 2\phi)^{1/2}. \tag{8.120}$$

Both solutions are graphically represented in fig. 8.55. Corresponding collapse

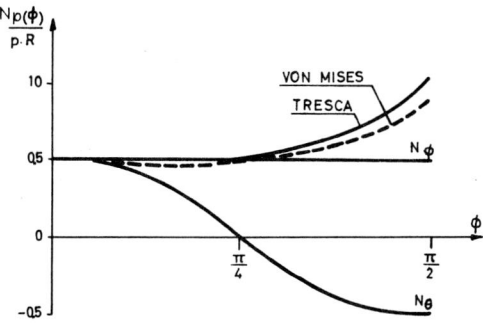

Fig. 8.55.

mechanisms, as well as indications on the uniform-strength design of a shell of revolution subjected to both axial forces and bending moments, can be found in ref. [8.65]. The uniform-strength design of membranes of revolution has been studied in detail by Issler in ref. [8.66], where the solutions (8.118), (8.119), (8.120) can also be found. Both yield conditions of Tresca and von Mises are used. Two kinds of problems are treated: (1) shells with negligible dead weight, for which general expressions are given for N_ϕ, N_θ, t, and applied to spherical shell examples; and (2) shells loaded by their own weight: because of the symmetry of revolution, the state of stress at the vertex of the shell is represented by point A fig. 8.56 (or A', if the shell is suspended instead of being supported). For a given shape of the midsurface, two uniform-strength designs exist, corresponding to two different stress profiles lying on both sides of the yield curve with respect to point A (fig. 8.56).

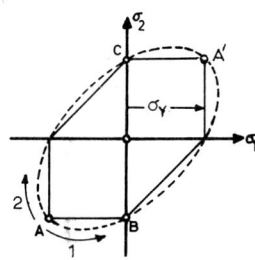

Fig. 8.56.

If γ is the weight per unit volume of the shell material, the two equilibrium equations for a spherical shell are

$$\frac{d}{d\phi}(t\sigma_\phi \sin\phi) - t\sigma_\theta \cos\phi + R\gamma t \sin^2\phi = 0, \qquad (8.121)$$

$$\sigma_\phi + \sigma_\theta + R\gamma \cos\phi = 0. \qquad (8.122)$$

At the vertex, $\sigma_\theta = \sigma_\phi = -\sigma_Y$ (point A, fig. 8.56) and $\phi = 0$. We obtain $R = 2\sigma_Y/\gamma$ from eq. (8.122), whereas eq. (8.121) is identically satisfied. Hence, the thickness t_o at the vertex remains arbitrary. The function $t(\phi)/t_o$ is obtained from integration of eq. (8.121), using eq. (8.122) and the considered stress profile 1 or 2 (see fig. 8.56) on the Tresca hexagon or on the von Mises ellipse. The four solutions are given in fig. 8.57. It is seen that stress profiles 2 give lower weights. Curves 1 and 2 corresponding to the Tresca condition were first obtained by Ziegler [8.67].

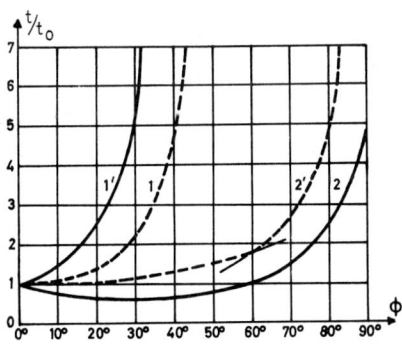

Fig. 8.57.

The uniform-strength condition is sometimes formulated [8.68, 8.69, 8.12] as

$$\sigma_1 = \sigma_2 = \sigma_Y, \qquad (8.123)$$

or

$$-\sigma_1 = -\sigma_2 = \sigma_Y. \qquad (8.124)$$

394 METAL SHELLS [Ch. 8

Both principal stresses must be of same sign and equal to the yield stress at every point of the shell. Because condition (8.123), or (8.124), completely defines the stress tensor when the principal directions are known, the two equilibrium equations will give both the meridian curve and the thickness distribution.

8.8.2.2. *Curved pressure vessel heads*

Minimum-weight design of pressure vessels is a problem of great practical importance. Assuming pure membrane behavior, Hoffman [8.53] has considered curved heads of cylindrical vessels. The yield condition of Tresca is used. The head is made of a spherical cap and a knuckle joining the cap to the cylinder (figs. 8.58 and 8.59), or is a ellipsoidal shell (fig. 8.60). The "Biezeno head" [8.52] has a knuckle with varying principal radii of curvature. *For heads with constant thickness*, the condition that the maximum shear stress be equal to $\sigma_Y/2$ throughout the head, together with continuity conditions in slope and thickness at the knuckle-sphere junction, define the head

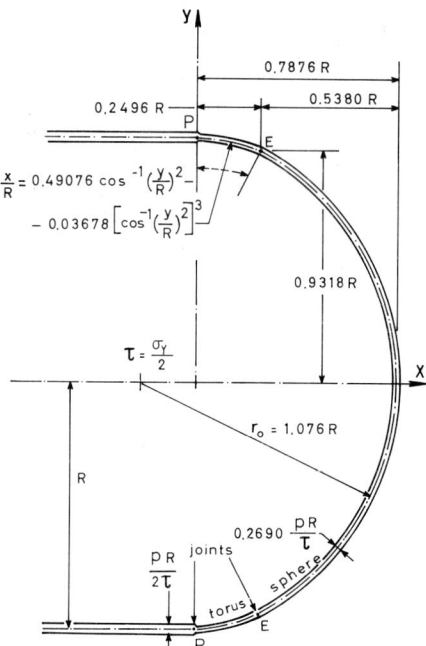

Fig. 8.58. Minimum-weight dimensions of constant-thickness Biezeno heads.

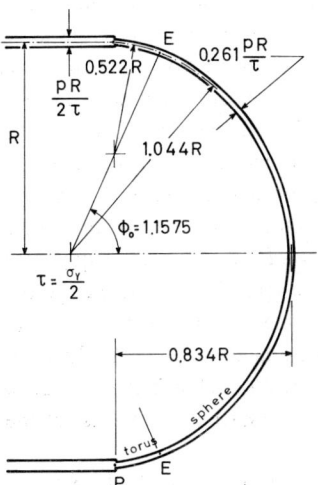

Fig. 8.59. Minimum-weight dimensions of torispherical heads of constant thickness.

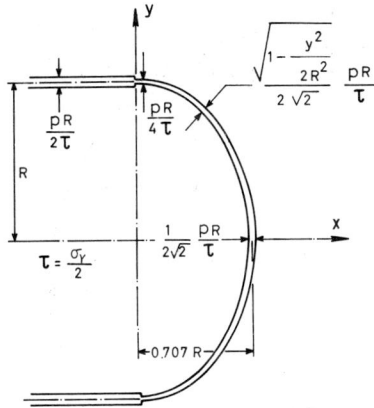

Fig. 8.60. Minimum-weight dimensions of ellipsoidal heads of varying thickness [8.53].

geometry except for a parameter β which is the ratio of the head thickness to the cylinder thickness. The value of β is obtained from the minimum-weight condition. The resulting "optimum" Biezeno head is shown in fig. 8.58. Its weight is about 90% of that of the hemispherical cap with a thickness equal to half the cylinder thickness.

In a torispherical head with constant thickness, the meridional curvature of the knuckle is constant and the maximum shear stress therefore varies in the knuckle. Its highest value is attained at the knuckle-sphere junction (see table 8.6). Equating this value to $\sigma_Y/2$ and expressing geometric continuity conditions at the junction of the torus and the sphere, the head geometry is obtained except for the angular parameter ϕ_o that fixes the location of junction E (fig. 8.59). The value of ϕ_o is obtained from the minimum-weight condition. The resulting head is shown in fig. 8.59. Its weight is only about 0.5% larger than the optimum constant-thickness Biezeno head.

If the head thickness is allowed to vary, supplementary economy in weight can be achieved. In a torispherical head, adequate variation of the knuckle thickness will enable the maximum shear stress to be equal to $\sigma_Y/2$ throughout. The minimum-weight head in this family is shown in fig. 8.61. If the meridional curvature of the knuckle is in turn allowed to vary, a head with smaller weight is obtained asserting the stress regime C in the whole knuckle (fig. 8.62, insert).

Various designs are compared in table 8.8. The weight of the hemispherical head with $\sigma_1 = \sigma_2 = \sigma_Y$ has been regarded as unity for comparison purposes. The thickness of the cylinder is equal to pR/σ_Y. The ideal Biezeno

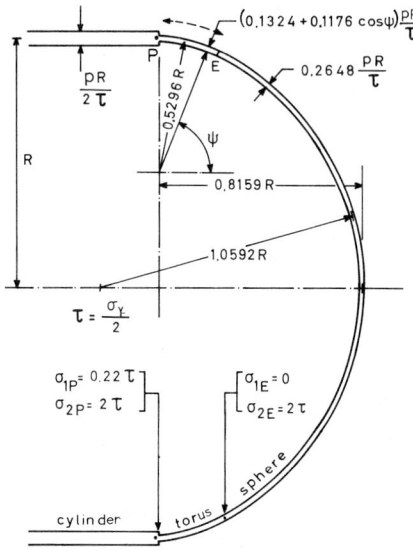

Fig. 8.61. Minimum-weight dimensions of torispherical heads with varying thickness.

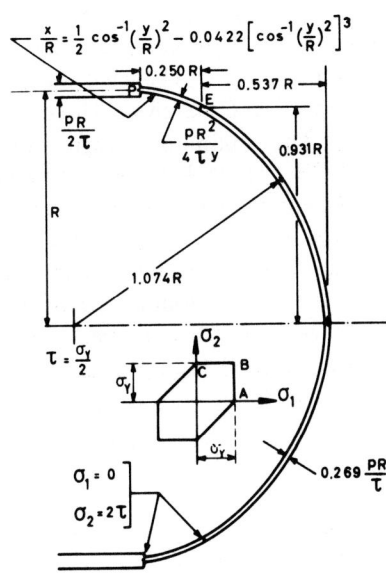

Fig. 8.62. Minimum-weight dimensions of variable-thickness, constant-shear head.

Table 8.8. Design properties of pressure vessel heads.

Head shape (arranged by weight)	Head weight	Knuckle thickness ($\times pR/\sigma Y$)	Depth of head ($\times R$)
Ideal Biezeno	1.425	1	0.515
Hemispherical	1	0.5	1
Torispherical, constant thickness	0.903	0.522	0.834
Constant shear, constant thickness	0.898	0.538	0.788
Torispherical, varying thickness	0.895	Increases from 0.500 at P to 0.530 at E	0.816
Constant shear, varying thickness	0.888	Increases from 0.500 at P to 0.538 at E	0.788

head has the same thickness as the cylinder. It can be seen that the torispherical head with constant thickness, while being a practical design, is only about 1.5% heavier than the lightest of the heads considered.

8.8.2.3. *Cylindrical vessels with flat heads*

Consider the cylindrical vessel with either two flat heads or one flat head, the other end being built-in (fig. 8.13). The yield stress σ_Y, the limit pressure p_l and the containment capacity C are given. The design unknowns are the dimensions R, L, t, and t'. The "optimum" design is the design with largest economy coefficient $k = C/V$, where V is the steel volume. Using Hodge's results [8.28] in the slightly modified form of eqs. (8.31) and (8.32), it can be shown that maximum k is always attained for $L = \infty$. In practical problems, L must remain finite. Accordingly, the maximum possible value of L is taken. The radius R is then obtained from $C = 2L\pi R^2$. The remaining design parameter is $\alpha = t'/t$. The values α_o of α corresponding to the largest k for various $l = L/R$ are given in fig. 8.63 [8.70, 8.71]. They are given by the following equations:

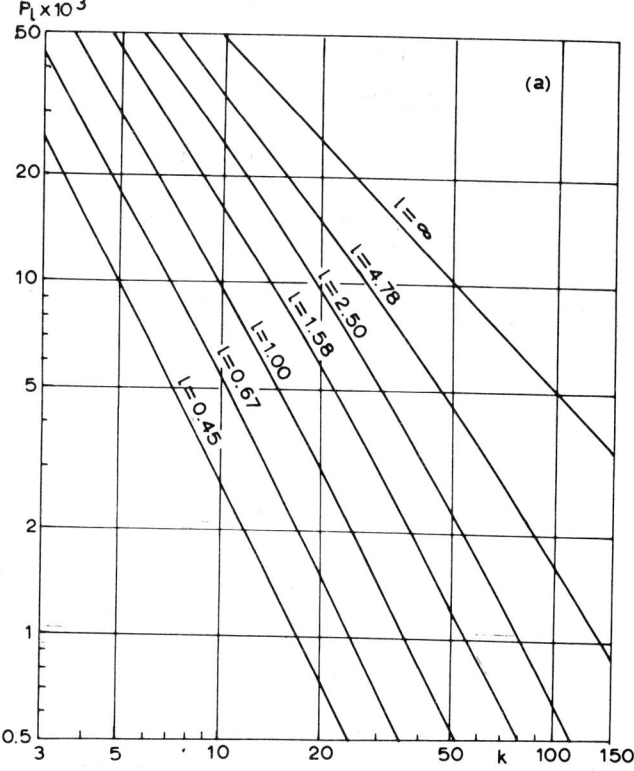

Fig. 8.63(a).

8.8] MINIMUM-WEIGHT DESIGN 399

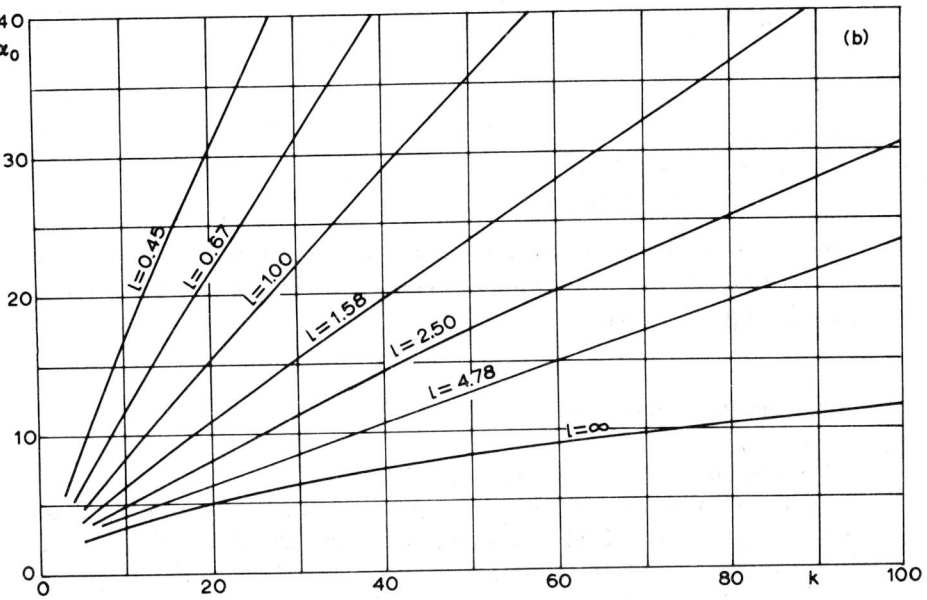

Fig. 8.63(b).

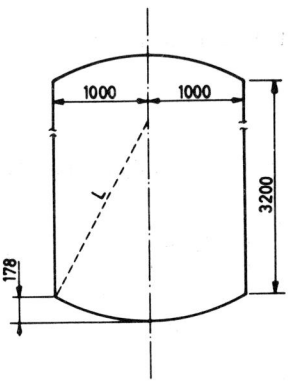

Fig. 8.64.

$$\alpha_o = \left(\frac{2}{3}\frac{p_l^r}{h^2} - 0.94\right)^{1/2}, \tag{8.125}$$

with

$$h = p_l^r \frac{2}{1 + (1+4p_l/l^2)^{1/2}}, \tag{8.126}$$

where $h = t/R$ and $p_l^r = p_l/\sigma_Y$. Information concerning the experimental bursting pressures of plastic optimum designs based on fig. 8.63 is found in refs. [8.70] and [8.71].

8.9. Examples of application

8.9.1. *Cylindrical vessel with reinforcing rings*

Consider the cylindrical vessel with spherical heads, the median surface of which is shown in fig. 8.64. The junction of the cylinder to the sphere will be reinforced by a heavy ring on which both parts can be regarded as built-in. The internal service pressure is p_1 = 150 psi. With a load factor (safety factor) of 2, the limit pressure is p_l = 300 psi.

The cylindrical shell has a limit pressure given by eq. (8.27):

$$p_l = \frac{\sigma_Y t}{R}\left(1 + \frac{1}{1+\omega^2}\right). \tag{a}$$

A first approximation to the limit pressure is

$$p_l = \frac{\sigma_Y t}{R} \tag{b}$$

for a cylindrical membrane. The resulting thickness is

$$t = \frac{p_l R}{\sigma_Y} = \frac{300 \times 100}{36000} = 0.833 \text{ in.}, \tag{c}$$

with σ_Y = 36,000 psi. We then obtain

$$\omega^2 = \frac{l^2}{2tR} = \frac{(320)^2}{2 \times 0.833 \times 100} = 614. \tag{d}$$

It is readily verified that the limit pressures given by eqs. (a) and (b) will practically coincide for this value of ω (long cylinder). If a smaller shell thickness is desirable (for some technological or financial reason), reinforcing rings must be placed to reduce the magnitude of ω^2 to, say, 10. Suppose we place the reinforcing rings 40 in. apart. If the thickness of 0.833 in. was maintained, we would obtain

$$\omega^2 = \frac{40}{2 \times 0.833 \times 100} = 9.6 < 10 \,. \tag{e}$$

The necessary thickness is then obtained introducing $\omega = 9.6$ and the given values of R and σ_Y into the condition

$$p_l \equiv \frac{\sigma_Y t}{R}\left(1 + \frac{1}{1+\omega^2}\right) \geqslant 300 \tag{f}$$

derived from relation (a). The resulting value of t is

$$t = 0.77 \text{ in.} \,, \tag{g}$$

and the corresponding ω is such that

$$\omega^2 = 10.37 \,. \tag{h}$$

The economy on the thickness is only $0.833 - 0.770 = 0.063$ in. The minimum cross-sectional area of a typical reinforcing ring is given by eq. (8.28):

$$\frac{4\omega^2}{4+\omega^2} \leqslant \gamma \,, \tag{i}$$

with

$$\gamma = \frac{l}{2M_p R}\sigma_{Ya} A_a \,. \tag{j}$$

In relation (j), l is twice the distance between two adjacent reinforcing rings, A_a the cross-sectional area of a ring with yield stress σ_{Ya}, whereas M_p and R are the plastic moment and the radius of the cylinder, respectively. One successively obtains

$$\omega^2 = \frac{(80)^2}{2 \times 0.77 \times 100} = 41.48,$$

$$\frac{4\omega^2}{4+\omega^2} = 3.65,$$

and

$$A_a \geqslant 3.65 \frac{2M_p R}{l\sigma_{Ya}} = 1.35 \text{ in}^2.,$$

with $\sigma_{Ya} = \sigma_Y = 36{,}000$ psi. It is seen that relatively small rings are sufficient.

The spherical caps have a radius L such that $17.8(L-17.8) = (100)^2$, whence $L = 290$ in. The cap angle is $2\alpha \cong 40°$. It is seen in fig. 8.27 that the best available lower bound is $p^* = 1$. Hence, the thickness will be

$$t_{\text{sphere}} = \frac{p_l L}{2\sigma_Y} = \frac{300 \times 290}{2 \times 36\,000} = 1.21 \text{ in.}$$

8.9.2. Torispherical head

Consider the same cylindrical vessel as in Section 8.9.1, but without reinforcing rings and with torispherical heads. We take

$$\frac{r}{D} = \frac{20}{200} = \frac{1}{10}. \tag{a}$$

Eq. (8.82) then becomes

$$0.88 \frac{t}{L} + 22.6 \left(\frac{t}{L}\right)^2 - 0.0006 - \frac{300}{36000} = 0, \tag{b}$$

from which one obtains $t/L = 0.00845$, or $t = 1.69$ in. It is checked in fig. 8.35 that, for $t/L = 0.0085$, one obtains $p_- D/2\sigma_Y t = 0.47$, that is, $p_- \cong 300$ psi. We finally note that the height H of the cap is $H = 32.7$ in., larger than $H = 17.8$ in. used in Section 8.9.2.

8.10. Problems

8.10.1. Prove relations (8.28).

8.10.2. Obtain the following bounds for the limiting value of a ring of force $2F_o$ applied to an infinitely long cylinder:

$$(2F_o)_+ = 4N_p\sqrt{t/R},$$

$$(2F_o)_- = 2\sqrt{2}N_p\sqrt{t/R},$$

using the yield locus consisting of a rectangle inscribed in the polygon *ABCDEF* of fig. 5.24.

8.10.3. Obtain eq. (8.42) by detailed calculation.

8.10.4. Apply the kinematic theorem to the mechanism found in Section 8.3.2., eqs. (8.14), to verify that it gives the limit load (8.27).

References

[8.1] W.A.NASH, "Recent Advances in the Buckling of Thin Shells", *Applied Mechanics Reviews*, **13**, 3, March 1960.
[8.2] Y.C.FUNG, E.E.SECHLER, "Instability of Thin Elastic Shells", *Structural Mechanics; Proceedings of the First Symposium on Naval Structural Mechanics*, pp. 115-168, Pergamon Press, New York, 1960.
[8.3] S.B.BARDORF, "A Simplified Method of Elastic Stability Analysis of Thin Cylindrical Shells", *N.A.C.A. Tech. Rep.,* 874, 1947.
[8.4] S.C.BATTERMANN, "Plastic Buckling of Axially Compressed Cylindrical Shells" *A.I.A.A. Journal*, **3**, 2, 316, January 1965.
[8.5] A.F.KIRSTEIN, E.WENK Jr., "Observation of Snap-through Action in Thin Cylindrical Shells under External Pressure", *Proc. S.E.S.A.,* **XIV**: 1, 1956.
[8.6] W.A.NASH, "An Experimental Analysis of the Buckling of Thin Initially Imperfect Cylindrical Shells Subject to Torsion", *Proc. S.E.S.A.,* **XVIII**: 1, 1961.
[8.7] R.H.HOMEWOOD, A.C.BRINE, A.E.JOHNSON Jr., "Experimental Investigation of the Buckling Instability of Monocoque Shells",*Proc. S.E.S.A.,* **XVIII**: 1, 1961.
[8.8] H.SCHMIDT, "Ergebnisse von Beulversuchen mit doppelt gekrummten Schalenmodellen aus Aluminium", *Proc. Symp. Shell Research,* Delft, August-September 1961, North-Holland Publ. Comp., Amsterdam, 1961.
[8.9] W.R.OSGOOD, *Residual Stresses in Metals and Metal Construction*, Reinhold Publ. Corp., New York, 1954.
[8.10] M.E.SHANK, *Control of Steel Construction to Avoid Brittle Failure*, Plasticity Committee of the Welding Research Council, New York, 1957.

[8.11] J.T.P.YAO, W.H.MUNSE, "Low-cycle Fatigue of Metals-Literature Review", *Welding Journal*, 182, April 1962.

[8.12] W.FLUGGE, *Stresses in Shells*, Springer, 1960.

[8.13] C.MASSONNET, "Tensions dans les reservoirs sous pression", Technical note B. II, I of the C.E.C.M., Brussels.

[8.14] M.SAVE, "Le calcul à la flexion des coques de révolution chargées de façon quelconque", *Bulletin of the C.E.R.E.S.*, **XI**: Liège, 1960.

[8.15] R.L.CLOUD, "Interpretative Report on Pressure Vessel Heads", *Welding Research Council Bulletin*, 119, 1, January 1967.

[8.16] M.E.LUNCHICK, "Yield Failure of Stiffened Cylinders under Hydrostatic Pressure", *Proc. 3rd U.S. Nat. Cong. Appl. Mech.*, Providence, 1958.

[8.17] M.E.LUNCHICK, J.A.OVERBY, "Yield Strength of Machined Ring-stiffened Cylindrical Shell under Hydrostatic Pressure", *Proc. S.E.S.A.*, **XVIII**: 1, 1961.

[8.18] R.C.DEHART, N.L.BASDEKAS, "Yield Collapse of Stiffened Circular Cylindrical Shells", (report) Southwest Research Institute, San Antonio, Texas, September 1960.

[8.19] H.H.DEMIR, D.C.DRUCKER, "An Experimental Study of Cylindrical Shells under Ring Loading", *Progress in Applied Mechanics* (The Prager anniversary volume), pp. 205-220, Macmillan, New York, 1963.

[8.20] M.SAVE, "Vérification expérimentale de l'analyse limite plastique des plaques et coques en acier doux" (Experiments on Limit Loads of Mild Steel Plates and Shells), *Report MT 21*, C.R.I.F., 21 rue des Drapiers Brussels, 1966.

[8.21] B.PAUL, "Limit Loads of Clamped Shells with a Reinforcing Ring", *PIBAL Report*, 424, December 1958.

[8.22] B.PAUL, "Collapse Loads of Rings and Flanges under Uniform Twisting Moment and Radial Force", *J. Appl. Mech.*, **26**: Series E, 2, June 1959.

[8.23] B.PAUL, P.G.HODGE Jr., "Carrying Capacity of Elastic-Plastic Shells under Hydrostatic Pressure", *Proc. 3rd U.S. Nat. Cong. Appl. Mech.* 631, 1958.

[8.24] E.T.ONAT, "The Plastic Collapse of Cylindrical Shells under Axially Symmetrical Loading", *Quart. Appl. Math.*, **XIII**: 63, 1955.

[8.25] R.SANKARANARAYANAN, "Plastic Interaction Curves for Circular Cylindrical Shells under Combined Lateral and Axial Pressures", *J. Franklin Inst.*, **270**: 5, November 1960.

[8.26] P.G.HODGE Jr., J.PANARELLI, "Interaction Curves for Circular Cylindrical Shells According to the Mises or Tresca Yield Criteria", *J. Appl. Mech.*, **29**: 375, 1962.

[8.27] P.G.HODGE Jr., *Limit Analysis of Rotationally Symmetric Plates and Shells*, Prentice-Hall, Englewood Cliffs, N.J., 1963.

[8.28] P.G.HODGE Jr., "Plastic Design of a Closed Cylindrical Structure", *J. Mech. Phys. Solids*, **12**: 1, 1964.

[8.29] N.A.FORSMAN, "On the Carrying Capacity of Cylindrical Vessels with Flat Closures" (in Russian), *Isv. An. U.S.S.R. Mekh. Mashin.*, 3, 106, 1964.

[8.30] M.SAYIR, "Kollapsbelastung von rotationsymmetrischen Zylinderschalen", *Zeit. Ang. Math. Phys. Z.A.M.P.*, **17**: 353, 1966.

[8.31] M.JANAS, A.KONIG, "Limit Analysis of Shells: Shell Roofs and Vessels" (in Polish), Arkady, Warsaw, 1967.

[8.32] Y.V.NEMIROVSKY, Y.N.RABOTNOV, "Limit Analysis of Ribbed Plates and Shells", *Nonclassical Shell Problems*, pp. 786-807, North-Holland Pub. Co., Amsterdam, 1964.

[8.33] A.BIRON, A.SAWCZUK, "Plastic Analysis of Rib-Reinforced Cylindrical Shells", *Trans. A.S.M.E., J. Appl. Mech.,* **89**: Series E, 1, 37, March 1967.
[8.34] M.CAPURSO, A.GANDOLFI, "Sul collasso rigido-plastico dei gusci nervati di revoluzione" (Plastic Collapse of Ribbed Shells of Revolution) (internal report), Istituto di Tecnica delle Costruzioni, Napoli, 1967.
[8.35] M.CAPURSO, A.GANDOLFI, "Sul collasso plastico dei tubi circolari nervati soggetti a pressione e sforzo assiale" (Plastic Collapse of Ribbed Cylinders under Pressure and Axial Force) (internal report), Istituto di Tecnica delle Costruzioni, Napoli, 1967.
[8.36] G.EASON, R.T.SHIELD, "The Influence of Free Ends on the Load-carrying Capacities of Cylindrical Shells", *J. Mech. Phys. Solids,* **4**: 17, 1955.
[8.37] G.EASON, "The Load-carrying Capacities of Cylindrical Shells Subjected to a Ring of Force", *J. Mech. Phys. Solids,* **7**: 169, 1959.
[8.38] H.H.DEMIR, "Cylindrical Shells under Ring Loads", *Proc. A.S.C.E.,* **91**: ST3 (J. Struct. Div., Part I), 71, June 1965.
[8.39] W.OLSZAK, A.SAWCZUK, "Die Grenztragfähigkeit von zylindrischen Schalen bei verschiedenen Formen der Plastizitätsbedingung", *Acta Technica Academiae Scientiarum Hungaricae,* **XXVI**: 1, Budapest, 1959.
[8.40] B.PAUL, "Carrying Capacity of Elastic-Plastic Shells with Various End Conditions, under Hydrostatic Compression", *J. of Appl. Mech.,* **26**: Series E, 4, 553, December 1959.
[8.41] P.G.HODGE Jr., "Displacements in an Elastic-Plastic Cylindrical Shell", *J. of Appl. Mech.,* **23**: 73, 1956.
[8.42] P.G.HODGE Jr., S.V.NARDO, "Carrying Capacity of an Elastic-Plastic Cylindrical Shell with Linear Strain-hardening", *J. Appl. Mech.,* **25**: *Trans, A.S.M.E.,* **80**: 79, 1958.
[8.43] P.G.HODGE Jr., "Yield Conditions for Rotationally Symmetric Shells under Axisymmetric Loading", *J. Appl. Mech.,* **27**: *Trans. A.S.M.E.,* **82**: Series E, 323, June 1960.
[8.44] W.FLUGGE, T.NAKAMURA, "Plastic Analysis of Shells of Revolution under Axisymmetric Loads", *Ing. Archiv.,* **34**: 4, 238, 1965.
[8.45] D.C.DRUCKER, R.T.SHIELD, "Limit Analysis of Symmetrically Loaded Thin Shells of Revolution", *J. Appl. Mech.,* **26**: *Trans. A.S.M.E.,* **81**: Series E, 61, 1959.
[8.46] D.C.DRUCKER, R.T.SHIELD, "Limit Strength of Thin-Walled Pressure Vessels with an A.S.M.E. Standard Torsipherical Head", *Proc. 3rd U.S. Nat. Cong. Appl. Mech. A.S.M.E.,* 665, 1958.
[8.47] P.G.HODGE Jr., C.LAKSHMIKANTHAM, "Limit Analysis of Shallow Shells of Revolution", *Trans. A.S.M.E., J. Appl. Mech.,* **30**: Series E, 2, 215, June 1963.
[8.48] P.G.HODGE Jr., "Plastic Analysis of Circular Conical Shells", *J. Appl. Mech.,* December 1960.
[8.49] R.H.LANCE, E.T.ONAT, "Analysis of Plastic Shallow Conical Shells", *J. Appl. Mech.,* **30**: 199, 1963.
[8.50] R.W.KUECH, S.L.LEE, "Limit Analysis of Simply Supported Conical Shells Subjected to Uniform Internal Pressure", *J. Franklin Inst.,* **280**: 1, 71, July 1965.
[8.51] D.C.DRUCKER, R.T.SHIELD, "Design of Thin-walled Torsipherical and Toriconical Pressure Vessel Heads", *J. Appl. Mech.,* June 1961.
[8.52] R.A.STRUBLE, "Biezeno Pressure Vessel Heads", *J. Appl. Mech.* **23**: *Trans. A.S.M.E.,* **78**: 642, 1956.

[8.53] G.A.HOFFMAN, "Minimum-weight Proportions of Pressure-Vessel Heads", *Trans. A.S.M.E.*, **29**: Series E, 4, 662, December 1962.
[8.54] G.D.GALLETLY, "Torispherical Shells: A Caution to Designers", *J. Eng. Industry, Trans. A.S.M.E.*, **81**: Series B, 51, 1959.
[8.55] C.E.MASSONNET, M.A.SAVE, "Résistance limite d'une poutre courbe en double té à parois minces soumise à flexion pure" (Ultimate Strength of a I-curved Thinwall Beam Subjected to Uniform Bending), *Proc. Int. Ass. Bridge. Str. Eng.*, **23**: 245, 1964.
[8.56] C.E.MASSONNET, M.A.SAVE, "Résistance limite d'une poutre courbe en caisson soumise à flexion pure" (Ultimate Strength of a Curved Box-shaped Beam Subjected to Uniform Bending), *Amici et Alumni* (The Campus anniversary volume) Liege, 1964.
[8.57] P.G.HODGE Jr., "Full-strength Reinforcement of a Cutout in a Cylindrical Shell", *Trans. A.S.M.E., J. Appl. Mech.*, **31**: Series E, 4, 667, December 1964.
[8.58] M.D.COON, S.S.GILL, R.KITCHING, "A Lower Bound to the Limit Pressure of a Cylindrical Pressure Vessel with an Unreinforced Hole", *Int. J. Mech. Sci.*, **9**: 2, 69, 1965.
[8.59] C.RUIZ, S.E.CHUKWUJEKWU, "Limit Analysis Design of Ring-reinforced Radial Branches in Cylindrical and Spherical Vessels", *Int. J. Mech. Sci.* **9**: 1, 11, 1967.
[8.60] S.S.GILL, "The Limit Pressure of a Flush Cylindrical Nozzle in a Spherical Pressure Vessel", *Int. J. Mech. Sco.*, **6**: 1, 105, 1964.
[8.61] K.S.DINNO, S.S.GILL, "The Limit Analysis of a Pressure Vessel Consisting of the Junction of a Cylindrical and a Spherical Shell", *Int. J. Mech. Sci.* **7**: 1, 21, January 1965.
[8.62] K.S.DINNO, S.S.GILL, "An Experimental Investigation into the Plastic Behavior of Flush Nozzles in Spherical Pressure Vessels", *Int. J. Mech. Sci.*, **7**: 12, 817. 1965.
[8.63] W.OLSZAK, A.SAWCZUK, "Inelastic Behaviour in Shells", Noordhoff, Groningen, 1967.
[8.64] R.T.SHIELD, *Optimum Design Methods for Structures, Plasticity*, Pergamon Press, New York, 1960.
[8.65] W.OLSZAK, A.SAWCZUK, "Some Problems of the Limit Analysis and Design of Nonhomogeneous Axially Symmetric Shells", *Proc. 2nd Symp. Concrete Shell Roofs Structures*, Oslo, 1957.
[8.66] W.ISSLER, "Membranschalen gleicher Festigkeit", *Ing. Archiv,* **33**: 5, 330, 1964.
[8.67] H.ZIEGLER, "Kuppeln gleicher Festigkeit", *Ing. Archiv.*, **26**: 378, 1958.
[8.68] M.MILANKOVIC, *Arbeiten der Jugoslaw, Akad. der Wissensch.*, **175**, 1908.
[8.69] M.C.DOKMECI, "A Shell of Constant Strength", *Zeit. Ang. Math. Phys., Z.A.M.P.*, **17**: 4, 545, 1966.
[8.70] M.JANAS, M.A.SAVE, "Investigation of Optimal Design of Cylindrical Vessels for Assigned Bursting Pressure", *Proc. I.A.S.S., Symp.*, Weimar, May 1968 (to be published).
[8.71] M.JANAS, M.A.SAVE, "Etude des réservoirs cylindriques à fonds plats" (Cylindrical Vessels with Flat Heads), Report MT58 of the "C.R.I.F.", 21 rue des Drapiers, Brussels, 1970.

9

Reinforced Concrete Shells

9.1 Introduction

Reinforced concrete shell structures, and particularly reinforced concrete shell roofs, have been increasingly used in the last twenty years, and their development is continuing. There are numerous publications on the subject; a large number of references can be found in the books [5.1] and [8.12]. Despite this wide use of reinforced concrete shells, their limit analysis is still in a relatively early stage of development [9.1]. In particular, no direct experimental support has been given of the yield surfaces used. Simple physical considerations on the yield conditions of steel and concrete [7.10,9.2,9.3], as well as the desire of obtaining polyhedrons [9.4] have been taken into account. Numerous tests to collapse, though of high engineering interest [9.5 to 9.9], have been conducted to investigate some peculiar points of the behavior of some reinforced concrete shells and, hence, cannot give much information on the applicability of limit analysis, though they have influenced the development of approximate methods to evaluate the collapse load [9.10 to 9.12]. These methods are approximate either because they introduce assumptions or concepts that are of dubious value from the point of view of limit analysis [9.10], or because they are purely kinematical and, hence, furnish only an upper bound [9.12,9.13].

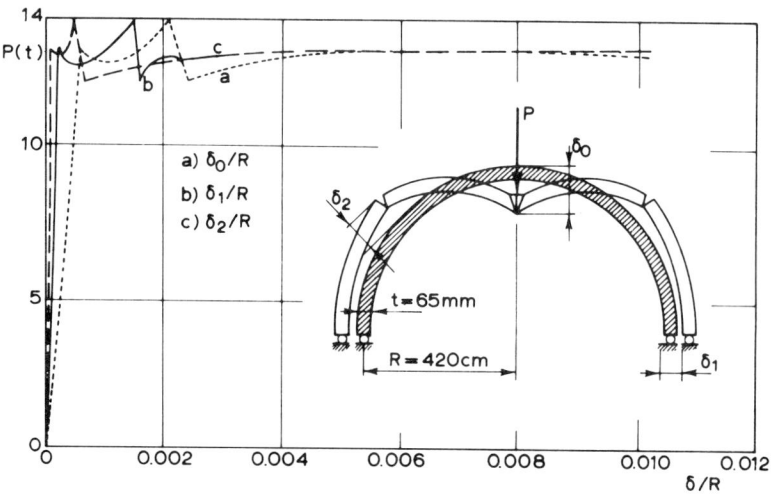

Fig. 9.1. Load-deflection relations for a reinforced concrete dome under concentrated load [9.18].

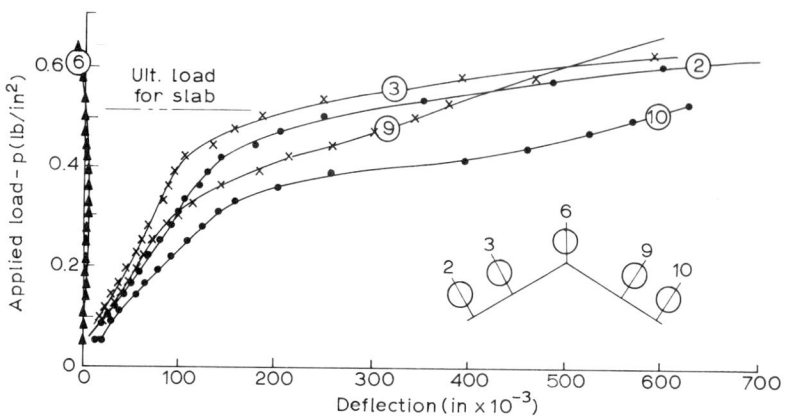

Fig. 9.2. Model structure 203(C1)3 free edges. Deflections of midspan cross section (collapse test) [9.15].

Available experimental data strongly indicates that collapse mechanisms and limit loads (in the sense of limit analysis) do exist, as can be seen in fig. 9.1 [9.14] and fig. 9.2 [9.15]. For the folded structures studied in ref.

[9.15], a safe estimate of the collapse load can be obtained by applying the yield line method to the plates. Experiments on cylindrical shell roofs also show [9.5] that plastic deformations are concentrated in "generalized yield lines" (see fig. 9.3) which will be used in Section 9.3. Because obtaining the right collapse pattern is often very difficult (at least as much as in plate problems) a sound trend is to combine experimental and analytical techniques, as strongly advocated by Sawczuk [9.4], Olszak and Sawczuk [8.63], and Haidukov [9.16]. From experimental collapse patterns, and with a sufficiently accurate yield surface, limit loads are obtained by the work equation of the kinematic theorem.

In thin shells, instability, and particularly creep-buckling, may be the relevant cause of collapse. Hence, taking account of the small amount of experimental support, limit analysis should be regarded as *one* theory to evaluate the safety with respect to a particular collapse mode, and should be used with adequate caution.

Fig. 9.3. Cylindrical panel after collapse [9.9].

In the following sections, we successively give some complete solutions for cylindrical tanks (Section 9.2) and various kinematic solutions (Section 9.3).

9.2. Circular cylindrical tank under axisymmetric loading

9.2.1. *Yield condition*

Longitudinal and circumferential directions are principal directions that we label x and θ, respectively. We assume that the concrete is perfectly plastic, with the yield curve in plane stress shown in fig. 9.4, where σ_r and σ'_r are the tensile and compressive yield stresses, respectively [9.2]. In this chapter, compressive forces and stresses will be regarded as positive, as is usual in reinforced concrete practice. Strain rates will be positive when corresponding to contractions.

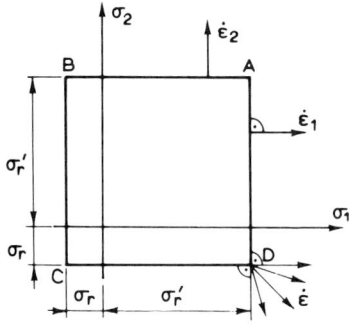

Fig. 9.4.

We assume that the amount of reinforcement subjected to compression is small enough to contribute in a negligible manner to the compressive strength, and that the reinforcement subjected to tension has a yield stress σ_Y and cross-sectional areas A_x and A_θ per unit of length. Because of the symmetry of revolution, the strain rates are distributed as shown in fig. 9.5 and we know (see Sections 5.1.4 and 8.3.1) that M_θ is a reaction. Applying the normality law to the strain rates of fig. 9.5, we conclude that, for the concrete, the stress point must be in C (fig. 9.4) for $-t/2 \leqslant z \leqslant \eta t/2$, and in D (fig. 9.4) for $-t/2 \leqslant z \leqslant \eta t/2$, with η either positive or negative. Taking into account that the reinforcement must be at yield in both axial and circumferential

9.2] CIRCULAR CYLINDRICAL TANK UNDER AXISYMMETRIC LOADING

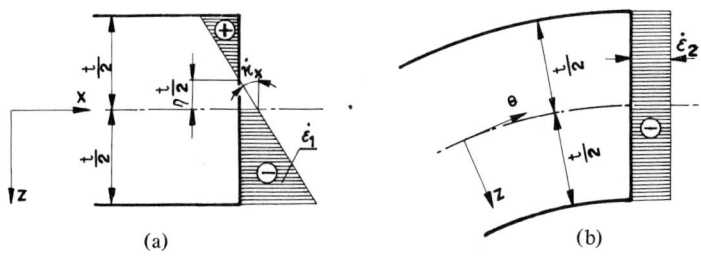

Fig. 9.5.

directions, the generalized stresses M_x, N_x, and N_θ can be evaluated for every value of the parameter η. Elimination of η from the expressions of M_x, N_x, and N_θ gives the yield surface. If we define the reduced generalized stresses

$$n = \frac{N}{\sigma'_r t},$$

$$m = \frac{M}{\sigma'_r t^2/4}, \qquad (9.1)$$

the reinforcement ratios

$$\mu_x = \frac{A_x}{2t},$$

$$\mu_\theta = \frac{A_\theta}{2t}, \qquad (9.2)$$

and the parameters

$$\alpha = \frac{\sigma_r}{\sigma'_r},$$

$$\beta = \frac{\sigma_Y}{\sigma'_r}, \qquad (9.3)$$

the yield surface of fig. 9.6 is obtained, formed of two parabolic cylinders with equations

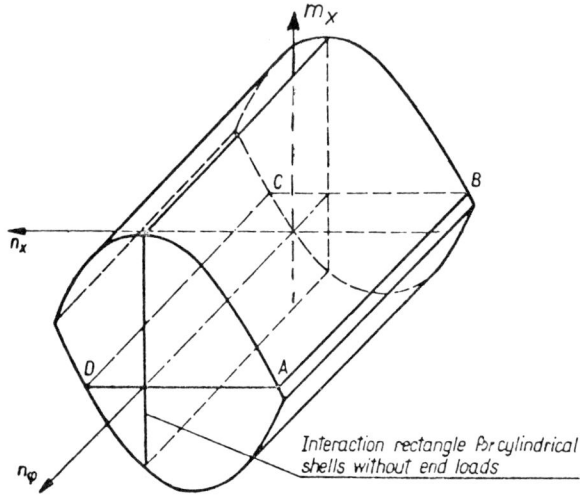

Fig. 9.6. Yield surface for reinforced-concrete cylindrical shells [9.17].

$$m_x(1+\alpha) + 2n_x^2 + 2n_x(2\beta\mu_x+\alpha-1) + 2\beta^2\mu_x^2 - 4\beta\mu_x - 2\alpha = 0, \quad (9.4)$$

$$-m_x(1+\alpha) + 2n_x^2 + 2n_x(\alpha-1) - 2\alpha = 0, \quad (9.5)$$

limited by the planes with equations

$$n_\theta = 1, \quad (9.6)$$

$$n_\theta = -\alpha - \beta\mu_\theta. \quad (9.7)$$

When the tensile strength of the concrete must be neglected, we simply set $\alpha = 0$. When there is no axial force N_x, intersection of the yield surface with

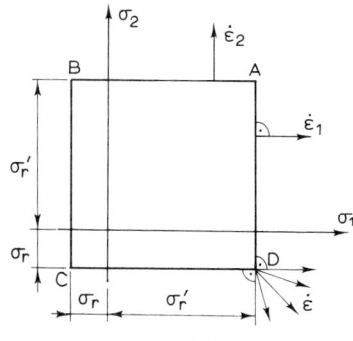

Fig. 9.7.

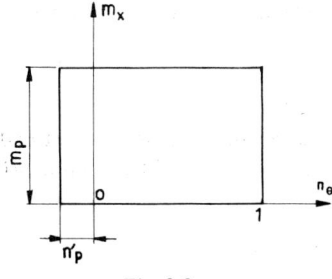

Fig. 9.8.

the plane $n_x = 0$ is the rectangle in fig. 9.7. In the absence of axial reinforcement at the external face of the cylinder, $m'_p = 0$, and the rectangle of fig. 9.7 reduces to that of fig. 9.8.

The values of m_p, m'_p, and n'_p can be evaluated from eqs. (9.4) to (9.7) by simply making $n_x = m_x = 0$, respectively.

9.2.2. Collapse mechanisms

We restrict ourselves to the case $n_x = 0$. The generalized strain rates $\dot{\phi}_x$ and $\dot{\lambda}_\theta$ corresponding to m_x and n_θ, respectively, are given in table 9.1. We have taken $\alpha = 0$ for simplicity (table 9.1), but this does not affect the conclusions regarding the possible mechanisms. From relations (8.1) and the normality law, it is easily seen that regimes AB and CD must be rejected and that regimes BC and DA furnish conical mechanisms with the general equation

$$\dot{w} = C_1 x + C_2 , \tag{9.8}$$

Table 9.1.

Stress profile fig. 9.7	Yield condition	Strain rates (to within a positive common factor)	
AB	$m_x = m_p \equiv -2\beta^2 \mu_x^2 + 4\beta\mu_x$ $-n_p \leqslant n_\theta \leqslant 1$	1	0
BC	$n_\theta = n_p \equiv -\alpha - \beta\mu_\theta$ $-m'_p \leqslant m_x \leqslant m_p$	0	-1
CD	$m_x = -m'_p \equiv 0$ $-n_p \leqslant n_\theta \leqslant 1$	-1	0
DA	$n_\theta = 1$ $-m'_p \leqslant m_x \leqslant m_p$	0	1

where $\dot{w}$ is the radial rate of displacement and C_1 and C_2 are integration constants.

9.2.3. Equilibrium equation

When compressive stresses and strains are regarded as positive, the fundamental equilibrium equation [eq. (8.10)] can be rewritten as

$$\frac{l^2}{C^2}\frac{d^2 m_x}{dx^2} - n_\theta - p^* = 0, \qquad (9.9)$$

where l is the length of the shell, $C^2 = 4l^2/Rt$, R is the radius of the median surface, t the effective thickness (see Section 7.7), and $p^* = p(x)R/t\sigma'_r$. The origin of the abscissa x is located at one end of the shell.

9.2.4. Cylindrical shell subjected to hydrostatic pressure
9.2.4.1. Nonuniform reinforcement: complete collapse

Consider a vertical cylindrical tank subjected to the hydrostatic internal pressure

$$p^*(x) = p_o^* \frac{x}{l}. \qquad (9.10)$$

The origin $x = 0$ is at the top cross section which is free, whereas the bottom end is built-in. The boundary conditions thus are (see fig. 9.9 for sign convention, all positive elements)

$$\dot{w}(l) = 0, \qquad \dot{w}(0) = -\dot{w}_o \qquad (\text{with } \dot{w}_o > 0),$$

$$m_x(0) = 0, \qquad v_x(0) \equiv \frac{4l}{t^2 \sigma'_r} V_x(0) = 0. \qquad (9.11)$$

Various collapse mechanisms can occur, depending on the geometry and the reinforcement of the shell. A complete discussion can be found in the paper by Sawczuk and Olszak [9.2]. Having in mind the "complete collapse" of the tank, corresponding to

$$w(x) = -\dot{w}_o \left(1 - \frac{x}{l}\right), \qquad \dot{w}_o > 0, \qquad (9.12)$$

we assume that (a) the axial reinforcement is uniform and exclusively at the

9.2] CIRCULAR CYLINDRICAL TANK UNDER AXISYMMETRIC LOADING 415

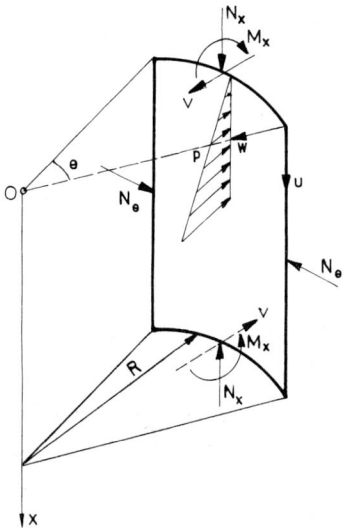

Fig. 9.9.

inner face: $m_p(x) = $ const, $m'_p(x) = 0$, and (b) the circumferential reinforcement is such that $n_p(x) = n_p x/l$. With the stress profile EB, fig. 9.8, that is, $n_x = -n_p(x)$, $0 \leqslant m_x \leqslant m_p$, integration of eq. (9.9) with due account of the boundary conditions (9.11), gives

$$m_x = m_p \left(\frac{x}{l}\right)^3 \tag{9.13}$$

and

$$p_0^* = \frac{6m_p}{C^2} + n_p . \tag{9.14}$$

It is easily seen that the mechanism (9.12) corresponds to the considered stress field. The expression (9.14) thus furnishes the exact limit load. The collapse mechanism and the stress field are shown in fig. 9.10.

If a uniform axial reinforcement at the external face of the cylinder ($m'_p \neq 0$) is added to the preceding reinforcement, the yield condition in fig. 9.7 must be used. For sufficiently short shells, complete collapse can still occur. Limiting values of the length parameter C, together with an auxiliary

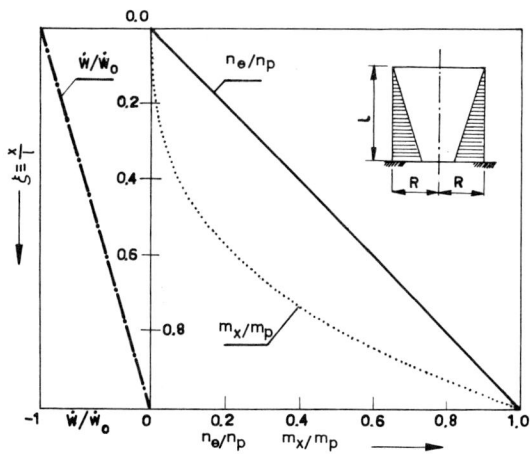

Fig. 9.10.

unknown x_1/l, are obtained from the conditions that, at $x = x_1$, $m_x = -m'_p$, and $dm_x/dx = 0$ (stress profile *ECEB* in fig. 9.7, with point *C* corresponding to abscissa x_1). In the present situation, it can also be shown [9.2] that the stress field remains admissible with nonvanishing circumferential reinforcement at the top of the cylinder. We thus have, in general,

$$n_p(x) = n_{po} + n_p \frac{x}{l}. \tag{9.15}$$

The limiting values C_l of C are given in fig. 9.11, for $0 \leqslant m'_p/m_p \leqslant 2$. When $m'_p = m_p = 1$ and $n'_p = n_p = 1$, we had found in Section 8.3.4.3 that $C_l^2 = 17.1$.

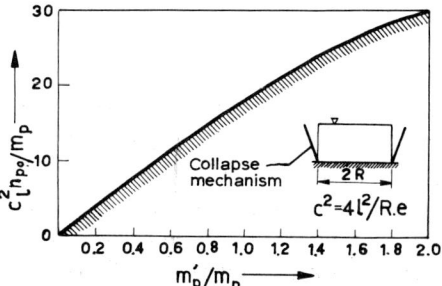

Fig. 9.11.

9.2.4.2. Uniform reinforcement: complete collapse

Consider the same loading as in Section 9.2.4.1, but a reinforcement with $m_p \neq 0, m'_p \neq 0$ in the axial direction, $n_p \neq 0$ in the circumferential direction, where all three values m_p, m'_p, and n_p do not depend on x. From what was seen in the preceding section, it can be expected that, with increasing length of the shell, a positive hinge circle at the bottom end will be accompanied by a negative hinge circle at a certain abscissa x_1, with possibly circumferential contraction of an upper region where $0 \leq x \leq x_o$ (fig. 9.12). At the boundary x_0 of this region, n_θ must jump from 1 to $-n_p$ because regimes AB and CD are not possible. Such a discontinuity is admissible (see Sections 5.6 and 8.6). We therefore consider the following stress field:

$$n_\theta = 1, \quad -m'_p \leq m_x \leq m_p \quad \text{for} \quad 0 \leq x \leq x_0,$$

$$n_\theta = -n_p, \quad -m'_p \leq m_x \leq m_p \quad \text{for} \quad x_0 \leq x \leq x_1,$$

$$n_\theta = -n_p, \quad -m'_p \leq m_x \leq m_p \quad \text{for} \quad x_1 \leq x \leq l, \quad (9.16)$$

with the boundary conditions:

$$m_x(l) = m_p, \quad m_x(x_1) = -m'_p, \quad m_x(0) = 0, \quad v_x(0) = 0 \quad (9.17)$$

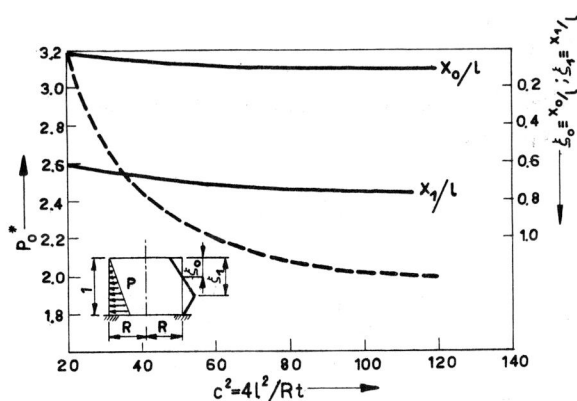

Fig. 9.12.

m_x	continuous in	$x = x_1$	and	$x = x_0$	
n_θ	discontinuous at	$x = x_0$	by the amount	$1 + n_p$,	
v_x	continuous at	$x = x_1$	and	$x = x_0$.	(9.18)

Integration of eq. (9.9) then gives

$$m_x = C^2 \left(\frac{x^2}{2l^2} + p_0 \frac{x^3}{6l^3} \right) \quad \text{for} \quad 0 \leqslant x \leqslant x_0. \tag{9.19}$$

$$m_x = C^2 \left(-n_p \frac{x^2}{2l^2} + p_o^* \frac{x^3}{6l^3} \right) + C^2 A \frac{x}{l} + C^2 B. \tag{9.20}$$

From conditions (9.18) at $x = x_0$, one obtains

$$A = \frac{x_0}{l^2} (1+n_p),$$

$$2B = -\frac{x_0^2}{l^2} (1+n_p). \tag{9.21}$$

The first two conditions of (9.17), together with the condition

$$\left(\frac{dm_x}{dx} \right)_{x=x_1} = 0$$

of an analytical extremum at $x = x_1$, give the following equations for the unknown p_o^*, x_1 and x_0:

$$\frac{6m_p}{C^2} + 3n_p - p_o^* + 6(1+n_p)\frac{x_0}{l} - 3(1+n_p)\frac{x_0^2}{l^2} = 0,$$

$$\frac{6}{C^2}(m_p+m_p') + 3n_p \left(1 - \frac{x_1^2}{l^2} \right) - p_o^* \left(1 - \frac{x_1^3}{l^3} \right) = 0,$$

$$\frac{2x_0}{l}(1+n_p) = 2n_p \frac{x_1}{l} - p_o^* \frac{x_1^2}{l^2}. \tag{9.22}$$

The corresponding mechanism is

$$\dot{w} = -\dot{w}_o \frac{x - x_0}{x_1 - x_0} \quad \text{for} \quad 0 \leqslant x \leqslant x_1,$$

$$\dot{w} = \dot{w}_o \frac{x - l}{l - x_1} \quad \text{for} \quad x_1 \leqslant x \leqslant l, \tag{9.23}$$

where $\dot{w}_o$ is the modulus of the radial displacement rate at $x = x_1$. The solution given above is statically and kinematically admissible and hence complete. It is valid provided the positive moment m_x in the vicinity of the upper edge does not attain m_p at a certain abscissa x_1. To study the limiting case where $m_x = m_p$ at $x = x_2$, we first remark that, to remain within the yield restriction, m_x must exhibit an analytical maximum at $x = x_2$, whereas expression (9.19) shows that m_x is monotonically increasing for $0 \leqslant x \leqslant x_0$. Hence $x_0 \leqslant x_2 \leqslant x_1$. The condition $m_x(x_2) = m_p$ and $(dm_x/dx)_{x_2} = 0$ give $x_2 = x_0 = 2n_p/p_o^* - x$. Substituting this expression for x_0 in eq. (9.20), using eqs. (9.21) and letting $m_x = m_p$, we finally obtain the critical value C_l at which eqs. (9.22) and (9.23) cease to apply. When $m_p = m_p' = n_p = 1$, the considered solution is valid for $17.1 < C^2 < 115$, and the corresponding values of p_o^*, x_0/l, x_1/l are given in fig. 9.12. If $C^2 < 17.1$ the mechanism described in Section 9.2.4.1 takes place. If $C^2 > 115$, partial collapse occurs, as will be seen in Section 9.2.4.3. Fig. 9.13 shows the stress profile, the stress field, and the yield mechanism (with a circumferentially compressed upper zone) when $m_p = m_p' = n_p = n_p' = 1$ and $C^2 = 75$.

9.2.4.3. *Uniform reinforcement: partial collapse*

For large values of C^2, only the bottom region of the tank will collapse,

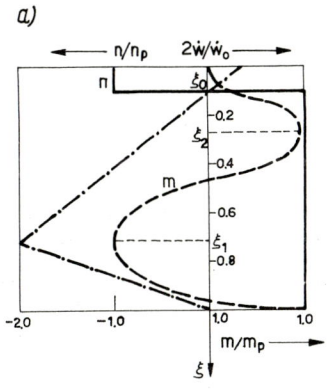
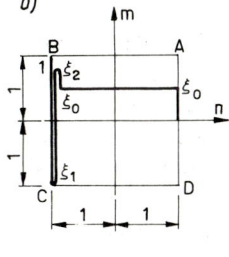

Fig. 9.13.

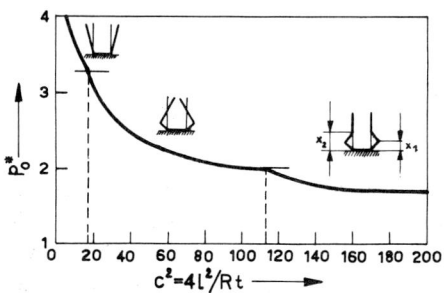

Fig. 9.14.

whereas the upper region will remain rigid. In the collapsing region, the stress profile BCB, fig. 9.7, will be valid, and the mechanism will be as shown in the insert at the right of fig. 9.14. Because the stress field can be continued in a statically admissible manner in the rigid region, the solution is complete. For detailed derivation, the reader is referred to the original paper [9.2]. The limit value of p_o^* is given in fig. 9.14 versus C^2, for the three types of mechanisms, when $m_p = m_p' = n_p = 1$.

9.2.5. *Cylindrical silo*

A vertical cylindrical silo, free at the upper edge, built-in at the bottom edge and containing a medium of density γ, with coefficient of friction μ on the wall and coefficient k of internal friction, has been studied by Sawczuk and König [9.17], in a manner similar to that used in Section 9.2.4. The limit load p_o and the collapse mechanisms are given in fig. 9.15, valid when $m_p = m_p' = n_p = 1$.

9.3. Upper-bound solutions

9.3.1. *Conical shell*

A truncated conical shell of the type used for foundations of high tower has been studied by S.Kaliszky [9.18]. Considering a conical mechanism bounded by a circumferential hinge circle, fig. 9.16, he assumes that the circumferential bending moment contributes in a negligible manner to the dissipation. The work equation then gives the limit load (see fig. 9.16 for notations):

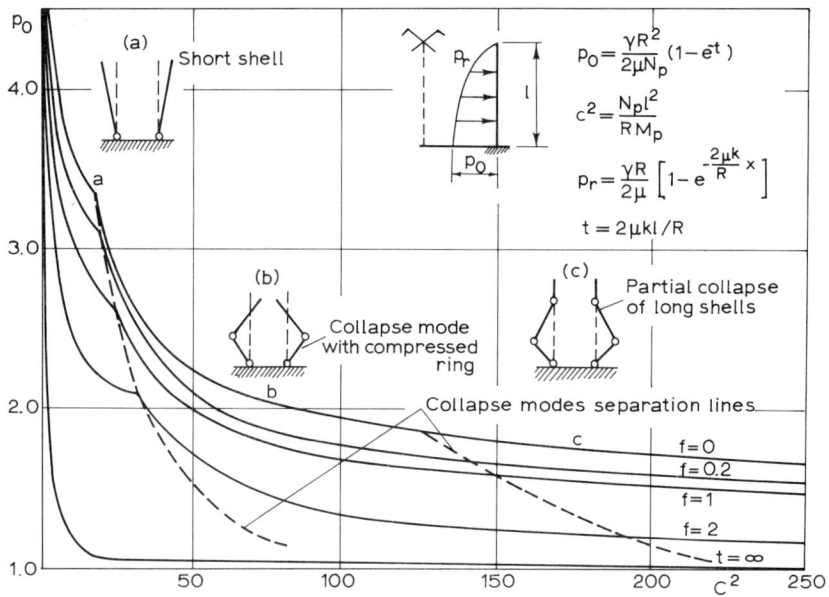

Fig. 9.15. Mechanisms of collapse and collapse pressure for nonuniformly loaded cylindrical shells [9.17].

$$p = \frac{3}{R^2} \frac{N_p R \cos\alpha (1-r_0)^2 + 2M_p r_0 \sin^2\alpha}{r_0^3 - 3r_0 + 2}, \tag{9.24}$$

where $r_0 = R_0/R$. The parameter R_0 is obtained from the minimum condition $dp/dR_0 = 0$, according to the kinematic theorem. It is given in fig. 9.17, as a function of the shell parameter k.

9.3.2. *Spherical cap*

A mechanism similar to that used in Section 9.3.1 is shown in fig. 9.18 in the case of a simply supported spherical cap subjected to vertical load uniformly distributed on the horizontal projection area. The meridional strip rotates with respect to lines tangent to the shell surface. For a unit vertical displacement rate at the boundary, we have (fig. 9.18) $\dot\phi = 1/(R-R_0)$, and the dissipation in the hinge circle is

$$D_1 = 2\pi R_0 M_p \dot\phi = \frac{2\pi M_p R_0}{R - R_0}.$$

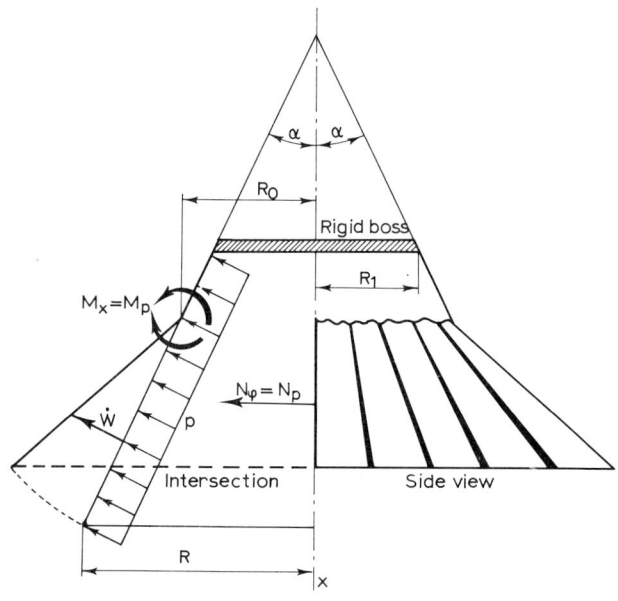

Fig. 9.16. Collapse mode for a truncated cone [9.18].

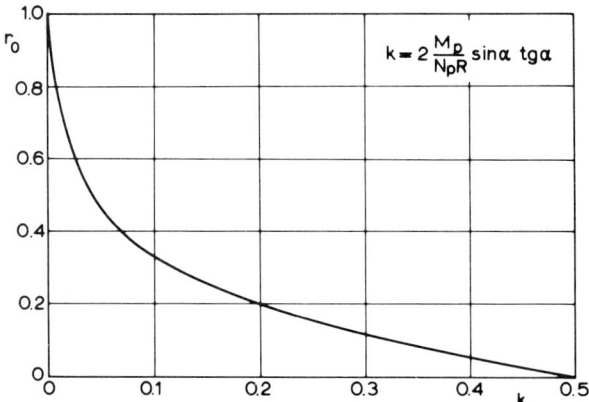

Fig. 9.17. Position of the circumferential hinge in a conical shell [9.18].

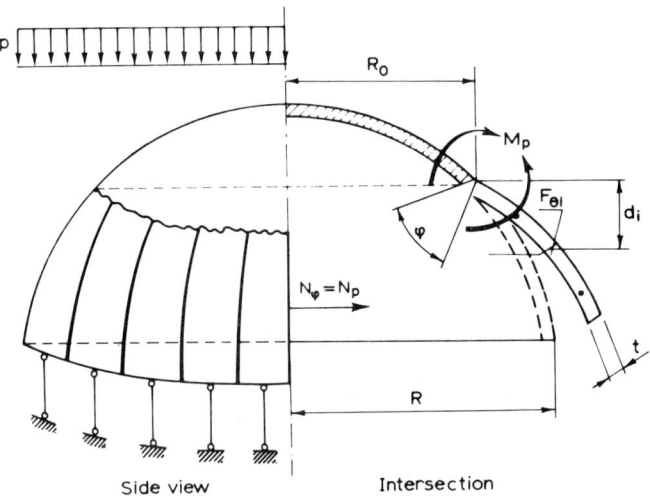

Fig. 9.18. Yielding zones in a spherical cap [9.14].

In the meridional yield lines, energy is dissipated by extension of the circumferential reinforcement. The corresponding dissipation is $D_2 = 2\pi\sigma_Y \Sigma F_{\theta i} d_i/(R-R_0)$, where d_i is the distance of the ith reinforcing bar with cross section $F_{\theta i}$ to the plane of the circumferential hinge circle (fig. 9.18). The upper bound p_+ to the limit load is then obtained from the work equation, where R_0 is evaluated from the condition $dp_+/dR_0 = 0$ (see [9.19, 9.20]).

9.3.3. Shallow shells

A simplified kinematic method for shallow shells has been developed by A.R.Rzhanitsyn (see [9.12, 9.13, and 8.63]). Consider, for example, a shallow shell (the rise f is small with respect to any span AD, BE, ...) with plane polygonal base, subjected to a concentrated force P, and hinged at the edges (fig. 9.19). A possible collapse mechanism is shown in fig. 9.19 (b). Lines of strain discontinuities OA, OB, OC, OD, OE, enable the various rigid parts to rotate about the edges AB, BC, etc. It is assumed that the dissipation in the yield lines is due exclusively to the axial circumferential force, with limit value N_p. The rate of circumferential extension $\epsilon_{\theta A}$ in line OA is (see fig. 9.19)

$$\dot{\epsilon}_{\theta A} = -\frac{\dot{w}_o}{l_{oA}} z (\cot\alpha + \cot\beta),$$

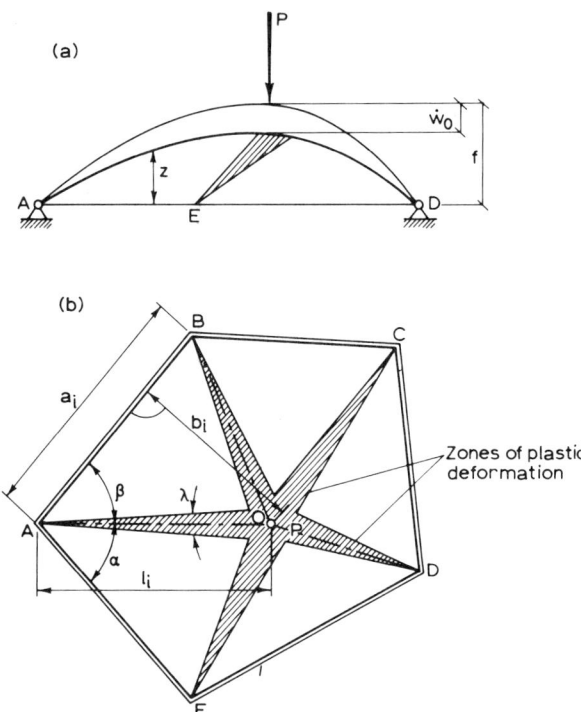

Fig. 9.19 (a). Fig. 9.19 (b). Collapse pattern for a shallow shell [9.12].

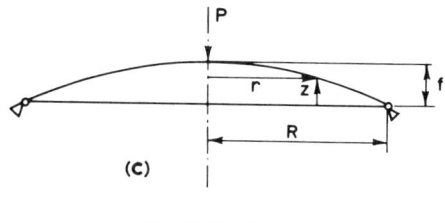

Fig. 9.19 (c).

where z is taken from the equation of the median surface of the shell. The total dissipation for the m yield lines is

$$D = N_p \dot{w}_o \sum_{i=1}^{m} \frac{\cot \alpha_i + \cot \beta_i}{l_i} \int_0^{l_i} z \, ds,$$

where s is the abscissa along the yield line i. The value of P is then obtained from the work equation, together with the minimum condition that gives the location of point 0.

For example, in the case of a shallow shell of revolution with equation $z = f[1-(r^n/R^n)]$, fig. 9.19 (c), the corresponding collapse load is

$$P_+ = 2\pi N_p f \frac{n}{n+1}. \qquad (9.25)$$

9.3.4. Cylindrical shell roofs
9.3.4.1. Introduction

Consider a cylindrical shell referred to a cylindrical system of coordinates (fig. 9.20). In the absence of symmetry of revolution, there are six generalized stresses: M_x, M_θ, $M_{x\theta}$, N_θ, N_x, $N_{x\theta}$. The shear forces V_x and V_θ are reactions. We also have $M_{\theta x} = M_{x\theta}$ and $N_{\theta x} = N_{x\theta}$ because, for all z such that

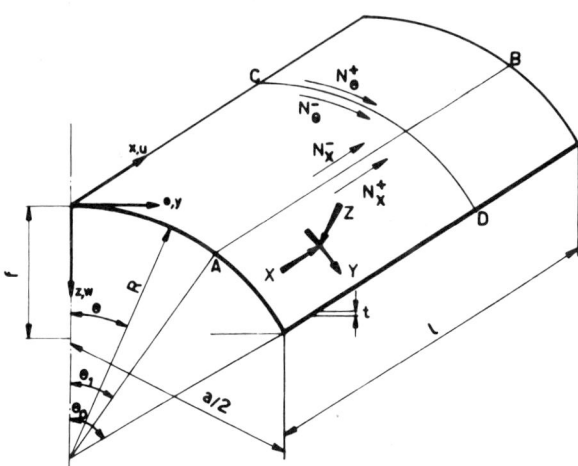

Fig. 9.20.

$-t/2 \leqslant z \leqslant t/2$, one has $z/R \ll 1$. Hence, the yield condition will be represented by an hypersurface in a six-dimensional stress space.

In a kinematic approach with lines of concentrated rate of deformation, the deflection rate $\dot{w}$ normal to the shell is bound to be continuous, whereas not only slope discontinuities are admissible but also discontinuities in the rates of tangential displacements $\dot{u}$ and $\dot{v}$. In order to evaluate these discontinuities, we shall need the following relations:

$$\dot{\epsilon}_x = \frac{\partial \dot{u}}{\partial x}, \qquad \dot{\epsilon}_\theta = \frac{1}{R}\left(\frac{\partial \dot{v}}{\partial \theta} - \dot{w}\right),$$

$$\dot{\kappa}_x = \frac{\partial^2 \dot{w}}{\partial x}, \qquad \dot{\kappa}_\theta = \frac{1}{R}\left(\dot{w} + \frac{\partial^2 \dot{w}}{\partial \theta^2}\right),$$

$$\dot{\gamma}_{x\theta} = \frac{1}{R}\frac{\partial \dot{u}}{\partial \theta} + \frac{\partial \dot{v}}{\partial x},$$

$$\dot{\kappa}_{x\theta} = \frac{1}{R}\left(\frac{\partial \dot{v}}{\partial x} + \frac{\partial^2 \dot{w}}{\partial x \partial \theta}\right). \qquad (9.26)$$

9.3.4.2. Collapse mechanisms

Collapse mechanisms formed of generalized yield lines [9.4,9.21,9.22,9.23] have been suggested by the experiments of Sawczuk [9.4], who tested cylindrical shells in reinforced mortar with the following characteristics: $\sigma'_r = 140$ kg/cm^2 (mortar); reinforcement: two orthogonal layers with square mesh of 13 mm side (78 wires per meter), at both the upper and lower face, made of wire of 1.2 mm diameter.

The geometry of the various shells is given in table 9.2. The ratio $t/R \simeq 0.023$ was chosen to avoid buckling failures. The support conditions are shown in fig. 9.21. There were no edge beams. The shells were subjected to

Table 9.2.

Type	θ_0	R, cm	f, cm	a, cm	L, cm	t, cm
A	30°	93	12.5	93	192	2.2
B	30°	93	12.5	93	95	2.0
C	17°	93	4.1	54.2	192	2.2
D	17°	93	4.1	54.2	95	2.2

9.3] UPPER-BOUND SOLUTIONS 427

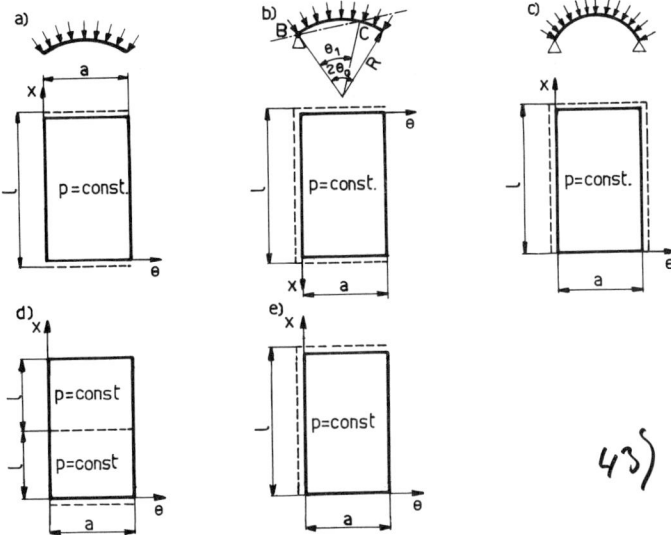

Fig. 9.21.

Table 9.3.

Type	a, cm	l, cm	p, kg/m^2	Type	a, cm	l, cm	p, kg/m^2
Aa	93	188	1,050	Cb	52	188	1,400
Ba	93	90	3,500	Db	52	90	1,600
Ca	54	188	700	Ac	89	190	1,250
Da	54	90	1,950	Bc	89	91	3,000
Ab	91	188	1,550	Ad	93	94	1,200
Bb	91	90	3,000	De	54	91	1,500

external radial pressure, applied by means of a rubber bag, and measured with an accuracy higher than 0.001 kg/cm^2. The collapse pressures given in table 9.3 are the maximum values that could be obtained, with wide opening of the cracks. Each value is the average of two or three measurements that do not differ by more than ±4%.

We now consider some of the observed collapse mechanisms:

1. Long shell, simply supported at two ends, types Aa and Da with $l/a \cong 2$,

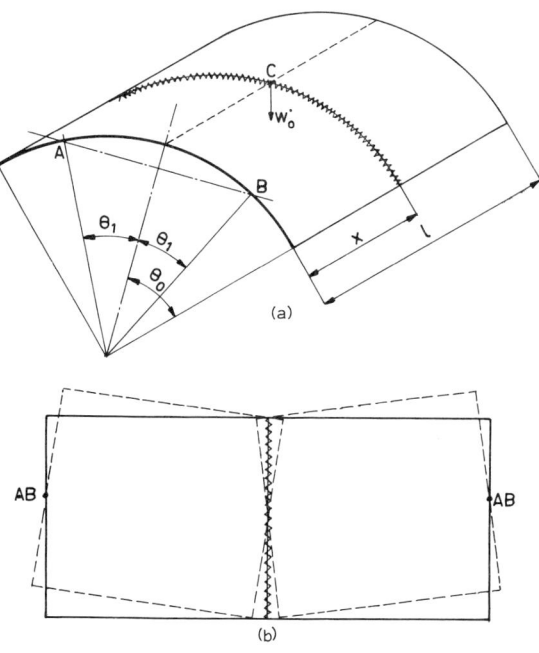

Fig. 9.22.

type Ca with $l/a \cong 3$, fig. 9.21. A "beam mechanism", with a transversal hinge circle at midspan, takes place. Each rigid portion on both sides of the hinge circle rotates with respect of an axis AB (fig. 9.22) located in the plane of the simple support. Consider the rigid part where $0 \leq x \leq l/2$. In relation (9.26) we use the conditions $\dot{\epsilon}_\theta = \dot{\kappa}_\theta = \dot{\gamma}_{x\theta} = 0$ and the boundary condition $\dot{u}(0,\theta,R) = 0$ that comes from the fact that points A and B do not move. We obtain

$$\dot{w} = x \frac{2\dot{w}_o}{l} \cos\theta, \tag{9.27}$$

$$\dot{v} = x \frac{2\dot{w}_o}{l} \sin\theta, \tag{9.28}$$

$$\dot{u} = \frac{2R\dot{w}_o}{l} (\cos\theta - \cos\theta_1), \quad 0 \leq \theta \leq \theta_0. \tag{9.29}$$

As both rigid parts are identical, it is easily seen that $\dot{v}$ is the same on both sides of the yield line, whereas $\dot{u}$ simply changes its sign as does $\partial \dot{w}/\partial x$. We therefore have the following discontinuities:

$$\dot{u}] = \frac{4R\dot{w}_o}{l}(\cos\theta - \cos\theta_1),$$

$$\left[\frac{\partial \dot{w}}{\partial x}\right] = \frac{4\dot{w}_o}{l}\cos\theta,$$

and the dissipation is

$$D = 2\int_0^{\theta_0}\left(N_x \dot{u}] + M_x\left[\frac{\partial \dot{w}}{\partial x}\right]\right)d\theta. \tag{9.30}$$

2. Shells simply supported on two short edges and on a third edge (Type b, fig. 9.21). When the third supporting edge is long, fig. 9.23 (a), there is complete collapse of the shell with the mechanism shown in the figure. Let w_o be the radial displacement rate of the segment DE. For part $ABDEA$ we have the following rates of displacement:

$$\dot{w} = \frac{\dot{w}_o}{\sin 2\theta_o}\sin\theta,$$

$$\dot{v} = \frac{\dot{w}_o}{\sin 2\theta_o}(\cos\theta - 1),$$

$$\dot{u}_o = 0, \tag{9.31}$$

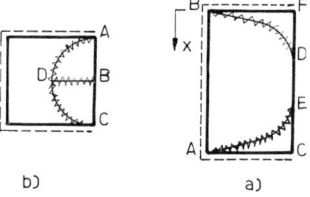

Fig. 9.23.

satisfying the boundary conditions $(\dot{w})_{\theta=0} = 0$, $(\dot{v})_{\theta=0} = 0$, $(\dot{w})_{\theta=2\theta_o} = \dot{w}_o$. Part *BFD* rotates with respect to an axis *BC* [fig. 9.21 (b)] in the plane $x = 0$. Point *C* is determined by the angle θ_1. The rates of displacement of this part are

$$\dot{u} = KR(\cos\theta - \cos\theta_1),$$

$$\dot{v} = Kx\sin\theta,$$

$$\dot{w} = Kx\cos\theta, \qquad (9.32)$$

where $K = \dot{w}_o/x_o \cos 2\theta_o$, x_o being the abscissa of point *D*, fig. 9.23. Because $\dot{w}$ must be continuous across the yield line *BD*, the following equation is obtained for this yield line

$$x = x_o \cot 2\theta_o \cdot \tan\theta. \qquad (9.33)$$

The following discontinuities thus occur in this yield line

$$\dot{u}] = \frac{\dot{w}_o R}{x_o \cos 2\theta_o}(\cos\theta - \cos\theta_1),$$

$$\dot{v}] = \frac{\dot{w}_o}{\sin 2\theta_o}\left(\frac{\cos 2\theta}{\cos\theta} - 1\right), \qquad (9.34)$$

$$\frac{\partial\dot{w}}{\partial x}\bigg] = \frac{\dot{w}_o}{x_o \cos 2\theta_o}\cos\theta,$$

$$\frac{\partial\dot{w}}{\partial\theta}\bigg] = \frac{1}{R}\frac{\dot{w}_o}{\sin 2\theta_o}\frac{\cos 2\theta}{\cos\theta}. \qquad (9.35)$$

The parameter θ_1 and x_o must then be adjusted to render the pressure *p* obtained from the work equation a minimum.

For the mechanism of fig. 9.23 (b), Sawczuk [9.4] has found the approximate relation

$$p_+ = 3M_p\left(\frac{1}{4R^2 \sin^2\theta_o/2} + \frac{8}{l^2}\right), \qquad (9.36)$$

assuming that bending is the prominent phenomenon and using approximate expressions for the slope discontinuities.

The simplest yield condition that can be imagined, in order to apply the method described in Sections 9.3.4.1 and 9.3.4.2, is

$$-Q_p \leqslant Q \leqslant Q'_p , \qquad (9.37)$$

where Q stands for any generalized stress, whereas Q'_p and Q_p are the corresponding plastic values for positive and negative sign of Q, respectively. The yield hypersurface is thus made of twelve hyperplanes. In the case of a long shell (head line 1), we have, at $x = \frac{1}{2}l$,

$$N_x = N'_p \quad \text{for} \quad 0 \leqslant \theta \leqslant \theta_1 ,$$

$$N_x = -N_p \quad \text{for} \quad \theta_1 \leqslant \theta \leqslant \theta_0 ,$$

(tensile forces are negative).

In the midspan cross section, there is a discontinuity on N_x by the amount $N_p + N'_p$ across the generatrices corresponding to $\theta = \pm \theta_1$. The bending moment M is equal to M_p all along the yield line. From the work equation, with expression (9.30) for the dissipation, the following load is obtained:

$$p_+ = \frac{8 N_p R}{l^2} \left[\left(1 + \frac{N'_p}{N_p} \right) \frac{\sin \dfrac{\theta_o}{1 + N'_p / N_p}}{\sin \theta_o} - 1 \right]. \qquad (9.38)$$

In the case of the tested shells N'_p = 30.8 t/m, N_p = 17.8 t/m, θ_o = 30°, R = 93 cm, l = 188 cm. Formula (9.36) then gives p = 1.56 t/m², to compare with the experimental value p_{ex} = 1.05 t/m². The same formula, applied to the shell Ca (1/a=3.5) gives p = 0.5 t/m² whereas p_{ex} = 0.7 t/m². Hence, formula (9.36) can be applied with confidence in the range $2.0 \leqslant 1/a \leqslant 3.5$.

For shells of the type Bb, formula (9.34) gives, with M_p = 0.1 t, $p_+ \cong$ 4 t/m², to compare with the experimental value p_{ex} = 3 t/m².

9.3.4.3. *Other work on the subject*

The method sketched above has been developed by M.Janas [9.3, 9.22] and used by him for the evaluation of the collapse load of a cylindrical shell hinged along heavy edge beams and supported on flexible diaphrams [9.22]. Unfortunately, the results are not directly applicable to reinforced concrete shells because of the choice of the yield surface. The same remark applies to the work of Fialkow [9.23], who obtained upper and lower bounds to the limit radial pressure of cylindrical roofs simply supported on two end diaphragms (the obtained bounds differ from the average value by 2% to 25%).

Recent work by Capurso [9.24] follows the same general procedure but is basically directed toward applications to reinforced concrete.

References

[9.1] F.LEVI, "Methodes simplifiees de calcul des voiles minces a courbure gaussienne nulle" (Simplified calculation methods of shells with vanishing gaussian curvature), *Simplified Calculation Methods for Shell Structures*, pp. 445-446, North-Holland Publ. Co., Amsterdam, 1964.

[9.2] A.SAWCZUK, W.OLSZAK, "A Method of Limit Analysis of Reinforced Concrete Tanks", Simplified Calculation Methods for Shell Structures, pp. 416-437, North-Holland Publ. Co., Amsterdam, 1964.

[9.3] M.JANAS, "Limit Analysis of Nonsymmetric Plastic Shells by a Generalized Yield Line Method", *Nonclassical Shell Problems*, pp. 997-1010, North-Holland Publ. Co., Amsterdam, 1964.

[9.4] A.SAWCZUK, "On Experimental Foundations of the Limit Analysis Theory of Reinforced Concrete Shells", *Shell Research*, pp. 217-231, North-Holland Publ. Co., Amsterdam, 1961.

[9.5] A.L.L.BAKER, "Further Research in Reinforced Concrete, and its Application to Ultimate Load Design", *Proc. Inst. Civil, Eng.*, 2: 2, August, 1953.

[9.6] P.B.MORICE, "Research on Concrete Shell Structures", *Proc. 1st Symp. Shell Roof Constr.*, London, 1952, Cement and Concrete Ass. London, 1954.

[9.7] A.C.van RIEL, W.J.BERANEK, A.L.BOUMA, "Tests on Shell Roof Models of Reinforced Concrete Mortar", *Proc. 2nd Symp. Shell Roof Constr.*, Oslo, 1957.

[9.8] G.R.MITCHELL, "Shell Research at the Building Research Station", *Proc. 2nd Symp. Shell Roof Constr.*, Oslo, 1957.

[9.9] A.L.BOUMA, A.C.RIEL, H.VAN KOTEN, W.J.BERANEK, "Investigations on Models of Eleven Cylindrical Shells made of Reinforced and Prestressed Concrete",*Shell Research*, pp.79-101, North-Holland Publ. Co., Amsterdam, 1961.

[9.10] H.LUNDGREN, *Cylindrical Shells*, The Danish Technical Press, Copenhague, 1949.

[9.11] A.L.L.BAKER, "Ultimate Strength Theory for Short Reinforced Concrete Cylindrical Shell Roofs", *Mag. Conc. Research,* 10, 3, 1952.

[9.12] A.R.RJANITSYN, "The Design of Plates and Shells by the Kinematical Method of Limit Equilibrium", *IX Congres Int. de Mec. Appl., Actes,* **VI**: 331, Brussels, 1956.

[9.13] A.R.RZHANITSYN, "Calculation of Shallow Shells by the Limit Design Methods", *Simplified Calculation Methods of Shell Structures*, pp. 438,444, North-Holland Publ. Co., Amsterdam, 1961.

[9.14] A.M.OVETCHKIN, "Analysis of Reinforced Concrete Rotationally Symmetric Shells" (in Russian), Gosstroyizdat, Moscow, 1961.

[9.15] A.R.DYKES, "Experimental and Theoretical Studies of Folded Plate Structures", *Nonclassical Shell Problems,* pp. 941-976, North-Holland Publ. Co., Amsterdam, 1964.
[9.16] G.K.HAIDUKOV, "Limit Equilibrium Design of Shallow Shell Panels", *Nonclassical Shell Problems*, pp. 977-996, North-Holland Publ. Co., Amsterdam, 1964.
[9.17] A.SAWCZUK, J.A.KONIG, "Limit Analysis of Reinforced Concrete Silo" (in Polish), *Arch. Inz. Lad.,* **8**: 161, 1962.
[9.18] S.KALISZKY, "Untersuchung einer Kegelstrumpfschale auf Grund der Traglastverfahrens", *Acta Tech. Hung.,* **34**: 159, 1961.
[9.19] N.V.AKHVLEDIANI, "On the Calculation of Reinforced Concrete Shells of Revolution According to the Limit Equilibrium Method" (in Russian), *Soob. AN Gruz. S.S.R.,* **18**: 205, 1957.
[9.20] N.V.AKHVLEDIANI, "Analysis of Reinforced Concrete Domes According to the Limit Equilibrium Method" (in Russian), *Issled. Teor. Sooruzh.,* **10**: 123, Gosstroyizdat, Moscow, 1961.
[9.21] M.JANAS, A.SAWCZUK, "On Carrying Capacities of Arch Dams", *Symp. Arch Dams* (Southampton, 1964), Pergamon Press, Oxford, 1964.
[9.22] M.JANAS, "Limit Analysis of a Cylindrical Shell Roof" (in Polish), *Arch. Inz. Lad.,* **8**: 365, 1962.
[9.23] M.N.FIALKOW, "Limit Analysis of Simply Supported Circular Shell Roof", *J. Eng. Mech. Div., Proc. A.S.C.E.,* June 1958 (with errata in the October 1958 issue).
[9.24] M.CAPURSO, "On the Limit Analysis of Reinforced Concrete Shells" (in Italian), *Giornale del Genio Civile,* **104**: 83, February 1966, and 167, March 1966.

10

Plane Stress and Plane Strain

10.1. Introduction

In Section 6.1 we have defined a disk as a body with the geometry of a plate but subjected to forces acting in its median plan. We recall that the thickness must be small with respect to the dimensions in the median plane. In this situation, if we refer the disk to the orthogonal coordinate system x, y, z of fig. 10.1 where axes x and y lie in the median plane, stress components parallel to the z-axis may be neglected. We thus set

$$\sigma_z = \tau_{xz} = \tau_{yz} = 0 . \tag{10.1}$$

Eqs. (10.1) define a *state of plane stress*. The z-axis is obviously a principal axis of the stress tensor, which is plane and will be completely determined by the components $\sigma_x, \sigma_y, \tau_{xy}$. These three stress components are our generalized stresses Q_i, whereas the corresponding generalized strain rates $\dot{q}_i$ are $\dot{\epsilon}_x, \dot{\epsilon}_y, \dot{\gamma}_{xy}$. As noted in Section 2.4, the principal axis z of the stress tensor is also principal for the strain-rate tensor when the body is isotropic. Hence $\dot{\gamma}_{xz} = \dot{\gamma}_{yz} = 0$. As a rule, $\dot{\epsilon}_z \neq 0$, but on account of $\sigma_z = 0$, the strain rate $\dot{\epsilon}_z$ does not appear in the expression for the dissipation and its value is irrelevant. From rotational equilibrium about the axis z, we have (see Section 1.1.1),

10.1] INTRODUCTION

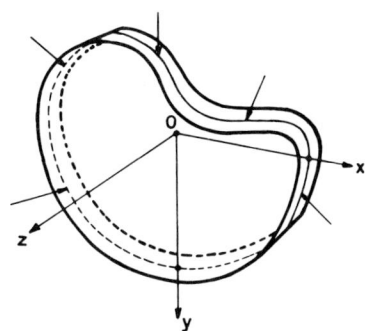

Fig. 10.1.

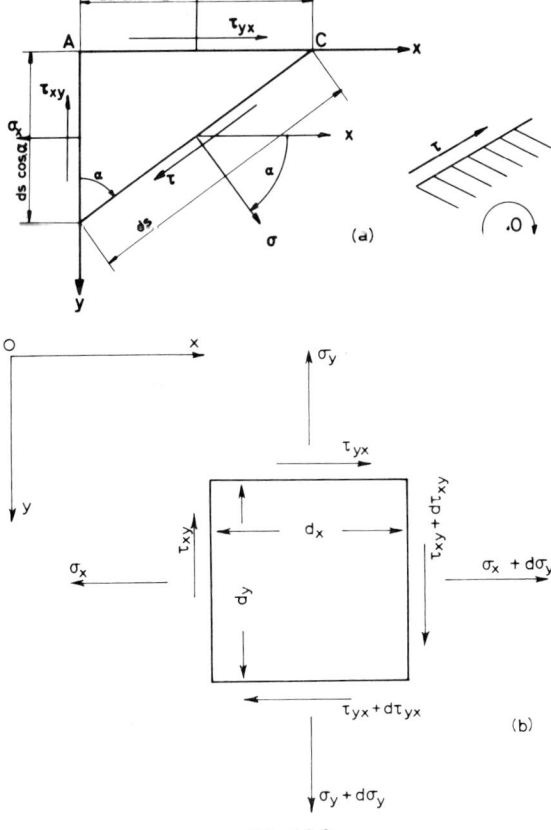

Fig. 10.2.

$$\tau_{xy} = -\tau_{yx} . \tag{10.2}$$

Positive stresses are shown in fig. 10.1 (a). Equilibrium of translation of the elementary parallelepiped of fig. 10.2 in the x- and y-directions give

(a) $\quad \dfrac{\partial \sigma_x}{\partial x} + \dfrac{\partial \tau_{xy}}{\partial y} + X = 0 ,$

(b) $\quad \dfrac{\partial \tau_{xy}}{\partial x} + \dfrac{\partial \sigma_y}{\partial y} + Y = 0 ,\tag{10.3}$

where X and Y are components of the body force per unit volume. The conditions of equilibrium at the boundary are

(a) $\quad l\sigma_x + m\tau_{xy} = \overline{X} ,$

(b) $\quad l\tau_{xy} + m\sigma_y = \overline{Y} ,\tag{10.4}$

where X and Y are the components of the surface traction per unit area, and l and m the direction cosines of the outward pointing normal at the considered point of the boundary.

With expressions (1.12) and (1.13) for the principal stresses σ_1 and σ_2 in the xy-plane, the yield condition (1.32) of Tresca can be written as:

$$\max \left[\left| \dfrac{\sigma_x + \sigma_y}{2} \right| + \sqrt{(\sigma_x - \sigma_y/2)^2 + \tau_{xy}^2} ; \quad 2\sqrt{(\sigma_x - \sigma_y/2)^2 + \tau_{xy}^2} \right] = \sigma_Y . \tag{10.5}$$

We have seen in Section 5.4.2, eq. (5.36) that the dissipation per unit volume is given by

$$D_V = \sigma_Y \cdot \max |\dot{\epsilon}_i| , \qquad i = 1, 2, 3 , \tag{10.6}$$

where $\dot{\epsilon}_i$ denotes a principal strain rate. The principal strain rates $\dot{\epsilon}_1$ and $\dot{\epsilon}_2$ in the xy-plane are related to the components $\dot{\epsilon}_x$, $\dot{\epsilon}_y$, $\dot{\gamma}_{xy}$ by eqs. (1.24) and (1.25), whereas $\dot{\epsilon}_3 \equiv \dot{\epsilon}_z$ is related to $\dot{\epsilon}_1$ and $\dot{\epsilon}_2$ by the incompressibility relation (1.26). Stress discontinuities are often used in plane stress problems. We recall from Section 5.6.2 that vectors $\boldsymbol{\sigma}'$ and $\boldsymbol{\tau}'$ normal to the discontinuity line AB (fig. 10.3) must be continuous across AB, whereas vector $\boldsymbol{\sigma}''$ parallel to AB may be discontinuous across AB.

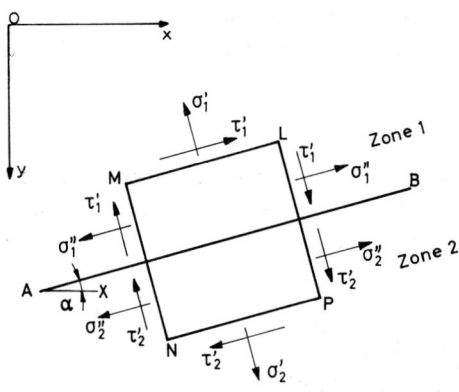

Fig. 10.3.

Discontinuity surfaces in the velocity field are also admissible, but it can be shown [3.9] that the component of the velocity vector **V** along the normal to the surface of discontinuity must remain continuous. Hence, the dissipation per unit area in the surface of discontinuity is

$$D = \frac{\sigma_Y}{2} |\mathbf{V}]| , \tag{10.7}$$

where $\mathbf{V}]$ is the jump in the tangential velocity and $\sigma_Y/2$ the yield stress in pure shear, σ_Y being the yield stress in simple tension.

Consider now the *state of plane strain*. It can be obtained practically, for example, with a very long solid that would be generated by the translation of the disk of fig. 10.1, with its loading, in the direction of the z-axis. Far from the end sections, all cross sections behave identically and, hence, the displacements of any point occurs in its own cross section. We thus have:

$$\epsilon_z = \gamma_{zx} = \gamma_{zy} = 0 , \tag{10.8}$$

and also

$$\dot{\epsilon}_z = \dot{\gamma}_{zx} = \dot{\gamma}_{zy} = 0 . \tag{10.9}$$

Eqs. (10.8) and (10.9) define a state of plane strain. From eq. (10.9) it is seen that the generalized stresses of the problem are $\sigma_x, \sigma_y, \tau_{xy}$ (as in plane stress) because the other stress components do not work. The z-axis is a principal axis of both the strain-rate tensor and the stress tensor. We thus have

$$\tau_{xz} = \tau_{yz} = 0, \tag{10.10}$$

whereas the magnitude of $\sigma_z \equiv \sigma_3$ depends on the state of stress in the xy-plane and on the yield criterion. With the yield condition of Tresca, it is easily seen in fig. 1.5 that for $\dot{\epsilon}_z \equiv \dot{\epsilon}_3 = 0$, the stress point must lie in one of the planes

(a) $\quad \sigma_1 - \sigma_2 = \sigma_Y$

(b) $\quad \sigma_1 - \sigma_2 = -\sigma_Y$ \hfill (10.11)

and σ_z may vary from 0 to σ_Y or from $-\sigma_Y$ to 0.

Eqs. (10.11) can also be written as

$$(\sigma_1 - \sigma_2)^2 = \sigma_Y^2, \tag{10.12}$$

or

$$(\sigma_x - \sigma_y)^2 + 4\tau_{xy}^2 = \sigma_Y^2. \tag{10.13}$$

With the yield criterion of von Mises, the condition $\dot{\epsilon}_3 = 0$ and the normality law give

$$\sigma_3 = \tfrac{1}{2}(\sigma_1 + \sigma_2). \tag{10.14}$$

Substitution of expression (10.14) for σ_3 into eq. (1.35) results in

$$(\sigma_1 - \sigma_2)^2 = \tfrac{4}{3}\sigma_Y^2, \tag{10.15}$$

that is,

$$(\sigma_x - \sigma_y)^2 + 4\tau_{xy}^2 = \tfrac{4}{3}\sigma_Y^2. \tag{10.16}$$

The similarity of relations (10.12) and (10.15) will be noted: they differ only by the magnitude of the constant in the right-hand side. In a (σ_1, σ_2) space they are both represented by two straight lines making 45° angles with the axes (see Problem 2.6.4, fig. 2.7). The dissipation per unit volume is, for the Tresca condition,

$$D_V = \sigma_Y |\dot{\epsilon}_1| \qquad \text{(with } |\dot{\epsilon}_1| = |\dot{\epsilon}_2|\text{)},\tag{10.17}$$

and for the Mises condition,

$$D_V = \frac{2\sigma_Y}{\sqrt{3}} |\dot{\epsilon}_1|.\tag{10.18}$$

10.2. Plane stress: perforated disks

10.2.1. *Square disk with a slit*

It is known from experiments that a perforated disk subjected to uniaxial tension and made of a ductile and negligibly work-hardening material flows plastically when the axial force attains the value

$$N^* = \sigma_Y A_n,\tag{10.19}$$

where A_n is the "net area" of the smallest cross section through the hole. This empirical concept of "net area" is at the basis of the pratical design methods for perforated or notched bars made of ductible metals. As will be seen in the following, plastic limit analysis gives a rigorous theoretical support to the concept of net area, while indicating its limits of applicability.

Consider the square disk of fig. 10.4 subjected to uniaxial tensile stress σ. The central slit has a length a larger than the thickness t but with negligible width h. From the empirical formula (10.19), the stress at the boundary is

$$\sigma^* = \frac{N}{A} = \frac{\sigma_Y(l-a)t}{lt} = \frac{l-a}{l}\sigma_Y.$$

In the absence of a slit, the limit stress would equal σ_Y. Hence, defining the "cutout factor" ρ as

$$\rho = \frac{\sigma^*}{\sigma_Y},\tag{10.20}$$

we immediately obtain

$$\rho = \frac{l-a}{l}.\tag{10.21}$$

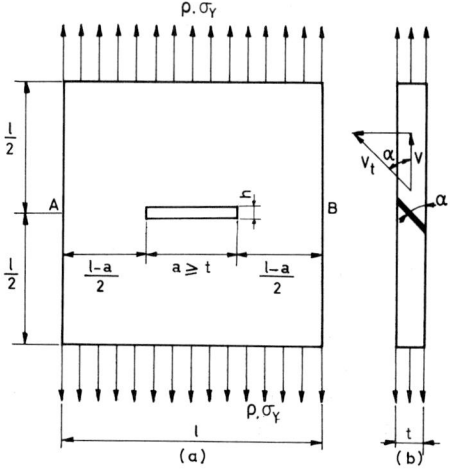

Fig. 10.4. (From *Plastic Analysis of Structures* by P.G.Hodge Jr. Copyright 1959, McGraw-Hill Book Company. Used by permission of McGraw-Hill Book Company, Inc.)

The empirical formula (10.21) has the following theoretical verification. Consider a collapse mechanism with a plane of discontinuity through line AB (fig. 10.4), inclined by an angle α with respect to the median plane of the disk. The area of this surface of slip is $A = [t(l-a)/\sin\alpha]$. If the upper and lower parts of the disk go apart with a relative velocity V in the direction of the applied load, the relative velocity tangential to the plane of discontinuity is $V_t = V/\cos\alpha$. The power dissipated thus is, according to eq. (10.7) for D,

$$D = \frac{Vt(l-a)}{\sin 2\alpha} \sigma_Y .\tag{10.22}$$

The power of the loads is

$$\mathcal{P} = \sigma l t V .\tag{10.23}$$

From the work equation $\mathcal{P} = D$, we obtain

$$\sigma = \frac{(l-a)}{l \sin 2\alpha} \sigma_Y .$$

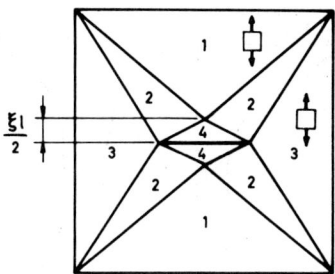

Fig. 10.5. (From *Plastic Analysis of Structures* by P.G.Hodge Jr. Copyright 1959, McGraw-Hill Book Company. Used by permission of McGraw-Hill Book Company, Inc.)

The minimum value of σ is obtained with $\alpha = \pi/4$, and the corresponding upper bound for the cutout factor is

$$\rho_+ = \frac{l-a}{l}. \tag{10.24}$$

A discontinuous statically admissible stress field is then constructed with regions of homogeneous state of stress (fig. 10.5). The parameter ξ defining the extent of the various regions is adjusted to maximize the applied stress σ without violating the yield condition. In this manner, one obtains [3.9]

$$\xi = 1 - \frac{a}{l} + \left(\frac{a}{l}\right)^2,$$

and the corresponding lower bound for the cutout factor is

$$\rho_- = \frac{l-a}{l}. \tag{10.25}$$

From comparison of eqs. (10.24) and (10.25) it is concluded that

$$\rho = \frac{l-a}{l}. \tag{10.26}$$

in accordance with the empirical expression (10.21).

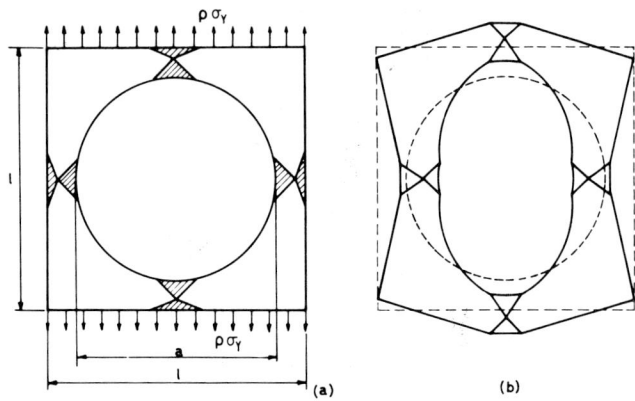

Fig. 10.6. (From *Plastic Analysis of Structures* by P.G.Hodge Jr. Copyright 1959, McGraw-Hill Book Company. Used by permission of McGraw-Hill Book Company, Inc.)

10.2.2. *Square disk with a central circular hole*

Consider the square disk of fig. 10.6, subjected to uniaxial tension. The mechanism of fig. 10.4 (b) can also be used in the present case, giving the upper bound (10.24). However, for increasing a, the disk is expected to collapse in the manner of a frame [fig. 10.6 (b)], with four plastic hinges in the weakest sections. With a mechanism of this kind, containing four plastified regions [in grey in fig. 10.6 (a)] similar to plastic hinges, the following upper bound is obtained [3.9]:

$$\rho_+ = \sqrt{2 - 4(a/l) + 3(a/l)^2} - \frac{a}{l}. \tag{10.27}$$

The upper bound (10.27) is smaller than the upper bound (10.24) when $\frac{1}{3} \leqslant a/l \leqslant 1$. With a discontinuous statically admissible stress field made of five kinds of regions, and maximization of the applied stress σ with respect to a parameter ξ, the following bounds are obtained [3.9]:

$$\rho_- = \frac{(l-a)^2}{2al}, \tag{10.28}$$

valid when $0.443 \leqslant a/l \leqslant 1$. When $0 \leqslant a/l \leqslant 0.443$,

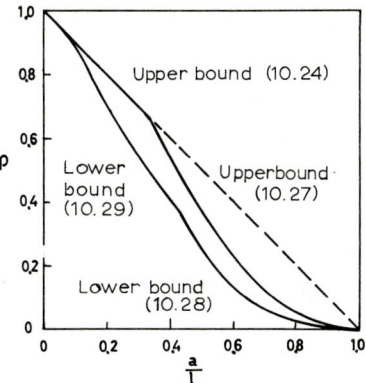

Fig. 10.7. (From *Plastic Analysis of Structures* by P.G. Hodge Jr. Copyright 1959, McGraw-Hill Book Company. Used by permission of McGraw-Hill Book Company, Inc.)

$$\rho_- = \frac{(1-a/l)(1-\xi)}{a/l}, \tag{10.29}$$

with

$$\frac{\xi(1-\xi)}{\sqrt{(3\xi-2)(2-\xi)}} = \frac{a}{l}.$$

The various bounds are represented in fig. 10.7. It is worth remarking that, in the present case, the concept of "net area" is valid for $a/l < \frac{1}{3}$ but ceases to be applicable for larger a/l.

A different situation arises when the tensile load is applied by a rigid pulling device that enforces uniform velocity at the upper and lower edge instead of uniform stress. The upper bound (10.27) must then be rejected whereas the upper bound (10.24) applies. On the other hand the stress field of fig. 10.8 is statically admissible for any a/l, and the corresponding lower bound is $\rho_- = (l-a)/l$ which thus turns out to be the exact value.

10.2.3. *Square disk with central circular hole, subjected to biaxial tension*

Consider the disk of fig. 10.9. Let $\rho_x \sigma_Y$ and $\rho_y \sigma_Y$ be the applied stresses that the disk can support separately at the limit state. For each ratio a/l, we want to determine, in a space with coordinates ρ_x, ρ_y the locus of the points corresponding to collapse. In the absence of a hole, this locus is identical to

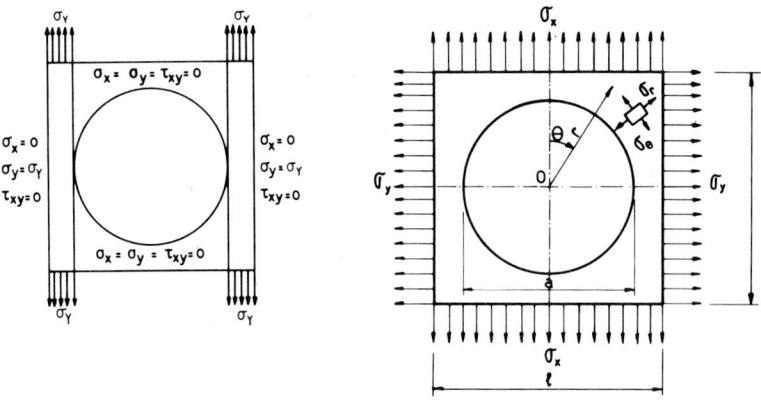

Fig. 10.8. Fig. 10.9.

the yield locus for $\sigma_Y = 1$. With the Tresca yield condition, the hexagon of fig. 10.10 is obtained. The desired curves have been bounded by Gaydon [10.1] for various values of a/l. They lie in the shaded regions in fig. 10.10. The bounds for ρ_x are given in fig. 10.11, when $\rho_x/\rho_y = -1$. In the case of equal tensile stresses ($\rho_x/\rho_y = 1$), eq. (10.24) again furnishes an upper bound. Consider now the following stress field, referred to the polar coordinate system of fig. 10.9:

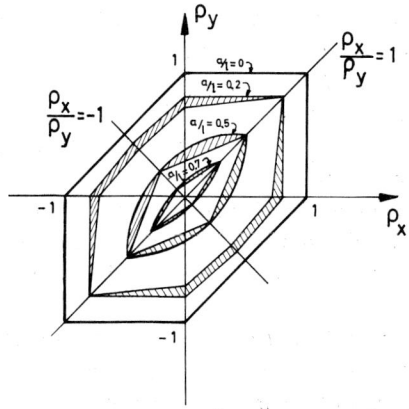

Fig. 10.10.

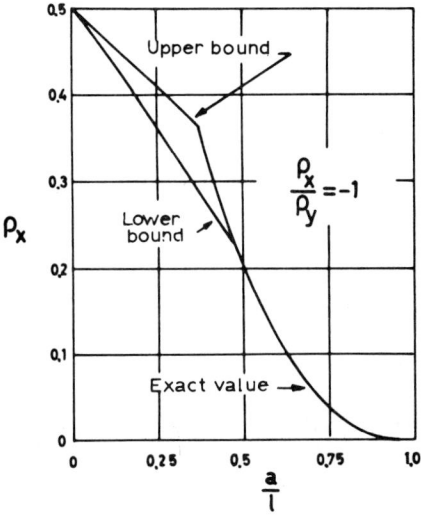

Fig. 10.11.

$$\sigma_r = \frac{\sigma_Y \rho}{1 - a/l}\left(1 - \frac{a}{2r}\right),$$

$$\sigma_\theta = \frac{\sigma_Y \rho}{1 - a/l}, \qquad \text{for} \quad \frac{a}{2} \leqslant r \leqslant \frac{l}{2}, \qquad (10.30)$$

$$\tau_{r\theta} = 0,$$

and

$$\sigma_r = \sigma_\theta = \sigma_Y \rho, \qquad \tau_{r\theta} = 0 \qquad \text{for} \quad \frac{l}{2} \leqslant r. \qquad (10.31)$$

The reader will easily verify that this field satisfies the equilibrium equation

$$\frac{d(r\sigma_r)}{dr} - \sigma_\theta = 0,$$

and the boundary conditions. The yield condition of Tresca is not violated as long as

$$\rho \leqslant 1 - \frac{a}{l}. \qquad (10.32)$$

Hence, when $\rho_x = \rho_y = -1$ the corresponding point of the interaction curve ρ_x versus ρ_y is given by

$$\rho_x = \rho_y = 1 - \frac{a}{l}. \qquad (10.33)$$

When the ratio σ_x/σ_y is not known beforehand, it is desirable to evaluate a "general cutout factor" ρ such that, if the disk is able to support the simultaneous stresses σ_x and σ_y it will be able, when perforated, to support simultaneously $\rho\sigma_x$ and $\rho\sigma_y$. We first assume that buckling does not occur. We next remark that the interaction curves ρ_x versus ρ_y are symmetrical with respect to the rays with eqs. $\rho_x = \rho_y$ and $\rho_x = -\rho_y$. Suppose that, for the considered ratio a/l, the points with coordinates $(\rho_x, 0)$, $(0, \rho_y)$ and the point corresponding to $\rho_x/\rho_y = 1$ are known (points 1, 2, and 3 in fig. 10.12, respectively). It is then possible to draw an hexagon inscribed in the unknown exact curve (dashed in fig. 10.12). The general cutout factor will be given by the smallest value of the ratio of the lengths of segments OA' and OB for all rays like OB, fig. 10.12. The smallest ratio $|\overline{OA}|/|\overline{OB}|$ will give a lower bound to ρ. Because segments $\overline{O1}$ and $\overline{O2}$ are equal, it remains to compare the

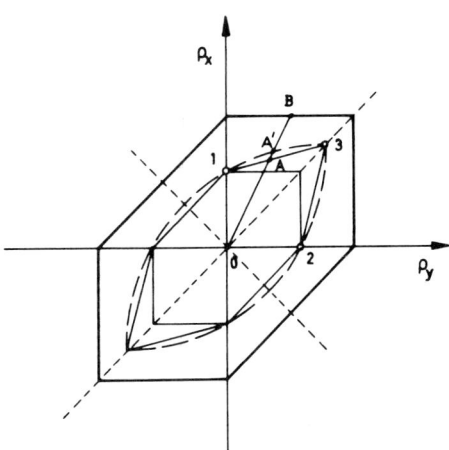

Fig. 10.12.

lengths of segment $O1$ with $|\overline{O3}|/\sqrt{2}$. From eq. (10.33) we have $|\overline{O3}|/\sqrt{2} = (1-a/l)\sqrt{2}$, larger than $|\overline{O1}|$ as given in fig. 10.7. Hence, the general cutout factor can be evaluated in this figure.

10.2.4. *Rectangular disk subjected to uniaxial tension and perforated by several holes*

Consider a rectangular flat bar pulled between rigid grips (as in a testing machine). When there is only one row of n holes with diameter a, the mechanism shown in fig. 10.4, together with the generalization of the stress field represented in fig. 10.8, give the exact cutout factor

$$\rho = \frac{l - na}{l}. \tag{10.34}$$

In analogy with the case of one single hole, relation (10.34) implies that $a \geqslant t$.

When there are two rows with n and $n' = n-1$ holes, respectively, the same mechanism remains applicable and gives

$$\rho_+ = \frac{l - na}{l}. \tag{10.35}$$

A more complicated discontinuous stress field is used [10.2], as shown in figs. 10.13 (b), (c), and (d). The dashed areas are in plane hydrostatic state of stress. The tractions on the longitudinal edges in fig. 10.13 (b) are eliminated by the field of fig. 10.13 (d). The parameters C, θ and γ are chosen to maximize the applied load without violating the yield condition. For the particular case $n = 5$, $n' = 4$, $a/l = 1/20$, Brady and Drucker [10.2] have found $\theta = 35°$, $\gamma = 19.9°$ and C as large as possible. The corresponding lower bound for the cutout factor is

(a) $\quad \rho_- = 0.70$, $\hspace{5em}$ (10.36)

whereas relation (10.35) gives

(b) $\quad \rho_+ = 0.75$. $\hspace{5em}$ (10.36)

In a similar manner, the following results have been obtained [10.2]:

1. With three rows of holes ($n = 5$, $n' = 4$, $n'' = 5$, $a = 0.5$ in., $l = 10$ in., distance between axes of rows: 1 in.),

$$0.67 \leqslant \rho \leqslant 0.72 \ . \tag{10.37}$$

2. With five rows of holes (same dimensions and distances),

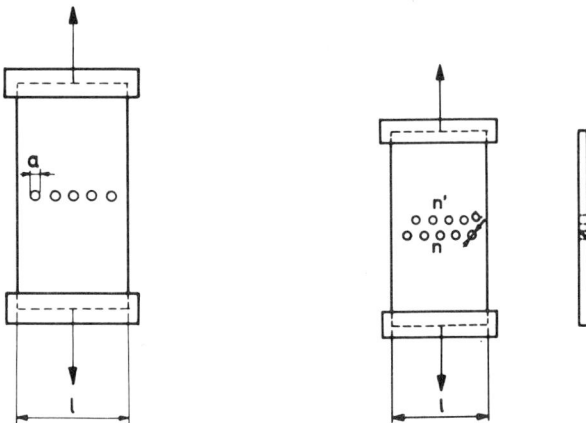

Fig. 10.13 (a).

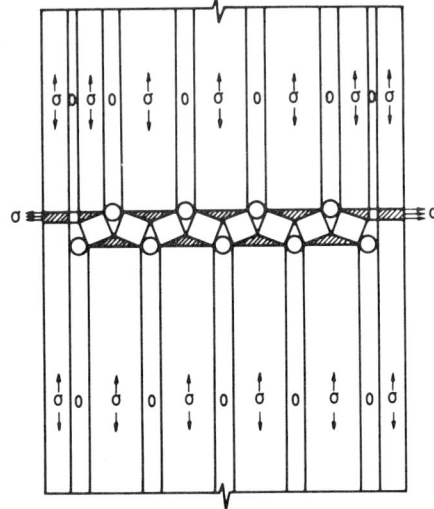

Fig. 10.13 (b).

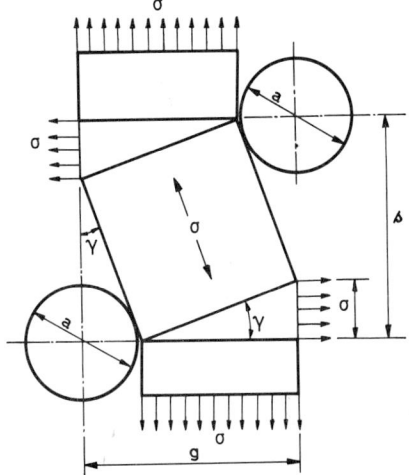

Fig. 10.13 (c).

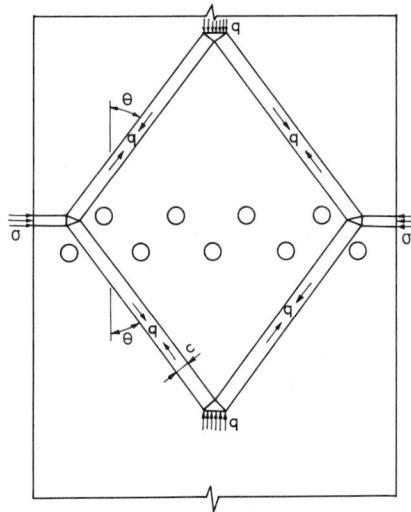

Fig. 10.13 (d).

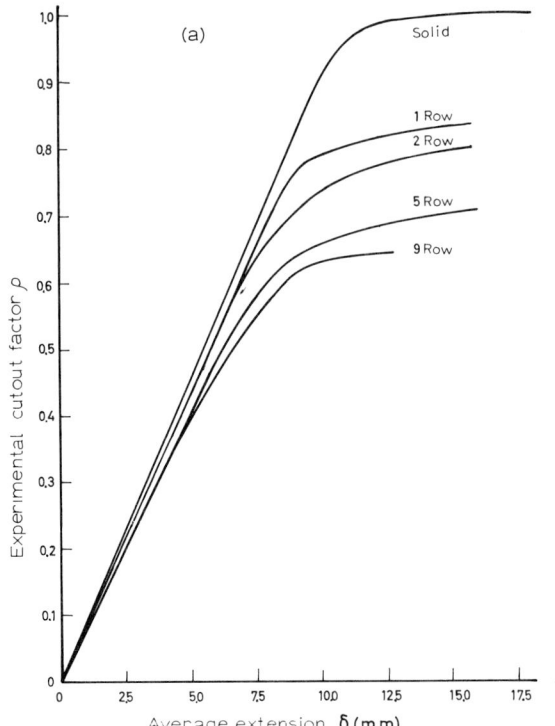

Fig. 10.15 (a). 61S-T6 Aluminium sheet.

In order to apply the "net area" method to the cases studied by Brady and Drucker, the smallest net area must be found, either cross-sectional or following a transversal broken line [10.3]. The geometrical parameters are l, a, n, n', and s and g (fig. 10.16). In the present situation, any segment like AB in fig. 10.16 is larger than $|CD| = |CE|/2$. Hence, the smallest net area is always a cross section through the centers of a row of holes. We thus have $A_n = \text{const} = l - 5a$, and the corresponding limit load is

$$P^* = \frac{\sigma_Y}{S}(l-5a),$$

where S is the safety factor. Application of the "$s^2/4g$" rule as prescribed by

$$0.58 \leq \rho \leq 0.69 \ . \tag{10.38}$$

3. With nine rows of holes (same dimensions and distances),

$$0.55 \leq \rho \leq 0.63 \ . \tag{10.39}$$

For more information on the subject, we refer the reader to the original paper [10.2] in which an experimental verification of the theory is also given. Figs. 10.14 and 10.15 summarize the experimental results. It is seen that, in accordance with the theory, the values of ρ are strongly influenced by the configurations of the holes. Experimental cutout factors are not, as a rule, smaller than the theoretical upper bounds probably because the Mises criterion applies better to the steel and aluminium used than does the Tresca criterion.

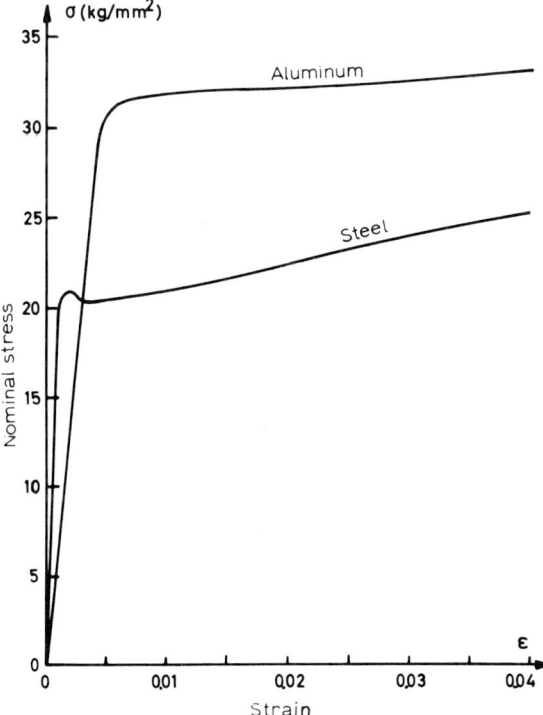

Fig. 10.14. Stress-strain curves for steel and aluminium sheet.

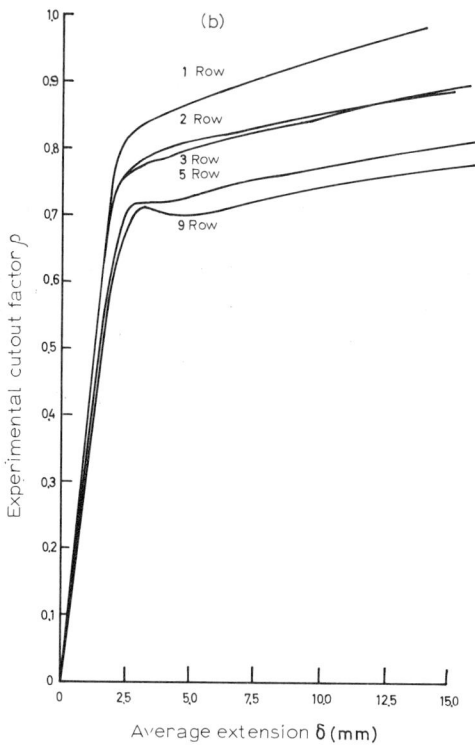

Fig. 10.15 (b). Typical hot rolled steel sheet results.

the A.I.S.C. specifications [10.4] would also result in a constant P^*. Because the limit load P_l (theoretical as well as experimental) is smaller than $\sigma_Y(1-5a)$ for more than one row of holes, the practical method gives a decreasing safety with an increasing number of rows. It must be remarked that the net area method is most often used for riveted or bolted connections, where forces are applied in the holes. In this latter case, the discrepancy seems smaller [10.2].

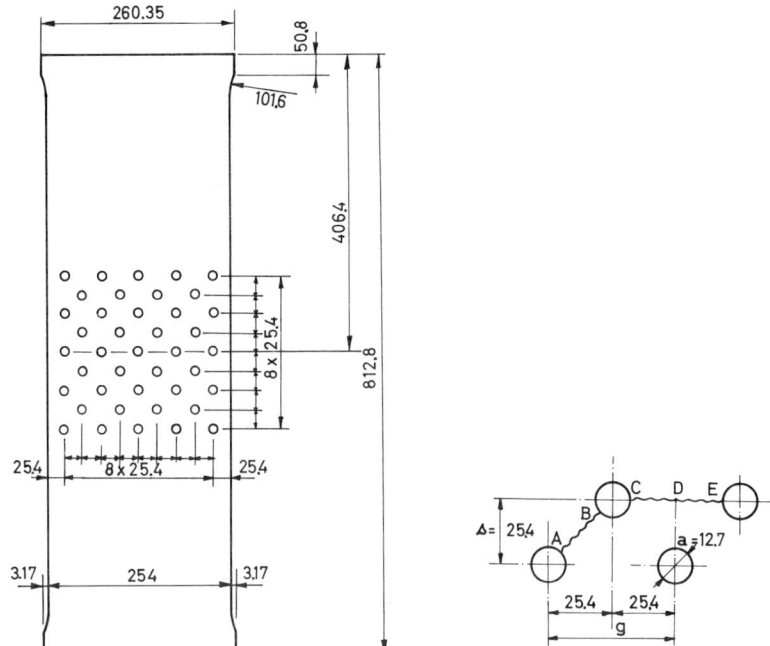

Fig. 10.16. Test piece configuration.

10.2.5. *Square disk with a central circular reinforced cutout*

In order to restitute, at least partially, the strength lost from the presence of a hole, the disk may be reinforced, for example, as shown in fig. 10.17. Whatever the type of reinforcement, the reinforced disk will be analyzed as formed of a basic disk and a hub of larger thickness. The mechanism shown in fig. 10.4 gives, in the present case,

$$\rho_+ = 1 - \frac{a}{l} + \frac{B}{lt/2E} \tag{10.40}$$

(see fig. 10.17 for notations). A lower bound to ρ will be obtained with a discontinuous stress field formed of regions of five types in the disk, and a special region in the hub (fig. 10.18).

The hub is loaded as shown in fig. 10.18 (b) and satisfies the interaction formula (see Com. V., eq. (5.4))

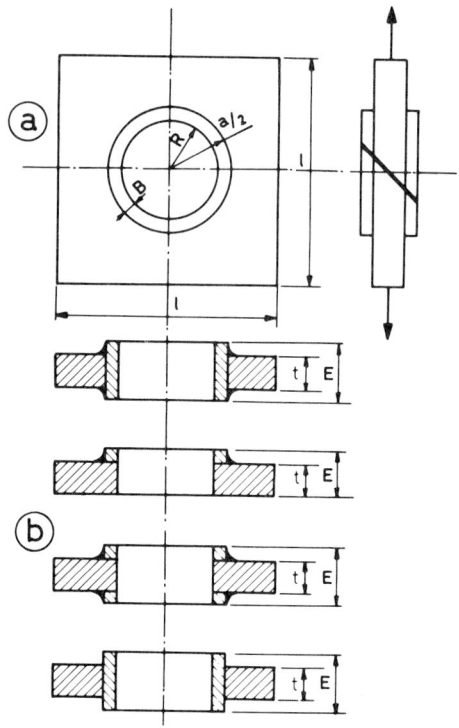

Fig. 10.17. (From *Plastic Analysis of Structures* by P.G.Hodge Jr. Copyright 1959, McGraw-Hill Book Company. Used by permission of McGraw-Hill Book Company, Inc.)

$$\frac{M}{M_p} + \left(\frac{N}{N_p}\right)^2 = 1 .$$

With the parameters $h = t/E$, $f = R/B$, $b = a/l$, $j = 2B/a$, one obtains [3.9]:

$$\rho_- = \min \, [\rho_1, \rho_2] \, , \qquad (10.41)$$

where

10.2]   PLANE STRESS: PERFORATED DISKS   455

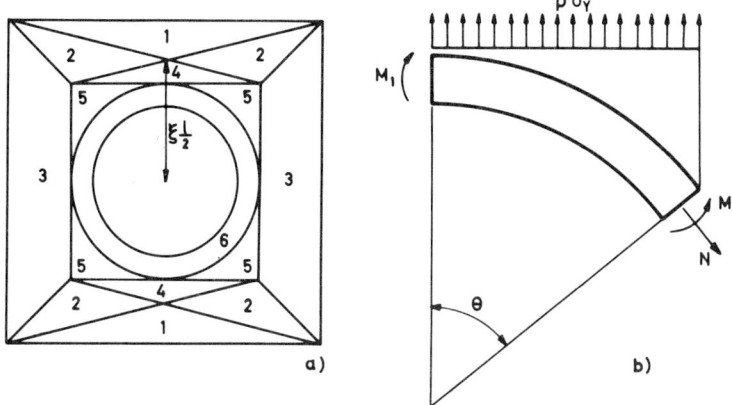

Fig. 10.18. (From *Plastic Analysis of Structures* by P.G.Hodge Jr. Copyright 1959, McGraw-Hill Book Company. Used by permission of McGraw-Hill Book Company, Inc.)

$$\rho_1 = \rho_h + \frac{(1-\rho_h)(1-b)^2}{b(2-\rho_h)}$$

$$\left[1 - \frac{b(\rho_2-\rho_h)}{1-b}\right]^2 = 4b^2 \frac{(\rho_2-\rho_h)^2}{(1-b)^2 - (\rho_2-b\rho_h)^2}, \quad (10.42)$$

with

$$\rho_h = \frac{-f + \sqrt{f^2 + 2}}{h(1+f)} \quad \text{if} \quad \tfrac{1}{2} \leq f,$$

$$\rho_h = \frac{1}{h(1+f)} \quad \text{if} \quad 0 \leq f \leq \tfrac{1}{2}. \quad (10.43)$$

When the reinforcement must be *designed* to ensure an assigned cutout factor ρ^* (for example, to give the full strength of the disk: $\rho^* = 1$), B and E will be chosen, from relation (10.40), to yield $\rho_+ > \rho^*$. It must then be verified, with eqs. (10.41) to (10.43), that $\rho^* \leq \rho_-$. In the opposite situation, larger E and B must be tried.

When the load is applied by rigid grips, the upper bound (10.40) remains valid, whereas the following lower bound can be obtained [3.9]:

$$\rho_- = 1 - \frac{a}{l} \sin \theta_1 , \qquad (10.44)$$

where θ must satisfy

$$\left[\frac{(1-\gamma f)^2}{2(2jf-h)} + 1 + \frac{h}{2} \right] \sin^2 \theta_1 - (1+jf+h) \sin \theta_1 + \left[jf + \frac{h}{2} - \frac{(1-jf)^2}{h} \right] \leqslant 0 .$$

Hodge and Perrone [10.5] have studied various types of reinforced holes and have compared their results with the experiments of Vasarhelyi and Hechtman [10.6]. The verification is very good for disks with rigid grips. Theoretical lower bounds assuming uniform stress at the boundary are always smaller than experimental values, most often by about 25%.

10.3. Notched bars in tension

Consider the V-notched bar shown in fig. 10.19. For very small thickness $(t/b \to 0)$ we are in plane-stress conditions, and the mechanism of fig. 10.19 (a), with slip planes at 45° of the median plane of the bar, gives the upper bound

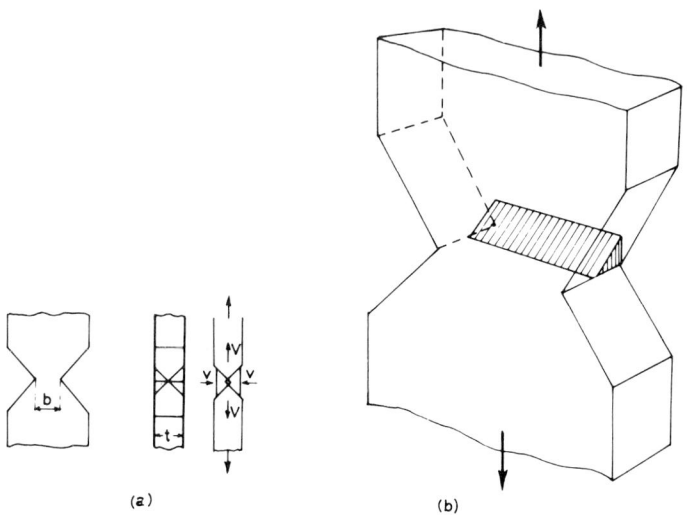

Fig. 10.19.

$P_+ = \sigma_Y bt$ for the limit load. A central band of width b in uniform traction with magnitude σ_Y [fig. 10.19 (d)] is a statically admissible stress field with the same load. Hence, the exact limit load is

$$P_l = \sigma_Y bt .\qquad (10.45)$$

For an arbitrary value of the ratio t/b, the mechanism shown in fig. 10.19 (a) gives the upper bound [10.7]

$$P_+ = \sigma_Y bt \left(1 + \frac{\sqrt{2}t}{4b} \right), \qquad (10.46)$$

the slip surfaces having the total area $2bt\sqrt{2} + t^2$, and the relative velocity of the sliding blocks in the slip planes being $\sqrt{2}v$. A fourth of the total sliding surface is shaded in fig. 10.19 (b).

In plane strain conditions ($t/b \to \infty$), the exact limit load is [10.8]

$$P_l = \sigma_Y bt \left(1 + \frac{\pi}{4} \right) = 1.785 \sigma_Y bt . \qquad (10.47)$$

For large but finite t/b ratio, the slip-line field of the plane strain state provides an upper-bound solution, with the load given by eq. (10.47). As shown in fig. 10.20, the upper bound (10.46) gives a higher load than eq. (10.47) (and, hence, should be rejected, according to the kinematic theorem) for $t/b \geqslant 2.22$. The experimental results of Findley and Drucker [10.9] show that: (a) the aluminium specimen tested seems to satisfy the yield criterion of von Mises better than Tresca's; the corresponding experimental limit loads, when divided by $2/\sqrt{3}$, come close to the points representing the limit loads of the steel specimen; and (b) a very large t/b ratio (larger than 6) is required to obtain plane strain conditions. The need of great experimental care (in particular for what regards axiality of the load) is emphasized in ref. [10.9]. The conclusion under (b) above is not confirmed by Sczcepinski and Miastkowski [10.10] who tested forty-eight mild steel flat-notched bars and concluded from their experimental results shown in fig. 10.21 that, for $t/b \geqslant 2$ the limit load is practically constant and equal to the plane-strain limit load.

Experiments on nonsymmetric notches are reported by Dietrich [10.11], whereas a survey on the problem of notched bars has been presented by Szczepinski [10.12], who cites important literature on the subject.

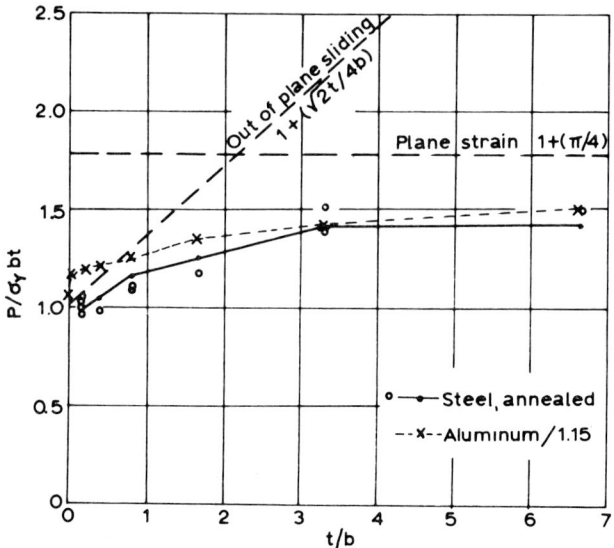

Fig. 10.20.

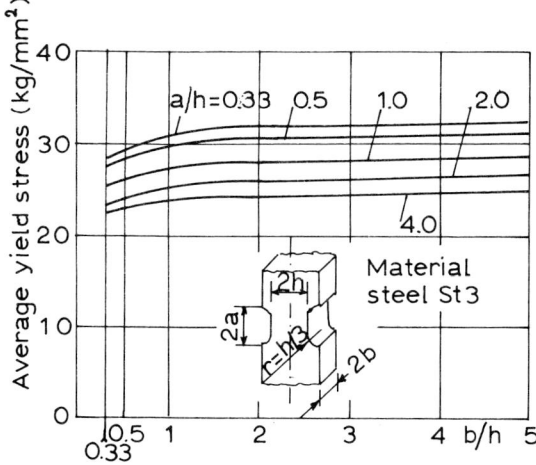

Fig. 10.21.

10.4. Thin rotating disks

10.4.1. *Introduction*

Consider a thin circular disk of uniform thickness t, rotating with an angular speed ω (rad/sec). Let γ be the weight per unit volume of the material. The force of inertia per unit volume has the value $F = \gamma \omega^2 r/g$, where r is the radial coordinate of the considered unit volume (fig. 10.22) and g the gravity acceleration. In elastic range, the principal* stresses σ_r and σ_θ (fig. 10.22) are given by [1.1]

(a) $\quad \sigma_r = \dfrac{3+\nu}{8} \dfrac{\gamma}{g} \omega^2 \left(b^2 + a^2 - \dfrac{a^2 b^2}{r^2} - r^2 \right)$,

(b) $\quad \sigma_\theta = \dfrac{3+\nu}{8} \dfrac{\gamma}{g} \omega^2 \left(b^2 + a^2 + \dfrac{a^2 b^2}{r} - \dfrac{1+3\nu}{3+\nu} r^2 \right)$. (10.48)

For thin disks, the third principal stress σ_z can be neglected (plane-stress conditions). The stress distribution (10.48) is shown in fig. 10.23 (a). Both principal stresses are positive throughout the disk. With the Tresca yield criterion, plasticity will first occur at $r = a$, when $\sigma_\theta = \sigma_Y$. With increasing speed ω,

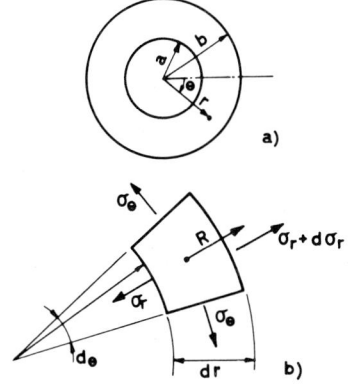

Fig. 10.22.

---

* Because of the symmetry of revolution.

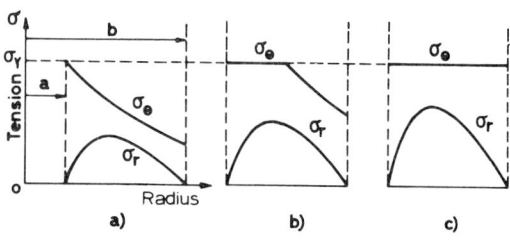

Fig. 10.23.

plasticity will spread in the disk and the stress distribution becomes the type shown in fig. 10.23 (b). As long as there remains an outer elastic annulus, displacements remain small (of the order of elastic displacements) and, hence, acceptable. At a certain value ω_l of the angular speed, the plastic zone spreads to the outer radius b and unrestricted plastic flow sets in, with the stress distribution of fig. 10.23 (c). If ω_s is the angular speed under service conditions, the safety factor against plastic collapse by overspeed will be $s = \omega_l/\omega_s$. Note that this safety factor is meaningful only if failure does not occur previously from other causes such as fatigue, excess elastic strains or creep. The influence of the temperature field on the elastic-plastic strains and on the local values of the yield stress σ_Y should also be considered.

At the plastic limit state, the stress field is statically determinate. Indeed, to the equilibrium equation in the radial direction

$$\frac{d}{dr}(r\sigma_r t) - t\sigma_\theta + \frac{\gamma}{g}\omega^2 r^2 = 0, \tag{10.49}$$

it is sufficient to add the yield condition

$$\sigma_\theta = \sigma_Y \tag{10.50}$$

to obtain the function $\sigma_r(r)$, taking into account the boundary conditions

$$\sigma_r = 0 \quad \text{at} \quad r = a, \tag{10.51}$$

$$\sigma_r = \sigma_{rb} \quad \text{at} \quad r = b. \tag{10.52}$$

The stress σ_{rb} vanishes when the outer edge is free, and is given by

$$\sigma_{rb} = \frac{T}{t_b} \qquad (10.53)$$

when the edge is loaded by radial tensile forces with magnitude T per unit of length (due to turbine blades, for example). It must also be verified that the obtained solution satisfies the inequality

$$\sigma_\theta > \sigma_r > 0. \qquad (10.54)$$

10.4.2. *Disk with constant thickness*

The equilibrium eq. (10.49) will be rendered nondimensional by dividing both sides by σ_Y. If we let

$$s_r = \frac{\sigma_r}{\sigma_Y}, \quad s_\theta = \frac{\sigma_\theta}{\sigma_Y}, \quad K = \frac{\gamma}{g}\frac{b^2\omega^2}{\sigma_Y}, \quad \rho = \frac{r}{b}, \qquad (10.55)$$

eq. (10.49) is rewritten, with the yield condition $s_\theta = 1$,

$$\frac{d}{d\rho}(s_r t\rho) = t(1-K\rho^2). \qquad (10.56)$$

Integration of (10.56) gives

$$s_r = 1 - \frac{1}{3}K\rho^2 + \frac{C}{\rho}. \qquad (10.57)$$

The integration constant C is determined from the boundary condition at the inner edge. In the absence of a central hole, $a = 0$ and, by symmetry, $s_r = s_\theta = 1$ at $\rho = 0$. Hence $C = 0$ and

$$s_r = 1 - \frac{K}{3}\rho^2. \qquad (10.58)$$

When there exists a central hole, $a \neq 0$, then $s_r = 0$ at $\rho = a/b \equiv \rho_a$, and the corresponding expression for s_r is

$$s_r = 1 - \frac{K}{3}\rho^2 - \frac{1}{\rho}\left(\rho_a - \frac{K}{3}\rho_a^3\right). \qquad (10.59)$$

To obtain the thickness t of the disk when ω_l is assigned, expression (10.55) for K is substituted into eq. (10.58) or eq. (10.59). If we let $\rho = 1$ in the resulting relation, and use the condition $s_{rb} = \sigma_{rb}/\sigma_Y$, we obtain t.

It is remarked that, in the absence of a central hole, the upper limit for K is 3, according to relation (10.58). On the other hand, when there is a central hole, the situation differs depending on the existence of edge forces T. If there are no such forces ($T=0$), use of $s_r = 0$ for $\rho = 1$ in eq. (10.59) yields

$$K = \frac{3}{1 + \rho_a + \rho_a^3}. \qquad (10.60)$$

From eq. (10.60) and (10.55) we see that neither K nor ω_l depends on the value t of the thickness. When $T \neq 0$ however, for given a, b, σ_Y, and ω_l, the thickness t is directly dependent on T. Because a uniform thickness may then become too large (see [10.13], p. 535), it is reasonable to consider disks with an inner part with constant thickness and an outer part with variable thickness.

10.4.3. *Practical shapes of rotating disks*

We restrict ourselves to disks such that $1 \leqslant K \leqslant 3$ (the usual range in turbine disks) and assume that the condition $s_{rb} = 1$ is to be satisfied. For more detailed analysis, in particular, when $s_{rb} < 1$, we refer the reader to Heyman [10.13].

Consider first a disk with a constant thickness for $0 \leqslant r \leqslant r_x$ and with a doubly conical outer part, with cone angle α, fig. 10.24. In this outer region

$$t = t_b + A(1-\rho), \qquad (10.61)$$

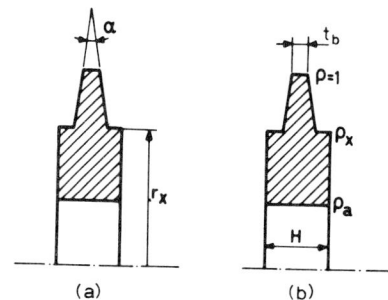

Fig. 10.24.

where A is a constant with the dimension of length. Substitution of expression (10.61) for t in eq. (10.56) and integration give

$$s_r[t_b+A(1-\rho)] = (t_b+A)\left(1-\frac{K\rho^2}{3}\right) - \frac{A}{2}\rho\left(1-\frac{K\rho^2}{2}\right) - \frac{C}{\rho}, \qquad (10.62)$$

where C is an integration constant.

The magnitude of A must be such that $(s_r)_{\max} \leqslant 1$, and the yield condition $s_\theta = 1$, $s_\theta \geqslant s_r \geqslant 0$ is not violated. Differentiating eq. (10.62), it is found that, for s_r to decrease with ρ decreasing from 1, it is necessary that

$$A \geqslant t_b K. \qquad (10.63)$$

The constant C is then obtained from the condition $s_r = s_{rb}$ at $\rho = 1$, and we finally have

$$s_r t = (t_b+A)\left(1-\frac{K\rho^2}{3}\right) - \frac{A\rho}{2}\left(1-\frac{K\rho^2}{2}\right) - \frac{1}{\rho}\left[(t_b+A)\left(1-\frac{K}{3}\right)\right.$$
$$\left. - \frac{A}{2}\left(1-\frac{K}{2}\right) - t_b\right]. \qquad (10.64)$$

If the radius r_x at which the profile of the disk changes is so chosen as to have the coefficient of A in eq. (10.64) vanish when $\rho = \rho_x$, the forces transmitted from the doubly-conical part to the central part will not depend on the angle α of the outer part. This condition is

$$3\rho^2 + 2\rho - \left(\frac{6}{K}-1\right) = 0, \qquad (10.65)$$

and its solution is

$$\rho^* = \frac{1}{3}\left[\left(\frac{18}{K}-2\right)^{1/2}-1\right]. \qquad (10.66)$$

If $\rho_x = \rho^*$, the radial force per circumferential unit of length transmitted across the section with equation $\rho = \rho_x$ is

$$(s_r t)_x = t_b\left(1-\frac{K\rho_x^2}{3}+\frac{K\rho_x}{3}\right). \qquad (10.67)$$

The thickness H of the constant thickness part is then obtained as explained in Section 10.4.2, regarding r_x as the outer radius with $T = (s_r t)_x \sigma_Y$.

From the preceding analysis it is seen that the thickness H will be independent of A if $\rho_x = \rho^*$. The smallest admissible value $t_b K$, relation (10.63), can then be given to A. Even more noticeable is the fact that, if the profile of the outer part is changed, the force $(s_r t \sigma_Y)_x$ transmitted in the section $\rho = \rho_x$ is practically independent of the chosen profile (and so is H), for $\rho_x \geqslant \rho^* = 0.7208$. This can be seen in fig. 10.25, and is discussed by Heyman in ref. [10.13]. In fig. 10.26 corresponding to $K = 1.5$, it can also be noted that, if $\rho_x \geqslant \rho^*$, the force $(s_r t \sigma_Y)_x$ depends very little on s_{rb} and, hence, the condition $s_{rb} = 1$ will have little influence on the magnitude of H.

For these reasons, the more easily feasible doubly-conical outer part will usually be preferred to other shapes. It will be connected by fillets to the central part with constant thickness. The curves in fig. 10.27 are good design aids. The radius a may be chosen for ρ_x to coincide with some critical non-dimensional radius ρ_c where s_r is a maximum in a disk with constant thickness (the dashed curve in fig. 10.27). Fig. 10.28 gives ρ_x versus K.

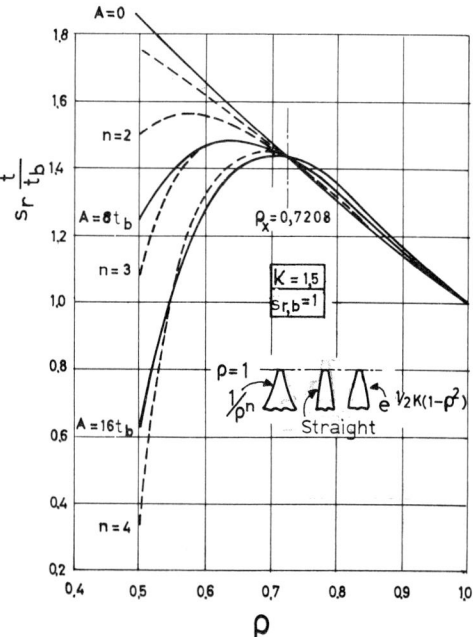

Fig. 10.25.

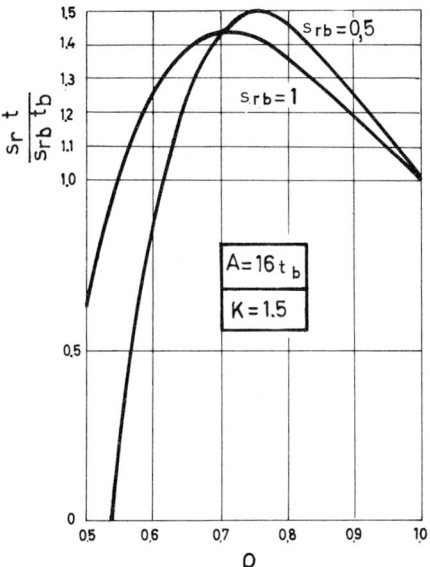

Fig. 10.26.

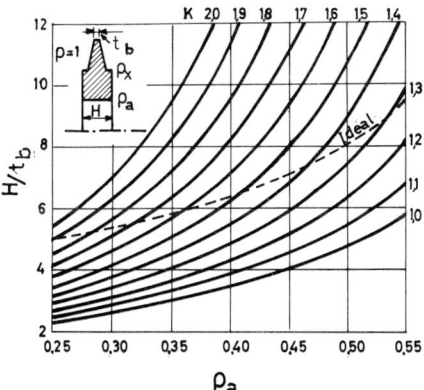

Fig. 10.27.

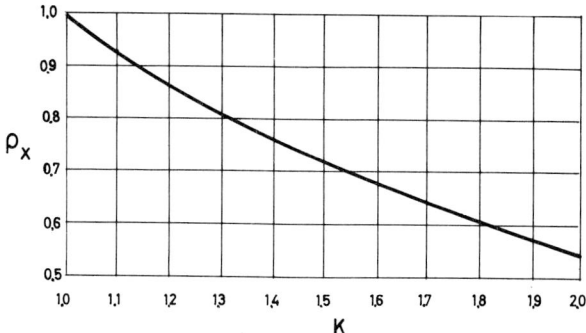

Fig. 10.28.

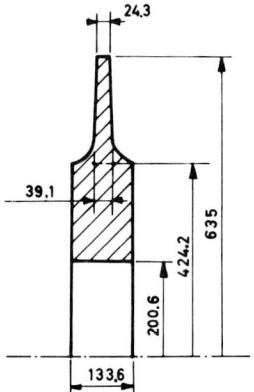

Fig. 10.29.

Consider, for example, the disk shown in fig. 10.29 [10.13]. The yield stress is $\sigma_Y = 60,000$ lb/in.^{2}. The design angular speed is 3,000 rpm. With a safety factor of 1.5, we have $\omega_l = 4,500$ rpm. At the bursting speed, the total edge loading is $2\pi bT = 9.10^6$ lb. We compute

$$K = \frac{\gamma}{g} \frac{b^2 \omega^2}{\sigma_Y} = \frac{0.283 \times (25)^2 \, (4500 \times 2\pi/60)^2}{32 \times 12 \times 60,000} = 1.706 \,,$$

and

$$t_b = \frac{T}{\sigma_Y} = \frac{9 \times 10^6}{50\pi \times 60{,}000} = 0.955 \text{ in.}$$

The dashed curve in fig. 10.27 gives $H/t_b = 5.5$ and $\rho_a = 0.315$. Hence, $a = 0.315 \times 25 = 7.9$ in. Then, $A = t_b K = 0.955 \times 1.706 = 1.628$. In fig. 10.28 we find $\rho_x = 0.64$. The thickness of the doubly-conical part at $\rho = \rho_x$ thus is

$$t_x = t_b + A(1-\rho_x) = 0.955 + 1.628(1-0.64) = 1.541 \text{ in.},$$

and the uniform thickness is $H = 5.5 \times 0.955 = 5.26$ in. Fillets will be machined at the junction of the two parts, as shown in fig. 10.29.

10.4.4. Disks of minimum weight

It is known from Sections 6.8.1 and 6.8.2 that to obtain a disk of minimum weight for given loads, it suffices to find: (a) a collapse mechanism with constant modified dissipation Δ/t, and (b) a corresponding statically admissible stress field. In the present case the only loads are the body forces due to the rotation. The modified dissipation per unit volume is

$$\frac{\Delta}{t} = \sigma_r \dot{\epsilon}_r + \sigma_\theta \dot{\epsilon}_\theta - \frac{\gamma}{g}\omega^2 r \dot{u}, \qquad (10.68)$$

where $\dot{u}$ is the radial displacement rate. Because

(a) $\quad \dot{\epsilon}_r = \dfrac{d\dot{u}}{dr},$

(b) $\quad \dot{\epsilon}_\theta = \dfrac{\dot{u}}{r},$ \hfill (10.69)

the condition $\Delta/t = \text{const} > 0$ obliges the stress point $(\sigma_r, \sigma_\theta)$ to be at A on the yield curve, fig. 10.30, for any ρ. Substitution of σ_Y for σ_θ and σ_r in eq. (10.49) and integration give

$$t = t_b \cdot e^{(1/2)K(1-\rho^2)}, \qquad (10.70)$$

with $t_b = T/\sigma_Y$.

The corresponding mechanism will be obtained upon integration of the equation

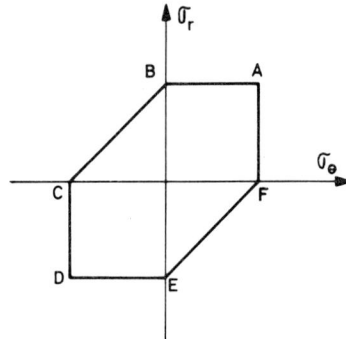

Fig. 10.30.

$$\frac{\Delta}{t} \equiv \sigma_Y \left(\frac{d\dot{u}}{dr} + \frac{\dot{u}}{r} \right) - \frac{\gamma}{g} \omega^2 r \dot{u} = \text{const} \tag{10.71}$$

with $d\dot{u}/dr \geqslant 0$ and $\dot{u}/r \geqslant 0$.

When the disk is pierced by a central hole, it is impossible to have $\sigma_\theta = \sigma_r = \sigma_Y$ at $r = a$. On the other hand, an upper limit H to the thickness is often assigned in practice. Drucker and Shield [6.49] have shown that, with this latter condition, the disk of minimum weight had the constant thickness H for $\rho_a \leqslant \rho \leqslant \rho_x$, and the exponential profile (10.70) for $\rho_x \leqslant \rho \leqslant 1$. The value of ρ_x is given by the relation

$$H \left[1 - \frac{K\rho_x^2}{3} - \frac{1}{\rho_x} \left(\rho_a - \frac{K\rho_a^3}{3} \right) \right] = t_b e^{(1/2)K(1-\rho_x^2)}, \tag{10.72}$$

obtained by the condition $\sigma_r t = \sigma_r H$ at $\rho = \rho_x$.

Instead of designing for minimum weight, it might also be desired to find the value of ρ_x corresponding to the smallest possible constant thickness H. The condition $dH/d\rho_x = 0$ gives, using (10.72),

$$(K\rho_x^2 - 1) = 0. \tag{10.73}$$

In practice, the condition of minimum H is often more important than that of minimum weight. The two conditions are incompatible. It is finally worth noting that very little economy in weight is achieved by the exponential profile over the tapered profile, as can be seen from numerical examples [10.14].

10.5. Other plane stress problems

Rectangular disks with various boundary conditions, made of a material with the von Mises yield condition, have been treated by Szmodits [10.15]. His lower bounds are obtained with discontinuous stress fields, and Mohr's circle is systematically used to deal with the discontinuity conditions. Discontinuous velocity fields are used in the kinematic analysis. Both types of fields are rather crude and give bounds that are far apart (lower bounds are roughly half of the upper bounds). Problems of the same type can be solved, using a general method of series expansions of the unknown functions that reduces the analysis to a linear programming problem where the coefficients of the expansions are the variables and the limit load is the cost function [10.16].

Minimum-weight design of various disks, mostly of theoretical interest, have been treated by Prager [10.17], and Hu and Shield [10.18].

10.6. **Plane strain: thick tube**

The state of plane strain has been widely studied in the mathematical theory of perfectly plastic solids. Prager and Hodge devote approximately half their book to the subject [3.13]. Plane strain occupies also an important part of the books by Hill [3.13], Nadai [1.4], and Sokolovsky [3.14].

Many of the problems treated do not, however, belong to the theory of structures but deal with various deformation processes: indentation, extrusion, etc., and therefore are not included in this volume. Only one structural plane-strain problem will be considered hereafter: the thick tube subjected to internal pressure. This subject is of important practical interest and has been widely studied experimentally [10.19]. For other plane-strain problems the reader is referred to the books cited above, where other interesting questions may also be found, e.g., torsion of prismatic and cylindrical bars with various cross sections, etc.

It is not possible to discuss here the numerous papers dealing with the plastic thick tube subjected to internal pressure. Such a discussion can be found in refs. [10.20] and [10.21]; autofrettage is treated in ref. [3.12]. We restrict ourselves to the following questions: (a) what is the magnitude of the pressure at which plasticity first occurs; and (b) what is the magnitude of the limit pressure, and what is the physical significance of it.

Let a and b be the internal and external radii of the tube, respectively, and p the internal pressure. The tube is referred to a cylindrical coordinate system,

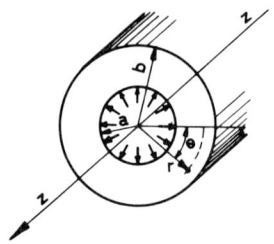

Fig. 10.31.

fig. 10.31. The longitudinal, radial, and circumferential directions are principal directions because of the symmetry of revolution. In elastic range, we have the well-known Lamé equations:

(a) $\quad \sigma_r = \dfrac{pa^2}{b^2 - a^2} \left(1 - \dfrac{b^2}{r^2}\right),$

(b) $\quad \sigma_\theta = \dfrac{pa^2}{b^2 - a^2} \left(1 + \dfrac{b^2}{r^2}\right).$ \hfill (10.74)

In plane strain $\epsilon_z = 0$ and Hooke's law gives

$$\sigma_z = \nu(\sigma_r + \sigma_\theta) = \dfrac{2\nu pa^2}{b^2 - a^2}.$$ \hfill (10.75)

In the case of a closed-end tube, eq. (10.75) must be replaced by

$$\sigma_z = \dfrac{pa^2}{b^2 - a^2} = \dfrac{1}{2}(\sigma_r + \sigma_\theta).$$ \hfill (10.76)

In both situations σ_z is the intermediate principal stress. With the Tresca yield condition, the maximum elastic pressure p_e will be given by the condition

$$(\sigma_\theta - \sigma_r)_{\max} = \sigma_Y.$$ \hfill (10.77)

From eq. (10.74) it is readily seen that $\sigma_\theta - \sigma_r$ is maximum at $r = a$. We obtain

$$p_e = \frac{\sigma_Y}{2} \frac{b^2 - a^2}{b^2}, \qquad (10.78)$$

an expression valid for both end conditions considered above. With the von Mises yield condition, the end conditions play a role because σ_z enters the yield condition

$$\sigma_R^2 \equiv \sigma_r^2 + \sigma_\theta^2 + \sigma_z^2 - \sigma_r\sigma_\theta - \sigma_\theta\sigma_z - \sigma_z\sigma_r = \sigma_Y^2. \qquad (10.79)$$

It turns out that σ_R is maximum at $r = a$ [3.12] and it is found that in the state of plane strain,

$$p_e = \frac{(\sigma_Y/\sqrt{3})[(b^2-a^2)/b^2]}{\sqrt{1+(1-2\nu)^2(a^4/3b^4)}}, \qquad (10.80)$$

and in the case of a closed end tube,

$$p_e = \frac{\sigma_Y}{\sqrt{3}} \frac{b^2 - a^2}{b^2} \qquad (10.81)$$

Note that, for an annulus in plane stress $[\sigma_z - 0, \epsilon_z = -2\nu p a^2/E(b^2\ a^2)]$ in elastic range], one would have

$$p_e = \frac{(\sigma_Y/\sqrt{3})[(b^2-a^2)/b^2]}{\sqrt{1+(a^4/3b^4)}} \qquad (10.82)$$

If we now increase p beyond p_e, a plastic annulus spreads from the internal surface toward the external surface. The state of restricted plastic flow, due to the presence of the external elastic annulus, ceases when the elastic-plastic interface reaches the outer surface, and plastic collapse occurs. The corresponding limit pressure is obtained in the following manner, assuming that the Tresca condition holds and that σ_z remains the intermediate principal stress. The stresses σ_r and σ_θ (fig. 10.32) satisfy the equilibrium equation

$$\frac{d\sigma_r}{dr} + \frac{\sigma_r - \sigma_\theta}{r} = 0, \qquad (10.83)$$

and the yield condition

$$\sigma_\theta - \sigma_r = \sigma_Y \quad \text{for} \quad a \leqslant r \leqslant b. \qquad (10.84)$$

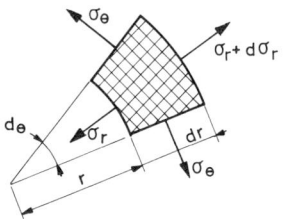

Fig. 10.32.

Integration of eq. (10.83), after substitution of σ_Y for $\sigma_\theta - \sigma_r$, gives

$$\sigma_r = \sigma_Y \ln r + C_1 . \tag{10.85}$$

The integration constant is obtained from the condition $\sigma_r = 0$ at $r = b$. We obtain $C_1 = -\sigma_Y \ln b$, and eq. (10.85) is rewritten

$$\sigma_r = \sigma_Y \ln \frac{r}{b} . \tag{10.86}$$

The limit pressure is given by $\sigma_r = -p_l$ at $r = a$, a condition that yields

$$p_l = \sigma_Y \ln \frac{b}{a} . \tag{10.87}$$

From eqs. (10.84) and (10.86) it is deduced that

$$\sigma_\theta = \sigma_Y \left(1 + \ln \frac{r}{b}\right). \tag{10.88}$$

The reader will verify as an exercise that σ_z is actually intermediate between σ_θ and σ_r, and that a mechanism corresponding to the stress field, eqs. (10.86) and (10.88) can be found.

In the absence of work hardening, the limit pressure (10.87) is a real collapse pressure because, when changes of geometry are taken into account, the pressure decreases with continuing plastic flow.

From a detailed elastic-plastic analysis by Hill, Lee, and Tupper [10.22, 10.23] where it is found that

(a) $\sigma_\theta = \sigma_Y \left(\dfrac{1}{2} + \dfrac{\rho^2}{2b^2} + \ln \dfrac{r}{\rho} \right),$

(b) $\sigma_r = \sigma_Y \left(-\dfrac{1}{2} + \dfrac{\rho^2}{2b^2} + \ln \dfrac{r}{\rho} \right),$

(c) $p = \sigma_Y \left(\dfrac{1}{2} - \dfrac{\rho^2}{2b^2} + \ln \dfrac{\rho}{a} \right)$ (10.89)

(ρ being the radius of the plastic region), it is also shown that σ_z tends very rapidly to the value $\frac{1}{2}(\sigma_r + \sigma_\theta)$. In a fully plastic state, it can thus be assumed with a good approximation (of the order of 1%) that $\sigma_z = \frac{1}{2}(\sigma_r + \sigma_\theta)$, and the condition of von Mises becomes formally identical to that of Tresca, except that σ_Y is replaced by $(2/\sqrt{3})/\sigma_Y$. Eq. (10.87) thus simply becomes

$$p_l = \dfrac{2\sigma_Y}{\sqrt{3}} \ln \dfrac{b}{a},$$ (10.90)

valid for both the closed-end tube and the tube in plane strain.

Bursting tests of approximately one hundred closed-end thick tubes have been performed by Faupel [10.19]. Various materials were used, among them mild-steel, brass, and allied steels, with fracture stresses ranging from 66,000 lb/in.² to 188,000 lb/in.², and average strains at rupture from 12% to 83%. The general shape of the stress-versus-strain curves of the tested materials is shown in fig. 10.33. The conventional highest elastic stress σ_e is chosen as the stress producing a residual strain of 0.01%, whereas the conventional yield limit $\sigma_{Y,0.2}$ corresponds to a plastic strain of 0.2%. The rupture stress σ_r is the highest attainable stress. Eq. (10.81), in which σ_e will be substituted for σ_Y, will give the highest elastic pressure. Because of work hardening of the

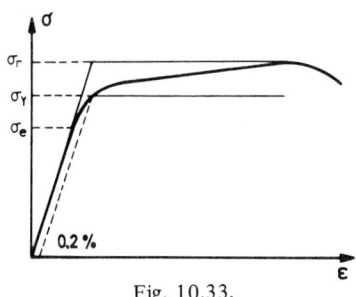

Fig. 10.33.

material, eq. (10.90) must be used with a "weighted" yield stress that depends on $\sigma_{Y,0.2}$ and on σ_r, to give the pressure at which continued plastic flow occurs and eventually burst under constant pressure. Faupel has suggested the formula

$$p_l = \frac{2\sigma_r}{\sqrt{2}} \left[2 - \frac{\sigma_r}{\sigma_{Y.02}} \right] \ln \frac{b}{a}, \qquad (10.91)$$

which is supported by his experimental results for ninety tests among one hundred, with 15% maximum discrepancy.

References

[10.1] F.A.GAYDON, "On the Yield-point Loading of a Square Plate with Concentric Circular Hole", *J. Mech. Phys. Solids*, **2**: 170, 1954.

[10.2] W.G.BRADY, D.C.DRUCKER, "An Experimental Investigation and Limit Analysis of Net Area in Tension", *Transactions A.S.C.E.*, **120**: 1133, 1955.

[10.3] F.SCHLEICHER, *Taschenbuch für Bauingenieure*, vol. 1, Springer, Berlin, 1955.

[10.4] *Manual of Steel Construction*, pp. 5-32, A.I.S.C., New York, 1964.

[10.5] P.G.HODGE Jr, N.PERRONE, "Yield Loads of Slabs with Reinforced Cutouts", *J. Appl. Mech.*, **24**: 85, 1957.

[10.6] D.VASARHELYI, R.A.HECHTMAN, "Welded Reinforcements of Openings in Structural Steel Members", *Weld. J. Research, Suppl.*, **16**: 182s, 1951.

[10.7] D.C.DRUCKER, "On Obtaining Plane Strain or Plane Stress Conditions in Plasticity", *Proc. 2nd U.S. Nat. Cong. Appl. Mech.*, pp. 485-488, A.S.M.E., 1954.

[10.8] E.H.LEE, "Plastic Flow of a V-notched Bar Pulled in Tension", *J. Appl. Mech.*, **19**: *Trans. A.S.M.E.*, **74**: 331, 1952.

[10.9] W.N.FINDLEY, D.C.DRUCKER, "An Experimental Study of Plane Plastic Straining of Notched Bars", *J. Appl. Mech.*, **32**: 493, 1965.

[10.10] W.SZCZEPINSKI, J.MIASTKOWSKI, "Experimental Limit Analysis of Flat Notched Bars Subjected to Tension" (in Polish), *Rozprawy Inzynierskie*, **13**: 637, 1965.

[10.11] L.DIETRICH, "Theoretical and Experimental Analysis of Load Carrying Capacity in Tension of Bars Weakened by Nonsymmetric Notches", *Bull. Acad. Pol. Sci., Serie des Sc. Tech.*, **14**: 7, 363, 1966.

[10.12] W.SZCZEPINSKI, "A Survey of Papers dealing with the Problem of Notched Bars Pulled in Tension" (in Polish), *Mechanika Teoretyczna i Stosowana*, **3**: 51, 1965.

[10.13] J.HEYMAN, "Plastic Design of Rotating Disc", *Proc. Inst. Mech. Eng.*, **172**: 531, 1958.

[10.14] J.HEYMAN, "Rotating Disks. Insensitivity of Design", *Proc. 3rd U.S. Nat. Cong. Appl. Mech.*, Brown University, Providence, R.I., 1958.

REFERENCES

[10.15] K.SZMODITS, "Scheibenmessung auf Grund der Traglastverfahrens", *Act. Techn. Hung,* **46**: 371, 1964.

[10.16] G.SACCHI, M.SAVE, "On the Evaluation of the Limit Load for Rigid Perfectly Plastic Continuum", Meccanica, 3: 3, 1968.

[10.17] W.PRAGER, "Dimensionnement plastique et économie des materiaux", *Bull. C.E.R.E.S.,* **X**: Institut du Génie civil, Liège, 1959.

[10.18] T.C.HU, R.T.SHIELD, "Minimum Volume of Disks", *Z.A.M.P.,* **XII**: 5, 414, 1961.

[10.19] J.H.FAUPEL, "Yield and Bursting Characteristics of Heavy-wall Cylinders", *Trans. A.S.M.E.,* **78**: 1031, 1956.

[10.20] P.G.HODGE Jr, G.N.WHITE Jr., "A Quantitative Comparison of Flow and Deformation Theories in Plasticity", *J. Appl. Mech.,* **17**: 180, 1950.

[10.21] D.N.de G.ALLEN, D.G.SOPWITH, "The Stresses and Strains in a Partly Plastic Thick Tube under Internal Pressure and End Load", *Proc. 7th Int. Cong. Appl. Mech.,* 403, Cambridge, 1958.

[10.22] R.HILL, E.H.LEE, S.J.TUPPER, "The Theory of Combined Plastic and Elastic Deformation with Particular Reference to a Thick Tube under Internal Pressure", *Proc. Roy. Soc.,* **A191**: 278, 1947.

[10.23] R.HILL, E.H.LEE, S.J.TUPPER, "Plastic Flow in a Closed End Tube with Internal Pressure", *Proc. 1st U.S. Nat. Cong. Appl. Mech.,* Chicago, 1951, 561, A.S.M.E., ed., 1952.

Subject Index

Adaptation, 80, 81, 107, 108
Affinity method, 236, 240, 277
   coefficient, 236
Arches, 60
Area, net, 439, 443, 451

Bar, 81
   notched, 456
Beam, 59, 88, 166, 203
   interaction with plate, 255, 259, 269, 271
   thin wall I, 375
   thin wall box, 385
Bounds, lower and upper, 25, 86, 89
Bursting, 473

Capacity, carrying, 40, 41, 115, 120, 157, 307, 353
Corner effects, 247
Cutout factor, 439, 446, 455

Dissipation, power, 14, 15, 70, 73, 170
   function, 15, 65

Dissipation, power (continued)
   with Tresca condition, 19, 436, 438
   with von Mises condition, 18, 439
   theorem of maximum, 26
   modified, 172
   for reinforced concrete plates, 216
Deformation theory, 36, 40
Design, 169, 172
   minimum weight, 169, 174
   uniform strength, 174, 175, 299, 392
Disk, 111, 434
   perforated, 439
   rectangular, 447, 469
   square, 452
   rotating, 459, 462
Energy, fictitious strain, 53
Enhancement factor, 301
Equilibrium, equations, 2, 3, 43, 67, 123, 150, 191, 336, 366, 381, 414, 436, 460, 471
   line, 52
   method, 263

Failure load, 115, 157, 165
  deflection, 119
Fan, 241, 243, 249, 250
  general, 259
Flow, incipient plastic, 13
  law (see normality law)
  plastic, 20
  restricted plastic, 31
  unrestrained plastic, 31, 36
Fore, axial, 296
  joint, 52, 54
  nodal, 262, 263, 265
Fracture line, 207, 225

Geometry, change of, 34, 297

Hinge line, 106, 207, 225
  circle, 337, 367, 421

Incompressibility, 20, 37, 436
Interaction, limited, 100, 102, 356, 361
  curve, MN, 299, 301

Limit state, 24
  analysis theorems, 25, 27, 28, 29
  load, 25, 48, 115, 204, 394
Linearization, 95, 97
Load, dead, 48
  live, 48
  failure, 115, 157, 165, 394
  circular line, 130
Loading parameter, 24
  in proportion, proportional, 24, 48, 58
  circular, 124, 129, 163, 184, 187
  annular, 128, 130, 186, 188

Mechanism, local flow, 15, 18
  of a body, 18
  collapse, 126, 161, 204, 240, 243, 254, 333, 413, 426, 427
  combined, 256, 258
Membrane, 117, 164, 175, 304, 305, 203–207, 367, 390
Moment, reduced, 72
  field, 126, 285
  volume, 285

Nonhomogeneous, 286
Normality law, 18
  rule, 107, 213

Orthotropy, 137, 207
  coefficient, 137, 215
  most economical coefficient, 292

Plate, 61, 68, 111, 178
  circular, 112, 113, 123, 157, 163, 165, 181, 243, 244, 245, 265, 284, 294
  annular, 112, 130
  rectangular, 112, 120, 144, 153, 166, 230, 257, 270, 273, 278
  square, 151, 153, 229, 248, 255, 267, 289, 293, 295, 297
  circular orthotropic, 119, 136
  polygonal, 156
  elliptic, 193, 261
  orthotropic in reinforced concrete, 207, 215, 236, 238, 277, 286, 292
Post-limit behavior, 161, 168

Reactions, 64, 65, 66, 80, 81, 87, 333
Regime, plastic, 132, 134
Rotation diagram, 227
  vector, 227, 233

Safety factor, 174
Sandwich, 83, 97, 174, 181, 190
Shake-down, 49, 54, 55, 57
Shear force, 66, 155, 308
Shell, 172, 325, 407, 409
  cylindrical, 62, 83, 91, 99, 326, 334, 350, 388, 389, 398, 414, 425
  metal, 326
  conical, 326, 361, 363, 420
  torispherical, 332, 364, 396
  toriconical, 332, 364
  short, 353
  long, 353
  rotationally symmetric, 356, 390
  spherical, 358, 421
  ellipsoid, 395
  shallow, 423
Silo, 420

Slenderness, 112, 325, 327, 355, 365
Solid, perfectly plastic, 23
Solutions, complete, 278
Stiffeners, 136, 327, 340, 344
Strain, tensor, 5
    plane, 5, 437, 469
    rate tensor, 6
    rate vector, 14
    rate intensity, 37
    rate discontinuities, 106
Strip method (Hillerborg), 290
Stress, tensor, 1
    components, 1
    principal, 3
    invariants, 3
    plane, 3, 434
    vector, 14
    joint, 14, 71
    reference, 11
    residual, 30
    constancy, 30
    deviator, 36
    intensity, 38
    reduced generalized, 77, 84
    generalized, 81, 213
    discontinuities, 104, 436
    profile, 124, 128, 137, 141, 327, 358
    concentration, 326
Superposition, in minimum-weight design, 195
    in reinforced concrete, 251

Tanks, 347, 410
Truss, 175
Tube, thick, 469
Twisting moment in yield line, 213

Uniqueness, 42, 279

Variables, generalized, 58, 63, 73
Velocity, 6, 437
    field, 27, 28, 30, 124, 129, 138, 142, 170

Weight function, 171
Work(s), theorem of virtual, 25, 42
    equation, 233
    dissipated, 235, 242
    hardening, 356, 472

Yield, conditions, 7, 67, 95, 207, 410
    condition of Tresca, 7, 72, 73, 86, 124, 145, 151, 181, 436, 438
    condition of von Mises, 9, 72, 84, 108, 129, 153, 190, 438
    surface, 14, 71, 77, 95, 212, 412
    point load, 32
    curve, 82
    stress, 96
    line, 106, 214, 215, 225, 251, 409
    condition of Johansen, 209, 223